Smith's Introduction to Industrial Mycology

Seventh Edition

A. H. S. Onions
Commonwealth Mycological Institute, Kew, Surrey

D. Allsopp
Biodeterioration Centre, University of Aston, Birmingham

H. O. W. Eggins
Biodeterioration Centre, University of Aston, Birmingham

Edward Arnold

© A. H. S. Onions, D. Allsopp and H. O. W. Eggins, 1981

First published 1938

by Edward Arnold (Publishers) Ltd,
41 Bedford Square,
London WC1B 3DQ

Second edition 1942
Third edition 1946
Fourth edition 1954
Fifth edition 1960
Sixth edition 1969
Seventh edition 1981

British Library Cataloguing in Publication Data
Smith, George
    Smith's introduction to industrial mycology. – 7th
    ed.
    1. Industrial microbiology
    I. Title   II. Onions, A. H. S.   III. Eggins, H. O. W.
    IV. Allsopp, D.
    660′.62   QR53

ISBN 0 7131 2811 9

Text set in 10/11 pt Linotron 202 Times, printed and bound in Great Britain at The Pitman Press, Bath

Preface to the Seventh Edition

The intention of this book is to produce a modern successor to Mr George Smith's book *An Introduction to Industrial Mycology*. As in the previous six editions, the aim is to assist non-specialist workers in the industrial field to identify the moulds and other fungi which are of importance in industry, either as useful organisms or as organisms causing damage. Many of the moulds described also cause food and grain spoilage and produce mycotoxins. It is hoped the book will help with the identification of some of these latter fungi, but we do not intend it to be a manual for the identification of mycotoxin fungi. This is only able to deal very superficially with the broader aspects of industrial mycology.

The first half of the previous volumes consisted of descriptions and illustrations of the moulds most regularly occurring in industrial products, with greater detailed consideration of the genera of most importance. The moulds involved have changed little over the years with perhaps a few additions. However, the study of taxonomy and nomenclature has increased considerably in the last few years. The treatment has been updated to follow these modern concepts and it is hoped we have included enough general mycology to enable the student to find and understand the more recent standard texts. The broad classification of Ainsworth (1966) (see References, p. 25) is followed as far as possible and the new approach to the identification of Hyphomycetes, based on methods of spore production, is introduced. Descriptions of laboratory methods are still included but not in such detail as in previous editions. Some chapters on other aspects have been reduced or omitted being now well documented elsewhere – for example the chapters on 'Mycology of Soil', 'Actinomycetes' and 'Microscopy' have been omitted. However, it was felt that the wood rotting Basidiomycetes should have a chapter in a book which includes spoilage fungi.

The chapter on industrial uses of fungi has been completely remodelled to cover modern appects of the subject including biochemical syntheses, harnessing of biodegradation and upgrading of materials, but this is now such a wide field that the treatment can only be an introduction. The chapter on control of mould growth has been rewritten and becomes 'Biodeterioration and its Prevention'.

G. Smith always maintained that figures from photographs were more suitable than line drawings for the use of beginners. However, line drawings have been introduced as it is felt that they often assist in interpretation of the photographs. Many of Smith's original photographs have been retained but some, through loss, damage or other causes, have been replaced and, as before, these replacements come from original photographs. For permission for their inclusion we are indebted to the Commonwealth Mycological Institute and in particular to the Institute photographer Mr D. W. Fry for his assistance with those of higher technical standard.

The authors wish to thank our many colleagues for help in producing this work: in particular Dr R. R. Davenport of Long Ashton Research Station for the chapter on Yeasts; Mr A. R. M. Barr of Catomance Ltd, Mr W. J. Irvine of Carreras Rothmans (Research Divison) and Dr B. Jarvis of the Food Research Association for advice on preservatives; Professor T. A. Oxley of the Biodeterioration Centre, University of Aston in Birmingham for contributing the section on water activity and Mr R. Young, also of the Biodeterioration Centre, for finalizing the chapter on industrial fungi.

In the taxonomy section we thank Dr A. F. Bravery of the Building Research Laboratories for examining the chapter on Basidiomycetes and are most grateful to our colleagues at the Commonwealth Mycological Institute who read and advised on much of the manuscript, especially Dr B. C. Sutton who examined many chapters, Dr D. L. Hawksworth the Ascomycetes, Mr P. Kirk the Zygomycetes, Dr P. M. Stockdale several chapters and all our other colleagues for support and interest, not least of all our ever patient typists, especially Mrs M. Rainbow who typed the taxonomic section.

1981

Agnes H. S. Onions
D. Allsopp
H. O. W. Eggins

Contents

viii   Contents

1

Introduction

We may rest assured that as green plants and animals disappear one by one from the face of the globe, some of the fungi will always be present to dispose of the last remains.

B. O. Dodge, *Rep. 3rd Int. Congr. Microbiol.*, 1940

1.1 Mycology

Mycology is concerned with the study of the Fungi, the term being derived from the Greek word *mykes*, meaning a fungus. The Fungi were, until comparatively recent times, regarded as members of the Plant Kingdom, and certainly, in general aspects, the majority of them bear a superficial resemblance to plants. Even at the present day nearly all the teaching in mycology in this country is carried out as part of courses in botany in our schools and universities, though there is an increasing number of courses in microbiology in which mycology forms a part.

1.2 The place of Fungi amongst living things, old and new theories

The supposed relationship of the Fungi to the various types of true plants has, until recently, usually been set out somewhat as in Table 1.1. In this system algae and fungi were regarded of the same phyletic origin and placed together as plants in the Thallophyta (Gr. *thallos*, a young shoot; *phyton*, a plant) which were plants which show no differentiation into root, stem, leaf etc., the vegetative structure being known as a *thallus*. This might be unicellular, as in some of the simplest fungi, or might show considerable specialization of structure with corresponding specialization of function.

The view that fungi and algae were derived from a primitive form of protozoa was originally put forward by Haeckel in 1866. Various classifications based on the derivation of fungi from this primitive kingdom have been favoured by many modern authors including Ramsbottom (1941), Langeron (1945) and Ingold (1959) and are fully discussed by Martin (1968).

However, it was still felt that fungi and algae were distinct and in 1969 Whittaker proposed his theory of classification of living things in which he separated the simplest prokaryotic organisms (Monera) from the eukaryotic organisms (Protista). From the latter he suggested three lines had developed with separation based on their nutrition: (1) the plant kingdom in which the

Table 1.1 The supposed relationship of the Fungi to the various types of 'true' plants and some other groups of organisms as usually set out before the more recent re-classifications.

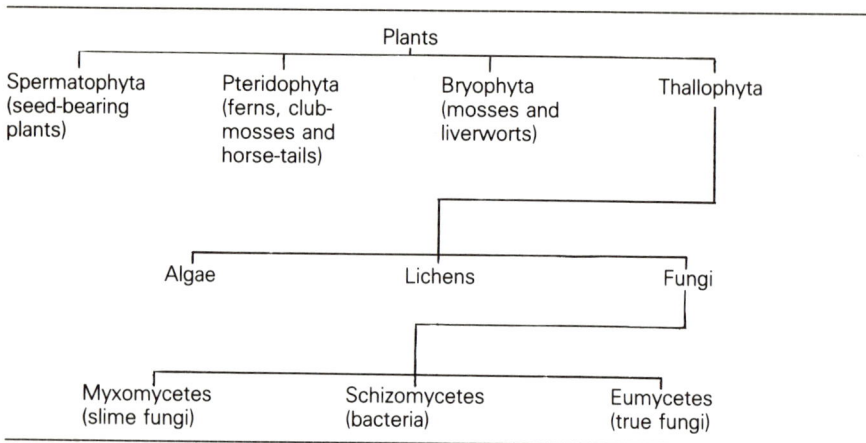

```
                                          Plants
   ┌─────────────┬─────────────────┬────────────────┬──────────────────┐
Spermatophyta    Pteridophyta      Bryophyta          Thallophyta
(seed-bearing    (ferns, club-     (mosses and
plants)          mosses and        liverworts)
                 horse-tails)
                              ┌──────────┴──────────┐
         ┌────────────────────┴────────────────────┐
        Algae                 Lichens               Fungi
                                                      │
   ┌────────────────────┬────────────────────┬───────┘
Myxomycetes           Schizomycetes         Eumycetes
(slime fungi)         (bacteria)            (true fungi)
```

organisms manufacture their own food by photosynthesis; (2) the animal kingdom characterized by ingestive nutrition, devouring organisms; and (3) the fungi which absorb nourishment from living or dead organisms or organic matter. This theory was adopted by Ainsworth (1973) who discusses the whole phylogenetic development in his book, *Introduction to the History of Mycology* (1976).

It is agreed, however, that whatever the origin of the fungi, they are a very distinct and separate group.

1.3   Separation of Fungi from other organisms

1.3.1   Algae

The fungi are distinguished from the algae in that they lack the green pigment which enables plants to bring about photosynthesis, i.e. the building up of complex organic compounds from carbon dioxide and water in the presence of sunlight.

1.3.2   Lichens

The lichens are compound organisms, consisting of algae and fungi in intimate association. Their study is a special branch of botany, since the alga-fungus association is so close that the lichens may be classified into genera and species just as if they were single organisms, and many of the fungi are unknown apart from their algal associates. Those who are interested in this group should consult the excellent text by Hale (1974).

1.3.3 Myxomycetes

The Myxomycetes (Gr. *myxa*, slime; *myketes*, pl. of *mykes*), the slime moulds, are a puzzling group of organisms and have been claimed by both mycologists and zoologists. However, Ainsworth (1973) included them with the fungi. No matter how they are classified their study does not come within the province of industrial mycology. The excellent monograph of the British Mycetozoa by Lister (1925) is still valuable and has only recently been replaced by the works of Gray and Alexopoulos (1968) on the biology, Martin and Alexopoulos (1969) on the taxonomy and a shorter key to the families and genera by Alexopoulos (1973).

1.3.4 Bacteria

The bacteria (Schizomycetes), comprising one of the three classes of fungi in Table 1 bear little resemblance to true fungi. Modern developments in microscopic techniques and molecular biology are presenting a clearer picture of their structure and nature, and the study of bacteria has now become a separate and important branch of science, with a technique of its own – their consideration is outside the scope of this book.

The only organisms listed in Table 1.1 which come within the province of the industrial mycologist are, therefore, the Eumycetes, or true fungi.

1.4 Numbers of Fungi

The number of species of fungi is a matter of speculation. Ainsworth (1968) reckoned that there were already 50 000 known species based on morphology and well over 100 000 specific names, and that this number was largely dependent on the number of working mycologists. He considered that there were at least 100 000 species and as there may be as many or more species of fungi than flowering plants this number could be as high as 250 000. The number of fungi depends on the understanding of an individual, but their range of habitat is wider than for flowering plants and many species are of world-wide distribution.

1.5 The activities of Fungi

As is to be expected of such a large group of living things, the fungi show great differences in size, structure, and metabolic activities. Some, such as yeasts, grow as loose aggregates of single detached cells, whilst others, such as the mushrooms and toadstools, form large fruit-bodies of complicated structure, with elaborate mechanisms for propagation. Some of the larger fungi are prized by the epicure whilst others are shunned as amongst the deadliest of poisons. The majority of the known fungi live on dead organic matter, performing a useful service in returning to the soil nutrients originally extracted by plants, but there is a large group of species which are inimical to man's activities through their habit of parasitizing plants which are grown for food and clothing, and a smaller group which are parasitic on animals, including man himself. Many fungi attack manufactured products of all kinds including foodstuffs, fabrics, leather, timber, cosmetics, pharmaceuticals, aviation fuel, and even glass. On the other hand a number of species are capable of synthesizing, under suitable conditions, substances useful to man, with an economy of effort which the chemist cannot emulate.

1.6   Fields of study

In mycology, as in other sciences, increased knowledge has resulted in complexity, and eventually the division of the science into a number of branches in which individual workers tend to specialize. What is usually termed pure mycology concerns the detailed structure, cytology, and modes of development of fungi. Field mycologists are interested in the fungi which are to be found in fields and woods, both the larger forms, known as mushrooms and toadstools, which grow on the ground or as parasites of forest trees, and the microscopic forms found on plant debris or as parasites of plants. The taxonomist studies structure with a view to classifying fungi, so as to show relationships and facilitate identifications by others. Although plant pathology is not a branch of mycology since it is concerned with the study and prevention of all kinds of abnormalities in cultivated plants, the plant pathologist nevertheless must have a good knowledge of mycology, for many diseases of plants are caused by fungi. Medical mycology deals with the fungi which cause disease in man. The study of toxins (mycotoxins) harmful to man and animals produced by fungi growing on deteriorating feedstuffs, in particular grains, is causing concern at the present time and is becoming an important branch of the science. Another somewhat specialized branch of the science is the study of the wood-destroying fungi, both those which attack standing trees and those which rot felled and worked timber.

1.7   Industrial mycology

The field of industrial mycology includes both the harmful activities of fungi in rotting or spoiling industrial raw materials and manufactured goods, and the uses of fungi in industrial fermentations. Their biodegradative activities are now being harnessed to assist in breakdown of organic waste and even upgrading of such waste for recycling or use as feedstuffs. The fungi concerned are commonly known as 'moulds'. They show considerable diversity of structure and are placed by the taxonomist in a number of widely separated groups. Although they are commonly visible to the naked eye, they all produce minute fruiting structures which cannot be studied without the aid of a microscope. One great advantage which industrial mycology has over some branches of the science is that the moulds are readily grown in the laboratory and can be studied quite independently of season or weather.

The various species of moulds differ in their responses to environmental conditions, in their abilities to attack various types of material, in toleration to preservatives and in synthetic activity. Hence little can be accomplished in the field of industrial mycology without a working knowledge of the moulds themselves, and the ability to recognize at least the more common species. Even in fermentation industries, which may use a single species of fungus, contaminants are likely to cause trouble unless they are recognized at an early stage. In addition, it is essential to be able to distinguish a highly active strain from other less useful strains of the same species, and to be able to recognize the species when searching for new and more active strains. A large proportion of this book is therefore devoted to descriptions and illustrations of most of the common species of moulds, and to the methods used in studying them for the purposes of identification.

The study of taxonomy has increased and undergone enormous change in approach and concept in the last 20 years. An attempt has been made to introduce the new ideas and to give enough new references to guide the reader to the extensive and expanding modern literature. However, descriptions of the fungi are based on culture characteristics and morphology and separation is by traditional keys – methods readily available to workers in industry. No effort is made to include the new, complex, and at present expensive, methods of electronmicroscopy, biochemical differentiation, electrophoresis, serology and numerical taxonomy.

The chapter on prevention of biodeterioration is given in such a way as to introduce the reader to the subject, as workers in this area are often new to industrial mycology, while it is now felt there have been such advances in the fermentation industries that workers will require no introduction to the technical aspects of the subject though they may still find information on fungi useful.

1.8 References

AINSWORTH, G. C. (1968). The number of fungi. In *The Fungi*, Vol. 3. AINSWORTH, G. C. and SUSSMAN, A. S. (Eds). pp. 505–14. Academic Press, New York.

AINSWORTH, G. C. (1973). Introduction and keys to higher taxa. In *The Fungi*, Vol. 4A. AINSWORTH, G. C., SPARROW, F. K. and SUSSMAN, A. S. (Eds). pp. 1–7. Academic Press, New York.

AINSWORTH, G. C. (1976). *Introduction to the History of Mycology*. Cambridge University Press.

ALEXOPOULOS, C. J. (1973). Myxomycetes. In *The Fungi*, Vol. 4B. AINSWORTH, G. C., SPARROW, F. K. and SUSSMAN, A. S. (Eds). pp. 39–60. Academic Press, New York.

GRAY, W. D. and ALEXOPOULOS, C. J. (1968). *Biology of the Myxomycetes*. The Ronald Press, New York.

HAECKEL, E. (1866). *Generelle Morphologie der Organismen*. Reimer, Berlin.

HALE, M. E. (1974). *The Biology of Lichens*. 2nd edition. Edward Arnold, London.

INGOLD, C. T. (1959). Fungi. In *Vistas in Botany*. TURRILL, N. B. (Ed). pp. 348–85. Pergamon Press, Oxford.

LANGERON, M. (1945). *Precis de Mycologie*. Masson & Cie, Paris.

LISTER, A. (1925). *A Monograph of the Mycetoza*. 3rd edition, revised by LISTER, G. British Museum Publication, London.

MARTIN, G. W. (1968). The origin and status of fungi. In *The Fungi*, Vol. 3. AINSWORTH, G. C. and SUSSMAN, A. S. (Eds). pp. 635–48. Academic Press, New York.

MARTIN, G. W. and ALEXOPOULOS, C. J. (1969). *The Myxomycetes*. University of Iowa Press, Iowa City.

RAMSBOTTOM, J. (1941). The expanding knowledge of mycology since Linnaeus. *Trans. Linnean Soc. London (Botany)*, **151,** 280–367.

WHITTAKER, R. H. (1969). New concepts of kingdoms of organisms. *Science*, **163,** 150–60.

2

A Brief Introduction to Morphology, Ecology and Physiology

2.1 Basic definitions and fungal morphology

When a mould begins to grow, whether it is on a natural substrate, on some manufactured material, or on a culture medium in the laboratory, there is an initial period during which nothing is visible to the naked eye. The time taken for visible growth to appear may vary from a few hours to many days, depending on a number of factors, the most important of which are the particular species of mould, the availability of any nutrient material present, and the relative humidity of the air. When the mould **colony** (as it is usually termed) has grown sufficiently to be readily seen, examination with a good hand-lens or low-power microscope will show the presence of a network of fine filaments. Each individual filament is termed a **hypha** (Gr. *hyphe*, a tissue), and the hyphae collectively are called **mycelium** (Gr. *mykes*, a fungus; *elos*, a wart). Some of the hyphae grow along the surface of the substrate, some may penetrate to a degree depending on the texture of the substrate, whilst others may stand above the surface, in some cases giving a hairy or fluffy appearance.

If the material on which the mould is growing is poor in nutritive value the mycelium spreads slowly and usually changes little in appearance, except that any aerial growth tends to collapse after a time. When, however, there is a reasonable amount of available food, and external conditions remain favourable, there is usually a gradual change in the appearance of the colony. Frequently there is a distinct change in colour, first manifested in the central and older parts of the mycelium. Microscopic examination at this stage will show the presence of fruiting structures, which are readily recognized as quite distinct from the ordinary hyphae. The individual reproductive bodies are called **spores** (Gr. *spora*, a seed). For example, the green mould usually found on leather (*Penicillium*) has more or less erect hyphae terminating in miniature broom-like structures, the individual bristles of the broom consisting of long chains of small roundish spores; the hairy-looking mould on bread (*Rhizopus*) bears small round, dark heads on erect hyphae, and if a head is crushed on a microscope slide it will be found to be full of small oval spores; a mould commonly found on jam (*Aspergillus*) produces spores of two kinds, one kind arranged in radiating chains, arising from the tops of erect hyphae, the other enclosed in yellow spherical bodies which are big enough to be seen without a lens. Other moulds form spores in a variety of ways, but, however they are formed, and whatever their appearance, they are almost always readily recognizable as distinct from the mycelium.

The spores of a fungus are designed for dissemination and reproduction, just as seeds are produced as a means of propagation by green plants. A spore, however, differs from a seed in that it never contains a germ, or plant embryo. It may consist of a single cell, or be a compound structure of several cells, each consisting of a mass of protoplasm surrounded by a firm containing wall. Spores of different species show considerable variation in size, ranging from a little over 1 μm, as in some species of *Penicillium*, to about 200 μm in greatest dimension in certain species of *Helminthosporium*. However, even the largest spores are light enough to be transported considerable distances by air currents, with the result that it is rare to find the atmosphere, either inside a building or outside, free from mould spores.

When a spore eventually alights it may remain dormant for a long period if conditions for growth are not favourable. As soon as the relative humidity of the air becomes sufficiently high (the actual figure varies for different species), germination occurs. It should be noted that although the term 'germination' is always used in reference to spores of fungi, the process has nothing in common with the process of germination of a plant seed. The first stage in spore germination is the absorption of water, with a consequent increase in size of the spore. One or more **germ-tubes** then emerge from points on the surface and rapidly elongate. Even a single-celled spore may give rise to more than one germ-tube, whilst in the case of many multicellular spores any cell may germinate separately. Figure 2.1 shows two spores of

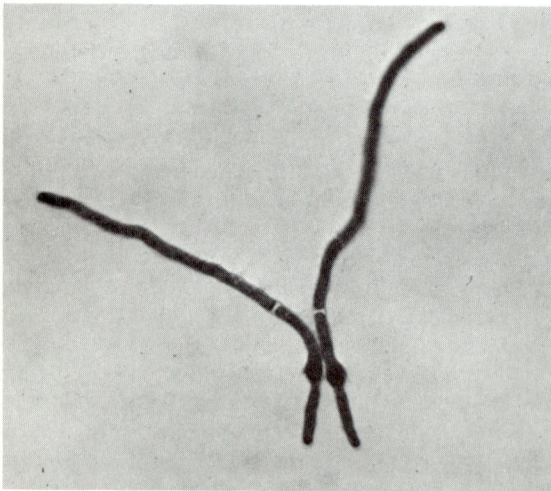

Fig. 2.1 *Penicillium notatum*   Germinating spores, ×250.

Penicillium notatum which have been allowed to germinate on cellulose sheet and subsequently stained (for the method see Chapter 13). Note that each spore has put forth two germ-tubes and that these have issued successively, not at the same time. If a supply of suitable nutrients, as well as sufficient moisture, is available, the germ-tubes rapidly increase in length and soon branch repeatedly, covering the surface of the substrate with a radiating network of hyphae. Frequently branches from one hypha fuse with other hyphae. This arrangement facilitates rapid transport of food material to the points where it is most required. The fusion of one hypha with another is termed **anastomosis** (Gr. *ana*, back; *stoma*, a mouth) (Fig. 2.2). When the

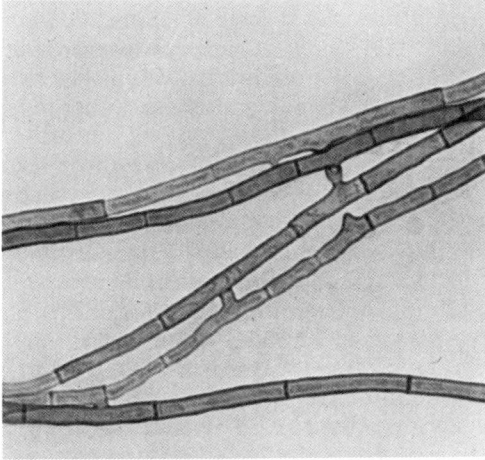

Fig. 2.2 *Uclocladium chartarum* Anastomosis, from slide culture, ×650.

mycelium has attained a certain size, that is, when the capacity for food intake has increased sufficiently, certain specialized hyphae begin to appear. Eventually these bear spores, ready to start the cycle over again. With many moulds spores appear first on the older central parts of the mycelium, the fruiting area gradually spreading outwards, but always remaining some little distance short of the edge of the mycelial area.

2.2 Ecology and physiology

Despite the many misconceptions which have arisen in recent years leading to misapplication of the word in general use, ecology is the study of the interactions between an organism and its environment. Ecology is a topic in its own right and is more fully discussed in Chapter 12. It is, however, useful to set the scene at this point, as it gives a more unified approach in considering the environment in which a mould grows, and also a warning that the various physiological aspects of growth, set out in more detail later (Chapter 12), must not be considered in isolation.

The environment in which any organism lives will contribute physical, chemical and biological factors which will have a bearing on the settlement, growth and development of an organism, and the organism in turn will modify the environmental factors. Interaction is, by definition, a two-way process.

The physical factors of main importance are those of temperature, water availability or humidity, O_2 and CO_2 levels, pH and perhaps pressure, light and other forms of radiation. The significance of these 'ecological' factors becomes obvious when considering control, for modification of these very factors may well be the first choice for prevention of growth (see Chapter 14).

The main factors in the chemical environment relate to nutrition and inhibition. An organism requires both **micro** nutrients (trace elements, growth factors, vitamins) in small quantities and **macro** nutrients (a carbon source for energy and assimilation, the 'food' in the case of saprophytic fungi) for growth. These growth requirements are discussed in Chapter 12, and have a bearing on choice of materials for culture media, discussed in Chapter 14. Chemical substances may also be present which inhibit or modify growth in some way. These compounds may be inherent in the material, may have been deliberately added to cause such an effect, as in the case of preservatives, or may come about as a result of waste secretion by organisms.

The physical and chemical tolerances and requirements of any organism are, of course, part of its physiological make up, and there are few aspects of fungal ecology which do not go hand in hand with physiological studies.

The biological factors in the environment include competition of all kinds from other organisms which may be present and also the various modifications in the physical and chemical environment which are brought about by organisms as they develop. Organisms compete for nutrients and space, some organisms are pathogens of others, and the waste products and secretions of organisms may build up, perhaps changing the pH or directly inhibiting growth by toxic action. Organisms create and maintain their own micro-climate. They may disrupt the substrate, and thus are able to modify not only the chemical but also the physical environment.

The environment is, therefore, seen to be both complex and dynamic, and for an organism to succeed it must be in balance, within fine limits, with the environment. If this balance can be disturbed to render it more hostile to the organism, less growth and therefore less damage will occur. This is the underlying reason why the industrial mycologist needs to think in ecological terms, and why ecology is of primary importance and direct relevance to applied microbiology.

2.3   Heterothallism

For many years it was a puzzle to mycologists why some species of the Mucorales regularly produced zygospores whilst others seemed to lose the power on continued cultivation, and in still other species, which were obviously nearly related, zygospores were quite unknown. Various theories attempting to explain the matter on a nutritional basis were not supported by experiment, and it was not until 1904 that the riddle was solved by Blakeslee (1904, 1920). He showed that the sporangiospores from a strain of *Mucor* producing abundant zygospores gave rise on germination to two distinct strains of mycelium, called by Blakeslee + and −. Each strain, by itself, was incapable of forming zygospores, but if both strains were grown in such a way that the two mycelia could come in contact, zygospores were produced along the line of contact. It was thus obvious that transfers of considerable masses of spores from a culture containing zygospores would usually result in a mixed culture of + and − strains and zygospores would again be produced, but that new cultures made from single spores would invariably lack zygospores and that sub-cultures made by transfer of a few spores would often result in a pure culture of one strain only. Such species, where sexuality resides in a whole thallus, are called **heterothallic**. In contrast to these, **homothallic** moulds produce zygospores on mycelia derived from single sporangiospores. Formation of gametangia is induced by contact of neighbouring hyphae, or even of branches from the same hypha. The illustrations of zygospore formation, Fig. 3.4 (p. 17), obviously depict a homothallic species. Both + and − strains of a number of heterothallic moulds are known and can be obtained from the various collections of type cultures, but, in a great many cases, only one strain of a species has so far been discovered and hence zygospores have never been observed.

Heterothallism is most frequently encountered in species belonging to the Mucorales, and it is to the numerous studies of such species that our knowledge of the phenomenon is chiefly due. However, in comparatively recent times it has been shown that heterothallism is not confined to this order. Many Ascomycetes have been found to be heterothallic and in certain of the Basidiomycetes a still more complicated state of affairs exists, the four spores from a single basidium being all of different mating types.

2.4   Heterocaryosis

In many fungi, the cells contain more than one nucleus. If these nuclei are genetically identical, the mycelium is said to be **homocaryotic**. If the nuclei differ in their genetic make-up the mycelium is said to be **heterocaryotic**. This latter situation may arise as a result of hyphal fusion or by mutation. This phenomenon is one more source of variation in the fungi which may be of evolutionary advantage.

2.5   Parasexuality

The mechanism for variation brought about by the recombination of genetic characteristics during sexual reproduction has long been recognized in the fungi as in other organisms. Recombination not associated with sexual reproduction has been recognized only more recently and is called **parasexuality**. Within a heterocaryon, dissimilar haploid nuclei may fuse, giving rise to diploid heterozygous nuclei. Recombination of genetic material may then take place, either during mitotic division of the diploid nucleus, or by random re-distribution of entire chromosomes when the nucleus breaks down to the haploid condition once more.

For a detailed account of these phenomena and sexual reproduction in fungi generally, the reader is referred to Chapters 14 and 15 in the book by J. H. Burnett (1976).

2.6 References

BLAKESLEE, A. F. (1904). Sexual reproduction in the Mucorineae. *Proc. Amer. Acad. Arts Sci.*, **40,** 205–319.

BLAKESLEE, A. F. (1920). Sexuality in the Mucors. *Science*, **51,** 375–82, 403–9.

BURNETT, J. H. (1976). *Fundamentals of Mycology*. 2nd edition. Edward Arnold, London.

3

Classification and Nomenclature

3.1 Systematics, classification and nomenclature

In all spheres of biological study it is useful to be able to identify the organisms one is dealing with. Man has, to a greater or lesser extent, been identifying organisms since time began for he quickly learnt which plants were poisonous and which animals were good to eat. Similarly, in the study of industrial mycology it is important to be able to recognize the fungi involved. If the wrong fungus contaminates an industrial process it will often result in the wrong product being produced. It is almost always necessary to be able to identify the organisms being handled in order to repeat a procedure, be it research or an industrial process. This is rendered possible since in nature organisms tend to fall into groups with many consistent characteristics in common.

The branch of biology in which the systematic arrangement of organisms into groups is studied is called systematics. The systematist must first study the organisms and from all the information available to him arrange them in recognizable consistent groups. The study involving the arrangement of organisms into groups is called classification. After organisms have been classified they then have to be named.

As classification consists of the arrangement of organisms into groups, mostly according to the subjective opinion of individual systematists, it is not surprising that many different systems of classification have been suggested. At any one time opinions differ as to the correct classificatory arrangement to adopt and this is to be expected if better systems of classification are to be made. On the other hand the name of an organism, which is ultimately based on an individual, should remain constant. This is not easy due to many causes, not the least of which is the lack of communication between systematists. In order to introduce stability into the naming of organisms, International Rules have been introduced. These are perforce quite complex. The study of the application of names to organisms is called nomenclature. Thus the organism has first to be classified and then named and it would be difficult to do one without the other.

3.2 Classification

There are some differences in detail between the schemes of classification propounded in the standard works on systematic mycology, but the scheme of Ainsworth (1966) is now fairly generally accepted.

The fungi are divided into two divisions, consisting of those that have a plasmodium, Myxomycota, and those in which a plasmodium is absent and in which the main assimilative phase is frequently filamentous, Eumycota (see Table 3.1).

The Myxomycota do not really concern the industrial mycologist and apart from *Plasmodiophora brassicae* (causing club root of brassicas) are of little economic importance. The common moulds fall in the Eumycota or true fungi. The subdivisions of the Eumycota are shown in Table 3.1.

Table 3.1 Classification of the fungi (following Ainsworth, 1966)

DIVISIONS

Division	Characteristics
Myxomycota	Plasmodium present
Eumycota	Plasmodium absent, frequently and typically filamentous

SUBDIVISIONS OF EUMYCOTA

Subdivision	Characteristics
Mastigomycotina	Motile cells (zoospores) present, perfect state (teleomorph) spores typically oospores
Zygomycotina	Perfect state spores zygospores
Ascomycotina	Perfect state spores ascospores
Basidiomycotina	Perfect state spores basidiospores
Deuteromycotina	No perfect state, reproducing asexually

The old scheme of classification

In many previous schemes of classification the Mastigomycotina and Zygomycotina were classed together as the Phycomycetes which were characterized by mycelium lacking cross walls (typically non-septate), and sexual reproduction was by oospores or zygospores. This concept of the Phycomycetes included a patently miscellaneous assemblage of organisms. Ainsworth (1966) retained the other main subdivisions, the Ascomycetes becoming the Ascomycotina, Basidiomycetes the Basidiomycotina and the Fungi Imperfecti the Deuteromycotina.

Separation in Ainsworth's scheme is still primarily based on the method of spore production of the perfect state except for the Deuteromycotina, which is an artificial subdivision included for the convenient accommodation of those fungi without a perfect state, but which may represent the sexual states of either ascospore or basidiospore producing forms.

Mastigomycotina

This subdivision, along with the Zygomycotina, is distinguished by having hyphae or filaments (where these are present) mostly without cross walls (septa) except in the fruiting organs and some of the older hyphae. Figure 3.1(a) shows vegetative hyphae of *Rhizopus stolonifer* consisting of a single enormously elongated branched cell, contrasted with Fig. 3.1(b) which shows

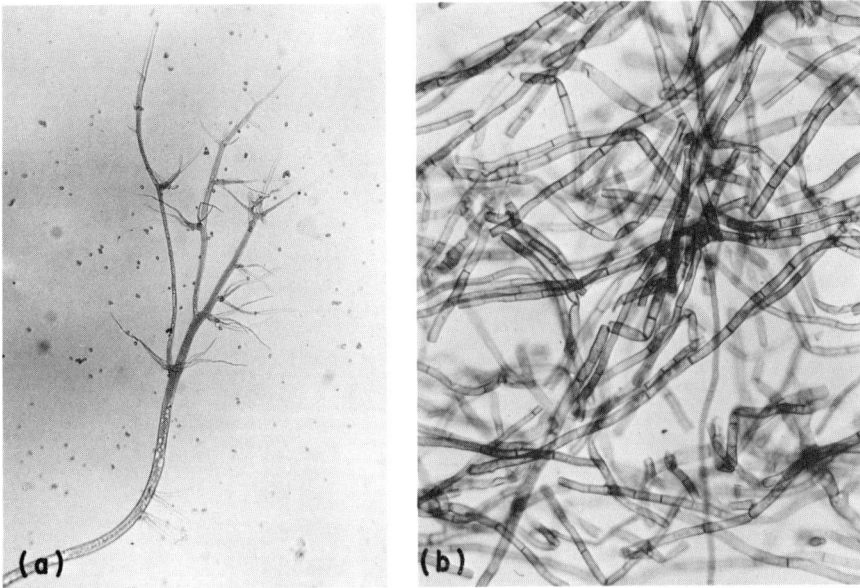

Fig. 3.1 (a) *Rhizopus stolonifer* Non-septate mycelium, ×60. **(b)** *Alternaria alternata* Septate mycelium, ×250.

septate hyphae of *Alternaria alternata* of the Deuteromycotina. If a young and vigorously growing culture of the common mould *Rhizopus stolonifer* is examined (with the low power of the microscope) it will be seen that the cell contents are streaming along both main hyphae and secondary hyphae. This rapid movement is possible only because of the entire absence of septa. It is true that it is possible to detect, by patient observation, the movement of protoplasm along septate hyphae but this streaming, which is very slow compared with the rate in non-septate hyphae, is possible because septa in living hyphae are not complete cell walls, but have small central pores. It is because of the free connection between all parts of the mycelium that most of the fungi of the Mastigomycotina and Zygomycotina which can be grown in the laboratory spread with extreme rapidity.

The sexual organs of the Mastigomycotina are readily distinguished as the antheridium, the male organ (Gr. *antheros*, flowery), and the oogonium, the female organ (Gr. *oon*, an egg; *gone*, seed, offspring). The oogonium, when ripe, shows most of the contained protoplasm aggregated into one or more oospheres (see Fig. 3.2). It is fertilized by the antheridium coming in contact with it or by motile male cells (spermatomoids) liberated by the antheridium. After fertilization the oospheres round up and develop cell walls to form the oospores, which lie free within the oogonium. Most also produce sporangia containing asexual zoospores.

The Mastigomycotina are divided into three classes – the Chytridiomycetes, Hyphochytridiomycetes and Oomycetes – according to the structure of the motile zoospores that they produce.

Some of the simpler forms of the Chytridiomycetes fail to produce or lack a

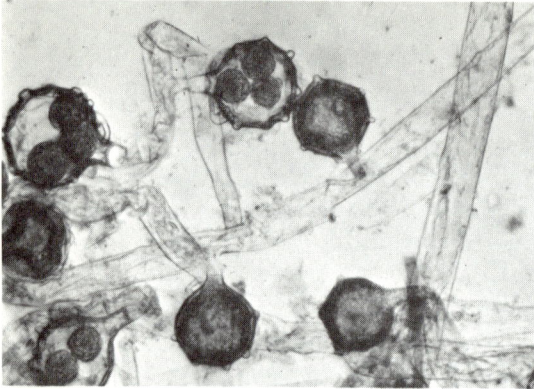

Fig. 3.2 *Achlya* sp.   Oogonia with oospheres, ×150.

vegetative structure. Of these, *Synchytrium endobioticum*, the cause of wart disease of potatoes, is of most economic importance.

The Oomycetes contain important plant pathogens such as various species of the genus *Pythium* which cause damping off diseases of seedlings and *Phytophthora* of which *P. infestans*, causing potato blight, is the best known. It also includes a few parasites of animals, the best known being species of *Saprolegnia* on fish and insects in water. However, none of the species is likely to be encountered in industrial work and therefore, their consideration will be limited to this brief account of their systematic position.

Zygomycotina
(Gr. *Zygon*, a yoke)

In this class the sexual organs are not distinguishable as antheridia and oogonia. They may be similar in size (usually the case) or markedly dissimilar, as in *Zygorhynchus*, but they are essentially alike in nature and function. They reproduce asexually by means of non-motile spores contained in a sporangium which may be violently projected. More usually the spores are dispersed passively by wind, rain or animals. The sporangia are round or pear-shaped and borne on simple or branched sporangiophores. Figure 3.3 shows the rounded sporangium of *Rhizopus stolonifer*. In a few genera sporangia are lacking and asexual reproduction is by conidia.

The zygospore results from the fusion of two cells delimited by septa from the sexual organs, and is thus exogenous. Two hyphae touch and at the point of contact develop protuberances, the progametangia, which gradually elongate and swell, remaining in contact themselves whilst pushing the parent hyphae further and further apart. Septa then appear in the progametangia, the cut off portions which remain in contact being the gametangia. The latter fuse, the cell walls in contact disappear, and the cell resulting from the fusion becomes rounded, usually thick-walled and dark coloured. Four stages in zygospore formation are shown in Fig. 3.4.

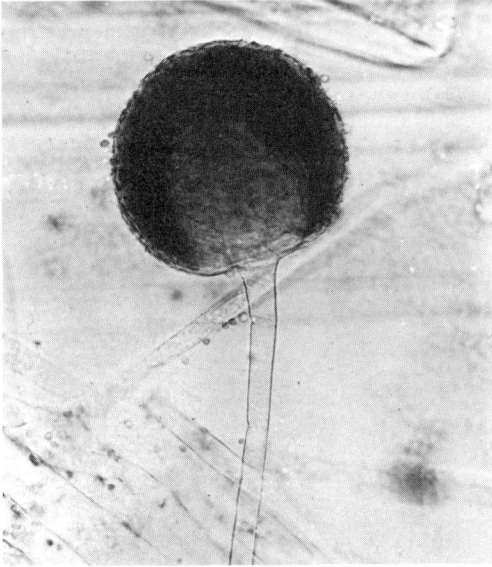

Fig. 3.3 *Rhizopus stolonifer* Sporangium (somewhat immature, but showing the essential points of structure), ×250.

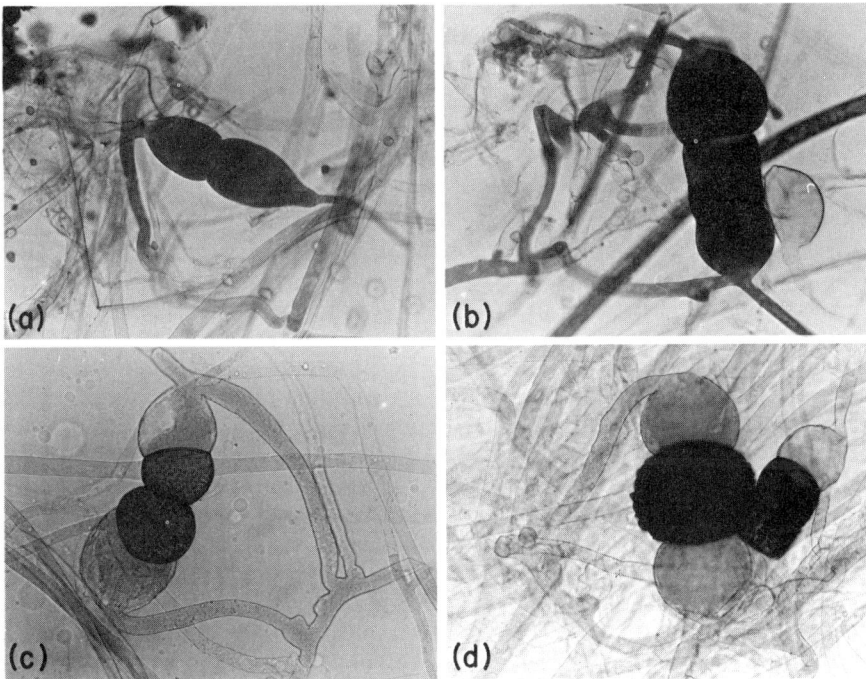

Fig. 3.4 *Rhizopus sexualis* **(a)–(d)** Successive stages in zygospore formation, ×120.

A variation of this process is found in *Piptocephalis*, a genus of moulds parasitic on other fungi belonging to the same class. In this case the zygospores are formed in a bud, put out from the cell produced by fusion of the gametangia. In some genera the zygospore is surrounded by stiff hairs or by a web of protecting hyphae, whilst in others it lies free amongst the aerial hyphae. In several genera which obviously belong here zygospores have never been seen.

The Zygomycotina include about 200 species, amongst which are a number of moulds of very common occurrence and considerable economic importance. These are discussed in more detail in Chapter 4.

Ascomycotina

In this class sexual spores are produced in asci (see pp. 50, 54). In most species each ascus contains eight spores. More rarely a larger or smaller multiple of two is produced, and only in some of the yeasts is it usual to find asci with an odd number of spores. The group shows a gradual transition from primitive forms which produce single, naked, globose asci to species which build up elaborate fruit bodies containing large numbers of club-shaped asci arranged in parallel series. The details of ascus formation are not readily observed and up to the present have been worked out for only a limited number of species. Asexual reproduction is by conidia, which show great diversity of form and arrangement, or by oidia. The class comprises about 15 000 species and includes both saprophytes and parasites, microscopic species and large fleshy fungi, some of which are edible. The genera which are of importance to the industrial mycologist are considered further in Chapter 5.

Basidiomycotina

The propagules which give the name to the subdivision are the basidiospores, borne exogenously on special organs, the basidia. In the typical species each basidium bears four spores. In the higher Basidiomycotina, the mushrooms and toadstools, the basidia are found in serrated ranks on the gills of the fleshy sporophores and there are specialized arrangements for ensuring the widespread distribution of the spores. The Basidiomycotina are divided into three classes. The Teliomycetes, which include the destructive parasites known as smuts and rusts, have a complicated life cycle. Most of these are obligate parasites and no attempt will be made to discuss their relationships and life cycles. These have been dealt with elsewhere. The Gasteromycetes, though of taxonomic interest, are not of economic importance. The species of the Hymenomycetes include mushrooms and toadstools, which are frequently responsible for rotting wood and similar materials. The Hymenomycetes are discussed in Chapter 7.

Occasionally when plating out materials Basidiomycetes appear on culture plates. A few species produce abundant conidia but in most cases the growths are sterile, and, in the present state of taxonomic knowledge, difficult to identify. Whether forming conidia or not it is often easy to recognize that such species belong to the Basidiomycotina by the presence in the mycelium of what are known as **clamp connections**. These peculiar structures are found only in the Basidiomycotina. A little distance from the tip of a growing hypha a very short branch appears; this curls backwards and reunites with the parent hypha close to the point of emergence; in the meantime a septum is formed midway in the hypha and another at the base of the hook. These structures are readily detected, as slight bulges, under the low power of a microscope (Fig. 3.5).

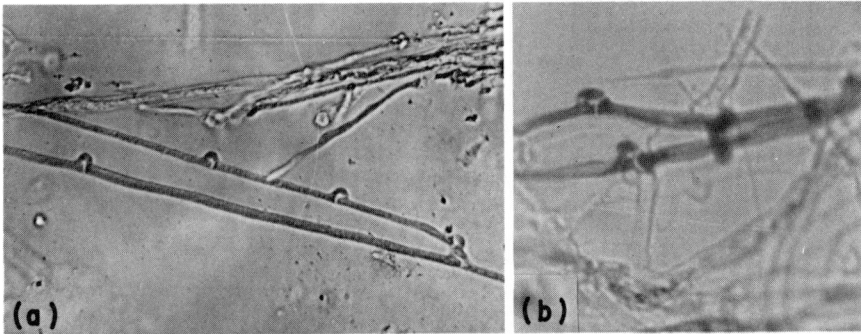

Fig. 3.5 *Polyporus* sp.   Clamp connections in mycelium of a Basidiomycete. **(a)** ×500; **(b)** ×800.

Deuteromycotina

This subdivision comprises a large number of fungi which produce neither ascospores nor basidiospores, but reproduce solely by means of conidia. (It should be noted that the moulds which produce sporangia are not included in the Deuteromycotina, but in the Zygomycotina, even though zygospores have never been seen.) It is safe to assume that in many cases their life cycles are incompletely known rather than incomplete. Most of the Deuteromycotina are probably Ascomycotina which produce perfect states (ascocarps) only under special conditions, which have not yet been discovered, or which have entirely lost the power of producing asci. It is now known with certainty in a few cases that heterothallism exists amongst the Ascomycotina and it is probable that this phenomenon accounts for many other cases of apparent loss of sexuality.

In the absence of perfect fructifications, it is necessary to classify the Deuteromycotina on characters of the asexual spores. According to the systematists, the genera thus set up are not true genera since it is known, on the one hand, that a single genus of Ascomycotina may include species with very different types of imperfect fructifications and on the other hand, that the conidial forms of species belonging to widely different genera may be similar. It follows, therefore, that a so-called 'form-genus', comprising a number of species with the same type of conidial fructification, may be a purely artificial group of fungi which are in fact quite unrelated. However, until the real relationships are discovered, it is necessary to have a convenient classification of the enormous number of Deuteromycotina, as otherwise identification of any form would be a hopeless task.

The great majority of the common fungi which are popularly known as moulds or mildew belong to this group and will be dealt with at greater length in Chapters 9, 10 and 11.

3.3   Nomenclature

In the early days of mycology there was no uniform method of naming fungi. To refer to any particular species it was necessary to use cumbersome descriptive phrases, for example 'Fungus campetris albus superne inferne rubens', literally 'the fungus of the fields, white on the top and reddening underneath', for the common mushroom. A great advance, not only for mycology but for the whole of biology, was made when the great Swedish botanist, Carl Linnaeus, in his *Species Plantarum* (1753), introduced the use of Latin binomials. In this system, soon to be universally adopted, the name of any organism consists of two words, a Latin noun, which is the name of the genus to which the organism belongs, followed by a specific epithet, also Latin and agreeing with the noun according to the rules of Latin grammar. The name is usually printed in italics, or underlined when written, the generic name with an initial capital letter and the specific epithet not capitalized. Each organism has its own name and the same name cannot legitimately be applied to two different species.

In principle this method of naming organisms is simple and straightforward, but, in the course of time, a number of difficulties have arisen. So long as there were comparatively few mycologists, and only a scant literature of the science, it was not difficult for any one worker to become acquainted with all the published descriptions of fungi, and to be in a position to decide whether any species under examination was new to science or had been described previously. However, the publication in the early part of last century of a number of fungus floras, notably Persoon's *Synopsis Methodica Fungorum* (1801) and Fries' *Systema Mycologicum* (1821–32), together with great improvements in the microscope, stimulated the search for fungi in all parts of the world, with a consequent enormous increase in the volume of literature. Inevitably there was a good deal of overlap, many species being described under two or more distinct names. In addition, many illegitimate names were bestowed in the sense that they had already been used for quite different organisms. Probably the greatest difficulty, however, has arisen from the fact that many of the descriptions in the older literature lack precision, the authors having been, of course, quite unaware of the enormous number of fungi yet to be discovered and of the necessity for detailed diagnosis. Many new names have been bestowed on fungi in good faith – names which were well known to the older mycologists. Subsequently, new interpretations of old publications have given rise to differences of opinion regarding the validity of many of these names, and the amount of attention which should be paid to the principle of priority. P. A. Saccardo, in his *Sylloge* (1882–1925) made an attempt to straighten out the confusion which, even at that time, had become serious, but eventually it was felt that only by international agreement was there any hope of putting nomenclature on a really sound basis.

International rules

The nomenclatural confusion was by no means restricted to the realm of mycology, but obtained throughout botany, so it became necessary for international agreement on a code of practice. A. de Candolle proposed a *Lois de la Nomenclature Botanique* at the International Congress held in Paris in 1867. This has been considered for amendment and revision at each subsequent International Botanical Congress and we now have the twelfth code in the series, the *International Code of Botanical Nomenclature*, adopted by the Twelfth International Botanical Congress, Leningrad, July 1975, prepared and edited by F. A. Stafleu *et al.*, Utrecht, 1978. No doubt there will be another after the next congress in 1980. Separate codes deal with Zoological Nomenclature and the Nomenclature of Bacteria.

The present code is presented in three languages – English, French and German. It is divided into three sections, dealing with principles, rules and recommendations with provision for modification. It contains six fundamental principles, and 75 articles or rules divided into six chapters, with supporting recommendations. There are also three appendices and a guide for the determination of types. The whole is very specific and deals with all the intricacies of nomenclature. It is the result of years of experience and discussion and is still being modified. Although every detail is covered concisely, 'Newcomers to systematic mycology frequently find the code itself rather a difficult document to interpret' (Hawksworth, 1974). There are, however, several useful studies which are helpful. One of the first of these was Bisby (1953), and later Ainsworth (1973) and Hawksworth (1974). Jeffrey (1977) has produced a useful book on Biological Nomenclature in which he deals with the problem of nomenclature as a whole, covering the Zoological and Bacteriological Codes as well. It is not proposed to discuss the whole code, the actual text of which must be carefully studied for full understanding, but some main concepts will be covered.

The principles

The principles in the code are fundamental to standardization of nomenclature. The code states that names must be determined by types, priority of publication and that each taxonomic group can have only one name and that name must be in Latin.

Ranks of taxa

The code designates names for the various ranks of taxa and recommends endings for the names (see Table 3.2). The last two ranks of genus and species are in Latin and consist of either a Latin word or some other word in Latin

Table 3.2 Ranks of taxa with endings recommended by the International Code of Botanical Nomenclature

Rank	Ending
Division	-mycota
Subdivision	-mycotina
Class	-mycetes
Subclass	-mycetidae
Order	-ales
Suborder	-ineae
Family	-aceae
Subfamily	-oideae
Genus, species, etc.	No standard suffixes but must be in Latin form

form. It is usual to use only the genus and specific name when referring to a species. The genus name is written with an initial capital letter, may be arbitrarily coined or derived but is usually given in the nominative singular. The species name is written with a small letter and agrees with the genus name, usually in the nominative but it can be in the genitive. It may be a noun or adjective. Names are usually descriptive but may indicate place of origin or be used to honour workers in the field of mycology. The rules governing the Latinization of words for nomenclature are fully described in the Code.

Typification

Each rank is indicated or designated by a **nomenclatural type**. This consists of the name of a single unit from the rank below until species level is reached and is then represented by a **Type** specimen. The type specimen (**holotype**) is the specimen on which the author based his original description of the species. Types are usually deposited in one of the older and better known herbaria. There are sometimes difficulties concerning the type. The original may have been lost, in which case an author who is classifying the group may designate a new type (**neotype**) to take the place of the original material. All new species described since 1st January 1958 must have a type, designated in the description of the species and deposited as dried material in a herbarium. As some specimens do not dry satisfactorily there is a move to allow the deposit of a living culture with a recognized culture collection to act as the type material. This may be introduced into the next version of the Code. The concept of names being based on a type is fundamental to all classification and nomenclature as, at least in theory, identifications are made by comparisons with type material. This may of course be at second or third hand as it is undesirable to be constantly handling the type material.

Publication

For a name to be effectively published it must have been published as printed matter which is distributed to the public, preferably in a recognized journal.

This publication must contain a description of the fungus which, from 1st January 1935, must be in Latin for the name to be validly published.

Priorities, name changes and starting dates

The rules lay down that each taxon shall bear one name only. However, it often happens that one fungus species is described in more than one place and thus has more than one name. This may be due to lack of communication or different concepts of taxa by different workers. When the problem does occur the oldest recognizable validly published name has priority and the other names are called **synonyms**. Where several different taxa have been given the same name they are called **homonyms**.

Nomina conservenda

A few exceptions to this rule of priority have been made for names of genera and taxa above the generic level. When a name in very common use has been found to be pre-dated by a little known name, for convenience the common name has been conserved, for example, *Alternaria* C. G. Nees ex Wallroth, 1833 in preference to *Macrosporium* E. M. Fries, 1832. These accepted conserved names are listed in an appendix to the rules. Although from time to time it has been suggested that some common species names might be conserved this has not yet been accepted.

Starting dates

To avoid problems resulting from finding and interpreting early literature definite starting dates have been designated and names introduced before these dates are not considered for purposes of priority. These dates vary according to the groups of fungi and are based on major works of famous early mycologists.

Starting dates for fungi
Lichenes
    1 May 1753, Linnaeus
Fungi: Uredinales, Ustilaginales and Gasteromycetes
    31 December 1801, Persoon
Fungi caeteri
    1 January 1821, Fries, *Systema Mycologicum*, Vol. 1, but this is treated as being the date of publication of the later volumes of the Systema and also of the *Eleachus Fungorum* (1828) and names published in other works between the dates of the first and last parts of the Systema do not affect the nomenclatural status of names used by Fries in this work.
Myxomycetes
    1 May 1753, Linnaeus

Note: The whole problem of starting dates came under review at the last International Mycological Conference at Tampa in 1977 and agreement of the most desirable starting dates for fungi is still under discussion. The above dates refer to the Code as it stands in the present edition (1978).

Author's names

It is customary, when mentioning the name of a genus or species, to append the name (often abbreviated) of the author who published the first description and bestowed the name. This custom makes for precision and is of special value when the same name has been used inadvertently by different authors for different genera or species, as it prevents confusion until such time as the more latterly described fungus can be renamed. When a species is transferred from one genus to another, the name of the original describer is put in brackets and followed by the name of the worker who made the transfer, for example *Scopulariopsis brevicaulis* (Saccardo) Bainier.

Name changes

Owing to advances in our knowledge of fungi it often becomes necessary to transfer the name of a species from one genus to another, or possibly to make it the type of a new genus. Such changes are quite in order, but with the provision that the specific epithet shall remain unchanged, except for inflection according to the rules of grammar. However often a species name is transferred, the correct specific epithet is always the first one which was applied to that particular species. For example, Bainier in 1907 decided that *Penicillium brevicaule* Saccardo is not a true *Penicillium* and made it the type of a new genus *Scopulariopsis*, the correct citation being *Scopulariopsis brevicaulis* (Sacc.) Bainier.

Pleomorphic forms

Some fungi produce more than one sort of spore, for example, Ascomycetes may produce an ascosporic state (**teleomorph**) and a conidial state (**anamorph**). It is quite possible for each of these to have been described under different names. Once it has been established that they both belong to the same fungus then, according to the rules, the earliest name for the perfect state (teleomorph) has priority and is the correct name. However, it is quite acceptable when referring to the imperfect state to use the imperfect name. Thus the many species of the imperfect genus *Aspergillus* are known to have perfect states, for example *Sartorya fisheri* the perfect state of *Aspergillus fumigatus*. When the ascosporic state is present with or without the conidial state the correct name is *Sartorya fisheri*. However, the conidial state is often seen without the perfect state and frequently shows no ability to produce the perfect state. In fact it is possible that it cannot do so and may even be a different fungus, although it appears the same as the imperfect state of *Sartorya fisheri*. Thus when the conidial state only is seen it is quite correct to refer to it as *Aspergillus fumigatus*.

Many of the species with *Aspergillus* conidial states have been described as *Aspergillus* species, but when the perfect states have been found they are of different genera. In a recent paper Subramanian (1972) found old names or gave new names to these perfect states with *Aspergillus* conidial states and these hold priority according to the rules, though workers in industry may still find it more convenient to think of this group of genera as the 'genus *Aspergillus*' and they are dealt with together in this book (Chapter 10).

Many other matters are dealt with in the Rules, all aimed at removing ambiguity and ensuring that a name shall have a precise meaning which can be understood everywhere, but these can not be summarized adequately here.

Note: Since the publication of the most recent Code and after the last International Mycological Congress in 1977 it has been noted that such terms as 'perfect', 'imperfect', and 'state' have become so familiar that they are used loosely. Hennebert and Weresub (1977) recommend that they be replaced and it seems likely that, in future, the following terms will be accepted and used.

Holomorph (holomorphic, etc.) for the fungus in all its forms, imperfect and perfect.

Teleomorph (teleomorphic, etc.) for the perfect state or the sexual morph or form involved in producing meiotically generated propagules.

Anamorph (anamorphic, etc.) for the imperfect state or the asexual morph or form.

In 1979 Weresub and Hennebert redefined these terms at greater length and with more precision.

3.4 References

AINSWORTH, G. C. (1966). A general purpose classification for fungi. *Bibl. Syst. Mycol.*, No. 1, 1–4.

AINSWORTH, G. C. (1973). Fungal nomenclature. *Rev. Plant Path.*, **52**, 59–68.

BISBY, G. R. (1953). *An Introduction to the Taxonomy and Nomenclature of Fungi.* 2nd edition. Commonwealth Mycological Institute, Kew.

CANDOLLE, A. DE (1867). *Lois de la Nomenclature Botanique.* Bailliere et Filo, Paris.

FRIES, E. M. (1821–32). *Systema mycologicum.* Griefswald, Lund.

HAWKSWORTH, D. L. (1974). *Mycologists Handbook.* Commonwealth Mycological Institute, Kew.

HENNEBERT, G. L. and WERESUB, L. K. (1977). Terms for states and forms of fungi, and their names and types. *Mycotaxon*, **6**, 207–11.

JEFFREY, C. (1977). *Biological Nomenclature.* 2nd edition. Edward Arnold, London.

LINNAEUS, C. (1753). *Species plantarum.* Stockholm.

PERSOON, C. H. (1801). *Synopsis methodica fungorum.* Göttingen.

SACCARDO, P. A. (1882–1925). *Sylloge fungorum omnium hucusque cognitorum.* Pavia, Italy.

STAFLEU, F. A. *et al.* (1978). *International Code of Botanical Nomenclature.* Adopted by the Twelfth International Botanical Congress, Leningrad, July 1975. Bohn, Schelema and Holkema, Utrecht.

SUBRAMANIAN, C. V. (1972). The perfect states of *Aspergillus*. *Current Science*, **41**, 755–61.

WERESUB, L. K. and HENNEBERT, G. L. (1979). Anamorph and teleomorph: terms for organs of reproduction rather than karyological phrases. *Mycotaxon*, **8**, 181–6.

4
Zygomycetes

4.1 Introduction

The essential features of the Zygomycotina have already been outlined in Chapter 3 and need not be further elaborated. Most of the common species are readily recognized as belonging to the group by their rapid rate of growth and their characteristic appearance, colonies usually being loosely floccose and of a grey or brownish-grey colour. Three genera, *Mucor, Rhizopus* and *Absidia*, include the great majority of the species which are normally encountered in the laboratory, but members of several other genera are found sufficiently frequently to justify descriptions being given here. Zygospores were originally found in comparatively few species (they have now been produced, at least in culture, for many species), and were not made the main basis of classification. However, even when species lack the usual fruiting stage, there has never been any question of placing these with the Fungi Imperfecti, owing to the very characteristic type of mycelium and imperfect fruiting stage which distinguish the subclass.

4.2 Classification of Zygomycotina

The Zygomycotina are divided into two classes, the Zygomycetes and the Trichomycetes (Ainsworth, 1973a). The Trichomycetes are a little known group mostly parasitic on arthropods.

Until recently the Zygomycetes were divided into two orders as follows:

Asexual spores occurring in sporangia	**Mucorales**
Asexual spores as conidia, forcibly shot away at maturity	**Entomophthorales**

Hesseltine and Ellis (1973) gave a comprehensive key to the groups and genera and accepted fourteen families in the Mucorales. However, Benjamin (1979), basing his work on phylogenetic relationships, has produced an entirely new classification of the Zygomycetes at ordinal level creating seven orders and 20 families, and included as one of the orders the Entomophthorales.

The Entomophthorales are mentioned here as one species, *Entomophthora muscae* Cohn, which is of common occurrence in the late summer, when it forms characteristic white halos round the bodies of dead flies, often stuck to the glass of windows. Waterhouse (1975) gives a simple key to the Entomophthorales and an introduction to the group.

However, only a few genera and species of the Zygomycetes are of interest to the industrial mycologist and most of these are to be found in the Mucorales.

4.3   Mucorales

The great majority of the species are saprophytic, occurring on a wide variety of organic substrates; others are parasitic on other members of the order. Typical colonies consist of coarse hyphae growing loosely, white in the early stages of growth and becoming grey or brownish with the production of fruiting structures. The usual mode of asexual reproduction is by spores produced in large numbers in globose sporangia, borne on sporangiophores which may be branched in various ways. The sporangial wall may be thin, in which case the spores are liberated by rupture or dissolution, or may be cutinized and shot off or broken off in one piece. In a number of families various modifications of the typical globose, many spored sporangium are found.

A number of different classifications of the Mucorales have been proposed but the total number of genera is not large and in the great majority of cases, the placing of a particular isolate in its correct genus is a fairly simple matter whichever key is used for the purpose.

Saccardo includes the group in his *Sylloge* (see p. 25), and there were monographs from Lender (1908) and Hagem (1910) and numerous papers on different genera. However, the literature was scattered and often difficult to follow, and as in other ubiquitous species there are minor differences between different isolates. Taxonomists who have handled comparatively small numbers of isolates have not always distinguished between true and spurious specific characters. Most original diagnoses have been drawn up in terms insufficiently broad to cover these isolate differences, with the result that workers trying to identify species have but rarely found published descriptions which tallied exactly with their own data.

However, the monograph by Zycha in 1935 considerably simplified the taxonomy by taking into account this natural variation. A further monograph by Naumov appeared in 1939 but Zycha remained the standard work for years. A modern version of Zycha's monograph was produced by Zycha, Siepmann and Linnemann in 1969. This comprehensive study is in German, but an English translation of the keys has been produced by Hanlin (1973). However, there are still difficulties and several workers are continuing studies of the group, many genera of which have now been monographed. Leaders in the field have been Hesseltine and his colleagues at Peoria, Illinois who have made a systematic and intensive study. They have produced general papers on the Mucorales and also monographs on separate groups and genera (Hesseltine, 1952, 1953, 1954, 1955, 1957, 1960; Hesseltine and Fennell, 1955; Hesseltine and Anderson, 1956, 1957; Hesseltine and Benjamin 1957; Hesseltine *et al.*, 1959; Hesseltine and Ellis, 1961, 1964, 1966, 1973; Ellis and Hesseltine, 1965, 1966, 1974; and Benjamin and Hesseltine, 1957, 1959).

M. Schipper, working at the Centraalbureau voor Schimmelcultures, is undertaking a critical study of the genus *Mucor* (Schipper, 1969, 1970, 1973, 1975, 1976, 1978). Other studies of interest include those of Gams (1969,

1977), Gams *et al.* (1972), and Chien *et al.* (1974) on *Mortierella*, Schipper *et al.* (1975) on *Zygorynchus*, Benjamin (1959) on *Syncephalastrum* and Nottebrock *et al.* (1974) on *Absidia*. Ingold's studies on spore liberation in the Mucorales are fascinating, in particular Ingold and Zoberi (1963).

As a simple introduction to the main groups Webster (1970), Talbot (1971) and Ingold (1978) are helpful as is Richardson and Watling's (1969) 'Keys to fungi on dung'.

Families and genera of Mucorales of most interest to industry are shown in Table 4.1.

Table 4.1 Families and genera of Mucorales of most interest to industry

Family	Genera
Mucoraceae	*Mucor, Rhizopus, Absidia, Zygorhynchus, Phycomyces*
Mortierellaceae	*Mortierella*
Syncephalastraceae	*Syncephalastrum*
Thamnidiaceae	*Thamnidium, Helicostylum*

4.4   The families of the Mucorales

Mucoraceae
Spores borne in a sporangium with the tip of the sporangiophore swollen and projecting as a **columella** into the sporangium. Its size and characteristics are of importance in determination of the genus. The spores are liberated by the rupture or dissolution of the sporangium wall. (See Fig. 4.1.) Genera of interest to industry are *Mucor, Rhizopus* and *Absidia*. Other interesting genera include *Zygorhynchus* and *Phycomyces*.

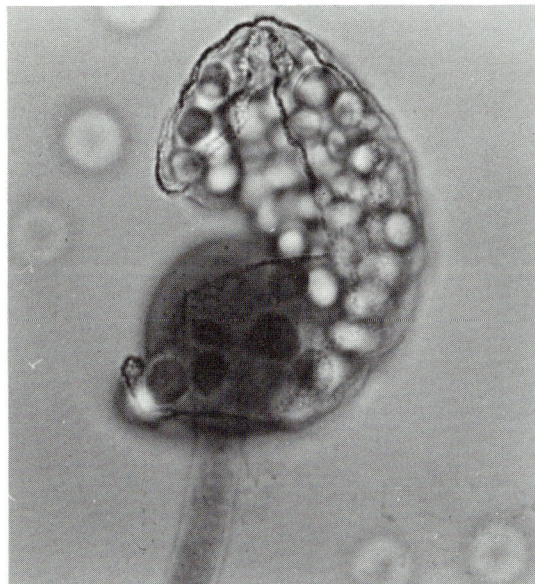

Fig. 4.1 *Mucor racemosus* Columella projecting into sporangium which is breaking up, a collarette is also present, ×650.

Mortierellaceae

Spores are produced in simple sporangia usually without a columella, on simple or branched sporangiophores. There may be one to numerous spores in the sporangium. Mostly soil organisms. The genus *Mortierella* is fairly common in soil. (See Fig. 4.2.)

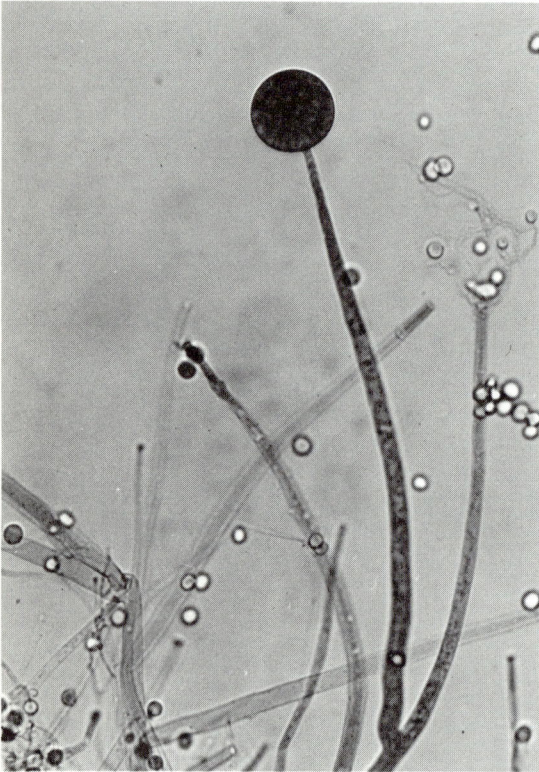

Fig. 4.2 *Mortierella* sp. Sporangium without columella, ×250.

Syncephalastraceae

Spores are produced in a single linear row in a tubular sporangium (which is called a merosporangium). The merosporangia are produced perpendicular to vesicular swellings at the apex of the sporangiophore, creating a radiate pattern (see Fig. 4.3 and refer to Benjamin, 1959). The one genus is saprophytic.

Thamnidiaceae

Members of this family always bear sporangiola, which are small sporangia with one to several spores. The sporangiola are usually borne in clusters well below a large *Mucor*-like sporangium or sterile point (Fig. 4.4). Benny and Benjamin (1975, 1976) have studied the group.

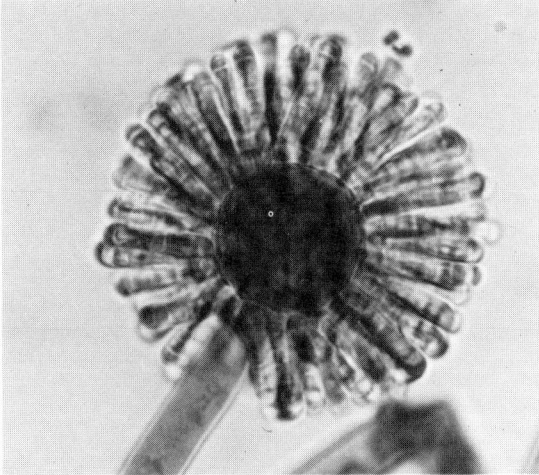

Fig. 4.3 *Syncephalastrum racemosum*   Columnar sporangium produced on vesicular swelling of sporangiophore, ×650.

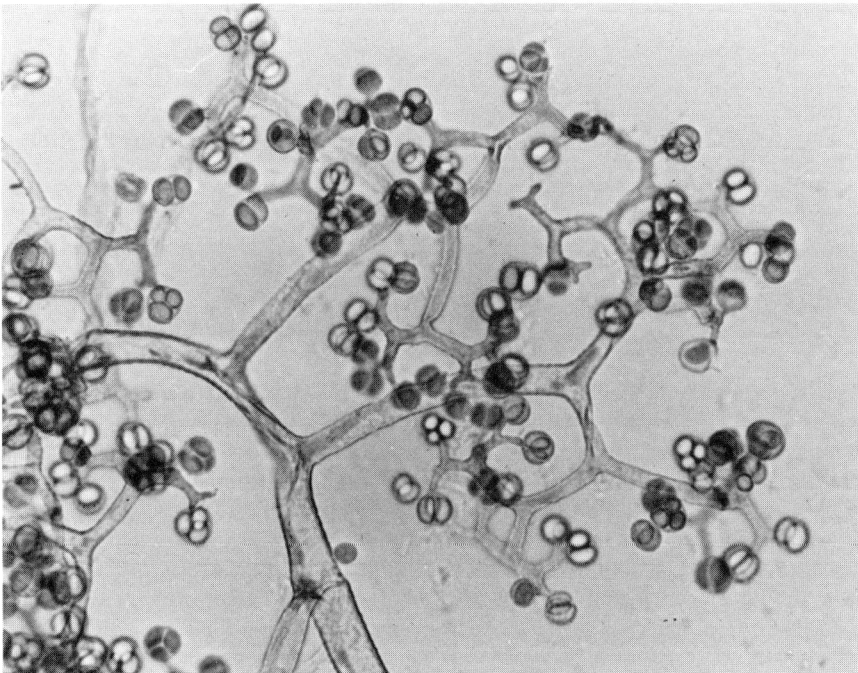

Fig. 4.4 *Thamnidium elegans*   Cluster of sporangiola, ×400.

A key to the common genera of Mucorales is given below, followed by descriptions of the genera.

Key* to the common genera of Mucorales

1.	Sporangia tubular, radiating from a vesicular swelling	*Syncephalastrum*
1'.	Sporangia globose or nearly so	2
2.	Many-spored sporangia and few-spored sporangioles both present	*Thamnidium*
2'.	Sporangia alone produced...	3
3.	Sporangiophores stiff, dark-coloured, metallic in appearance ..	*Phycomyces*
3'.	Sporangiophores otherwise	4
4.	Rhizoids and stolons present	5
4'.	Rhizoids and stolons absent	6
5.	Sporangia large, globose; sporangiophores arising from points of attachment of rhizoids..	*Rhizopus*
5'.	Sporangia small, pear-shaped; sporangiophores mainly as branches from the stolons	*Absidia*
6.	Homothallic; zygospores with very unequal suspensors	*Zygorhynchus*
6'.	Homo- or heterothallic; zygospores, when present, with approximately equal suspensors	*Mucor*

---

* In the dichotomous system, used for this key and others later in the book (also used by many other authors), all the essential data required for identifications are arranged as pairs of contrasted characters, the pairs being numbered consecutively on the left. Each member of a pair leads, on the right of the page, either to the name of a genus or to another higher number, i.e. to a further pair of contrasted features.

4.5   *Mucor* Micheli ex Fries
(Lat. *mucor*, a fungus) (Figs 4.5–4.9)

This is one of the largest genera of the order and includes a number of species which are of widespread occurrence and considerable importance. The criteria on which identifications are based are the mode of branching, if any, and the size of the sporangiophores; the structure of the sporangial wall; the size and shape of the columellae; the size and shape of the spores; the characteristics of the zygospores and chlamydospores, if present; and the general colony characteristics, such as colour and height of aerial growth and how they vary with temperature. About 150 species are listed in Saccardo's *Sylloge,* but many of these are inadequately described and a considerable number of the names are synonyms of common species. They have been discussed in the various works on the group and separation is still difficult. Schipper has carried out extensive studies of closely related species and has reduced many to synonymy. She stresses the importance of zygospore formation. Zygospores are formed between equal suspensors and are characteristically spiny or warted.

M. mucedo Linn. ex Fries (see Schipper, 1975) (Lat. *mucedo*, mucus). This is the type species of the genus. It occurs mainly on dung, and can usually be isolated, when required, if some horse-dung is incubated for a few days in a moist atmosphere. On its natural substrate it forms erect sporangiophores which may reach a height of 15 cm, with greyish sporangia, the wall of which is encrusted with crystals reported to be of calcium oxalate. In culture the sporangiophores are usually shorter, and bear at the base a number of short branches terminating in small sporangia. The terminal sporangia are 100–200 μm diam., with walls which disintegrate completely in fluid mounts; columellae pear-shaped to cylindrical, with orange-coloured contents; spores ellipsoid, 6–12 μm long or subspherical, 8–9 μm diam. There is no growth at 37°C.

M. piriformis Fischer (Lat. *piriformis*, pear-shaped) (see Schipper, 1975). Schipper treats this in the same group as *M. mucedo*. It is fairly common species particularly associated with rotting fruits. Colonies dense, cottony, 2–3 cm high; sporangiophores mostly unbranched, up to 50 μm diam.; sporangia globose and very large, 250–300 μm diam., with diffluent walls; columellae distinctly pyriform, pear-shaped (giving the name to the species), about 200 μm long; spores ellipsoid, 5–13 (Schipper gives 7–9.5) $\times$ 4–8 μm. No growth at 30°C, optimal growth at 10–15°C.

M. racemosus Fresenius (see Schipper, 1976; Lunn, 1977a) (Lat. *racemosus*, branched). (Figs 4.5, 4.6, 4.7. See also Fig. 4.1.) This is probably the most widely distributed of all the species of *Mucor*, and has frequently been

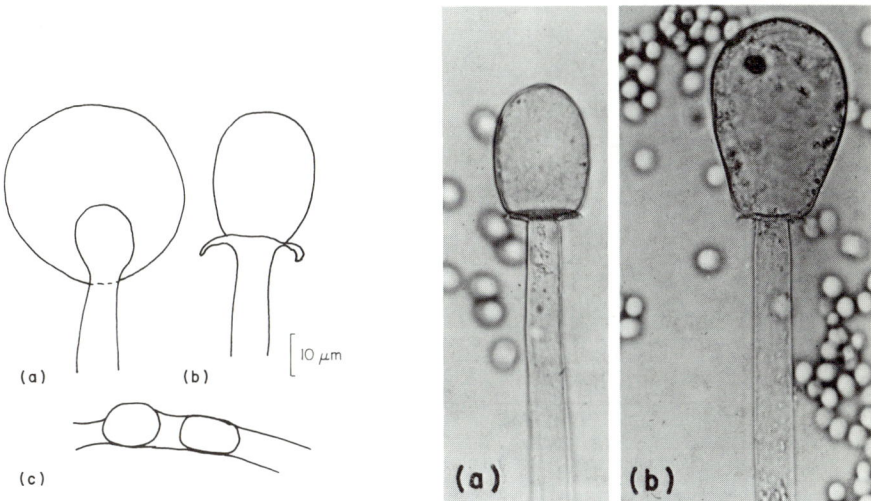

Fig. 4.5 (left)   *Mucor racemosus*   **(a)** Columella projecting into the sporangium; **(b)** columella and collarette; **(c)** chlamydospores.

Fig. 4.6 (right)   *Mucor racemosus*   **(a)** Oval columella with collarette, ×500; **(b)** pyriform columella with small collarette, ×500.

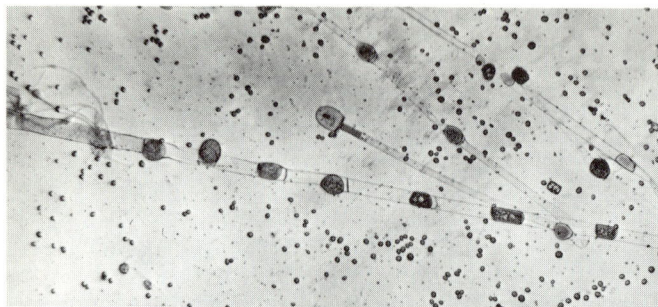

Fig. 4.7   *Mucor racemosus*   Chlamydospores in sporangiophore, ×100.

described under other names. It is found on almost every kind of damp material and particularly causing storage rot of fruit and vegetables. Colonies are grey or brownish grey, of loose texture and normally less than 1 cm high; sporangiophores simple at first, later becoming sympodially branched and very unequal in length; sporangia globose, very unequal in size but mostly small, 20–70 μm diam., with encrusted walls which break in pieces at maturity; columellae variable, globose or obovoid often with truncate base and collarette (a portion of the broken sporangial wall left *in situ*) (Fig. 4.6b); spores mostly ellipsoid, 6–10 × 5–8 μm. The most characteristic feature of the species is the production of abundant chlamydospores, which are formed in the sporangiophores, and occasionally in the columellae. They are of diverse shapes, colourless or yellow, smooth, and about 20 μm diam. (Fig. 4.7). Heterothallic but zygospores are rare in nature. No growth at 37°C. This species can grow submerged in liquid media and, like some yeasts, produces alcohol.

M. plumbeus Bonorden (Lat. *plumbeus,* leaden) (see Schipper, 1976). (Figs 4.8, 4.9.) Schipper groups this with *M. racemosus.* Two synonyms are often met with in the literature, *M. spinosus* van Tieghem and *M. spinescens* Lendner. Colonies at first white, then dull grey, and finally brownish grey, most frequently only a few millimetres high but varying considerably in different isolates; sporangiophores sympodially branched, about 1 mm long and 10 μm diam.; sporangia up to 80 μm diam., the wall encrusted and appearing spiny (Fig. 4.9a), rupturing at maturity; columellae obovoid or pyriform, with projections at the top (Fig. 4.9b); spores globose, 7–8 μm diam., occasionally smaller, verruculose. Heterothallic, no growth at 37°C. Of very common occurrence.

M. circinelloides van Tieghem (see Schipper, 1976). *M. circinelloides* with *M. hiemalis* is perhaps one of the commonest species of *Mucor* encountered. Colonies light yellowish brown, about 1 cm high and at 20°C in darkness produces two layers of growth. Sporangiophores sympodially branched, frequently with curved ends or occasionally monpodial; sporangia have slightly encrusted walls, larger ones deliquescent, smaller ones persistent and

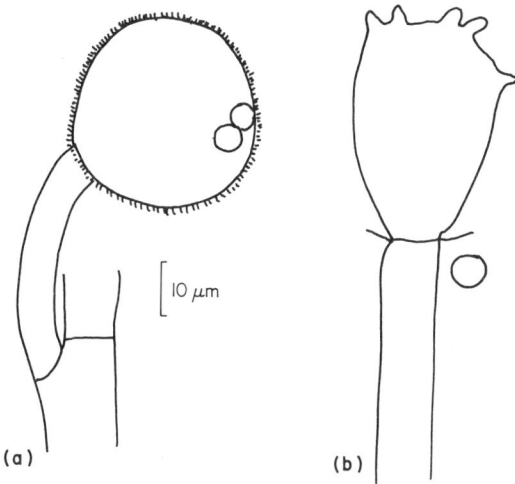

Fig. 4.8 *Mucor plumbeus* **(a)** Young sporangia showing spinescent walls; **(b)** columella with terminal spines.

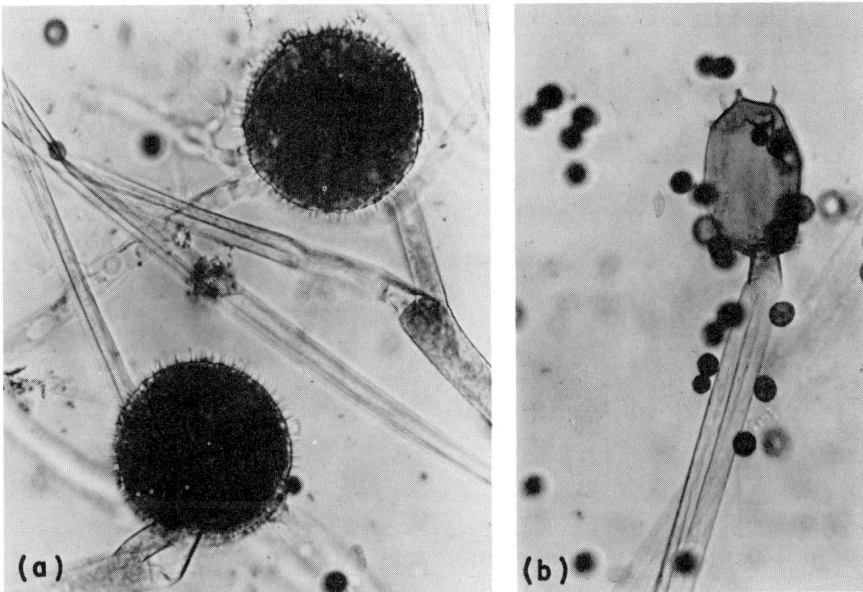

Fig. 4.9 *Mucor plumbeus* **(a)** Young encrusted sporangium, ×500; **(b)** columella with terminal spines, ×500.

rupturing at maturity; columellae ellipsoid, with a truncate base, to pyriform; spores broadly ellipsoid, 4.4–6.8 × 3.7–4.7 μm, smooth; numerous chlamydospores produced. Zygospores rarely produced in nature. Very poor growth at 37°C but good growth from 5°C to 10°C.

M. hiemalis Wehmer (Lat. *hiemalis*, pertaining to winter) (see Schipper, 1973). A common soil fungus and hence found on numerous soil contaminated products. Colonies yellowish or grey, 1–2 cm high; sporangiophores simple or sympodially branched; sporangia yellowish to greyish brown, 50–80 μm diam., with diffluent walls; columellae globose to obovoid, with a small colarette, up to 50 μm long; spores irregular in shape, but mostly ellipsoid in a mixture of large and small forms; chlamydospores formed in the mycelium but not so abundantly as in *M. racemosus*. Heterothallic. No growth at 30°C.

M. miehei Cooney & Emerson and *M. pusillus* Lindt (see Lunn, 1977b, c; recently transferred to the genus *Rhizomucor* by Schipper, 1976). Recently some mucoraceous fungi have been used to coagulate milk or to produce enzymes to coagulate milk. Frequently used are *Rhizopus* species and *Chlamydomucor oryzae* (Ellis *et al.*, 1974). However, two thermophilic *Mucor* species, *M. pusillus* and *M. miehei*, have also been used and are quoted in several 'Patent' processes. These species are thermophilic with poorly developed stolons and rhizoids and with dark coloured echinulate sporangia without apophyses. They both produce grey colonies. Sporangiospores are globose to subglobose, 3–5 μm. Columellae after liberating their spores sometimes invert to become umbrella shaped. *M. pusillus* is usually heterothallic and *M. miehei* is homothallic. Both have been reported as pathogens to man and animals, *M. pusillus* to a greater extent. Both cause storage rots and other spoilage problems.

4.6 *Mortierella* Coemans

Members of the genus *Mortierella*, although common soil organisms, are seldom found in the industrial environment. They have been discussed by Linneman (1941), Zycha *et al.* (1969). However, they are still a difficult group and Gams (1977) has produced a key to the species.

M. ramanniana (Möller) Linnemann (see Evans, 1971) (Figs 4.10, 4.11). This species is fairly common in soil and macroscopically looks very like *Mucor*. Colonies at first pinkish to brownish red, turning grey, only about 1 mm high and almost velvety in appearance; sporangiophores mostly unbranched, 2–6 μm diam.; sporangia reddish, small, 20–40 μm diam., with diffluent walls; columellae small, globose, 5–10 μm diam.; spores globose to short ellipsoid, 2–3 μm long. The species is transitional between *Mucor* and *Mortierella*, was described orginally as *Mucor* but transferred to *Mortierella* by Linnemann and with several other species forms a distinct group which was monographed by Turner (1963). Gams (1977) has proposed a subgenus of *Mortierella*, *Micromucor*, for these species with velvety, red or ochraceus colonies and very reduced or absent columellae.

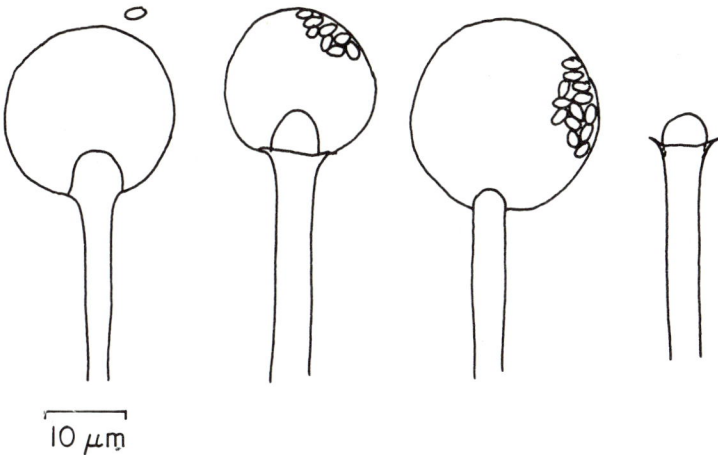

10 μm

Fig. 4.10 *Mortierella ramanniana* Sporangia and sporangiophores showing very small columella.

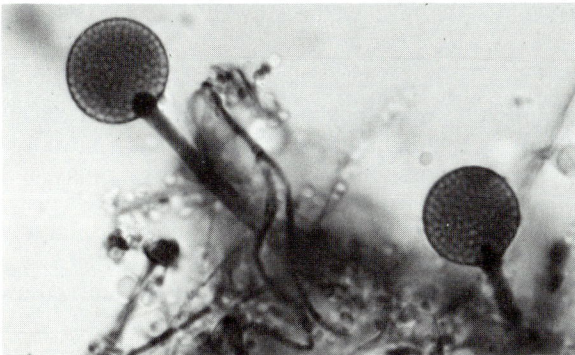

Fig. 4.11 *Mortierella ramanniana* Sporangia, ×500.

4.7 *Zygorhynchus* Vuillemin
(Gr. *zygon*, a yoke; *rhynchos*, a snout) (Figs 4.12, 4.13)

The genus is closely related to *Mucor*. It forms many spored sporangia of similar structure to those of *Mucor*. It differs in that all the species are homothallic, and the suspensors of the zygospores are very markedly unequal in size. The genus has been recognized by all students of the Mucorales except Lendner, who included the species in *Mucor*. Many species have been described. Hesseltine *et al.* (1959), in their monograph of the genus, conlcude that there were six species and one variety and Zycha *et al.* (1969) recognized only six species, the remaining names being reduced to synonymy. All the species are typically soil inhabitants, being found only rarely on other substrata. Two species are isolated fairly frequently, *Z. moelleri* being the more common of the two.

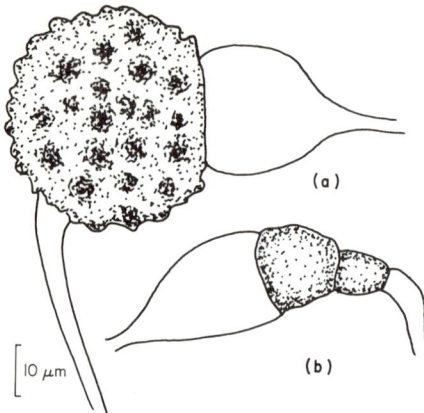

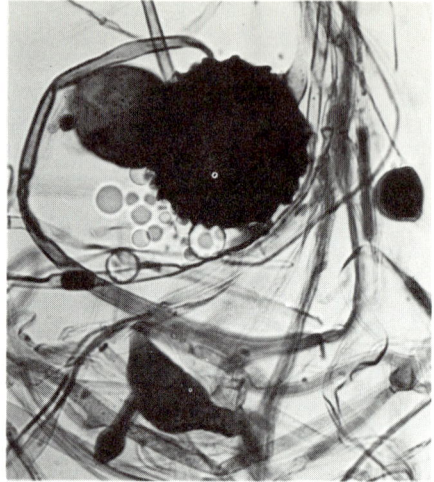

Fig. 4.12 (left) *Zygorhynchus moelleri* **(a)** Mature zygospore showing uneven suspensors; **(b)** developing zygospore.

Fig. 4.13 (right) *Zygorhynchus moelleri* Zygospore, ×500.

Z. moelleri Vuillemin (A. Möller, German mycologist). Grows well on all media, forming a loose felt only a few millimetres high; abundant zygospores formed in two to three days, with sporangia produced more sparingly and usually later; sporangiophores simple or irregularly branched; sporangia grey, usually slightly broader than long, mostly 48–50 μm; columellae definitely obovoid but broader than high, 25–30 μm high; spores ellipsoid, 4–5 × 3 μm; zygospores formed on bifurcated hyphae, spherical, 35–50 μm diam., dark brown to black and covered with short spines.

Z. heterogamus Vuillemin (Gr. *hetero-*, uneven; *gamos,* marriage, union). Somewhat similar to *Z. moelleri* but with larger sporangia and zygospores. Sporangiophores branched, often more or less verticillate with two to four branches; sporangia globose, 50–60 μm diam.; columellae globose; spores globose, 2–3 μm diam.; zygospores black, globose, warted, varying in size from 45–150 μm diam.; chlamydospores formed in the mycelium.

4.8 *Rhizopus* Ehrenberg ex Corda
(Gr. *rhiza*, a root; *pous,* foot) (Figs 4.14, 4.15)

Species of this genus occur on all kinds of material and are common as aerial contaminants in the laboratory. On most culture media they grow with extreme rapidity, spreading widely by means of their stolons. They completely fill culture tubes and Petri dishes with dense cottony masses of mycelium, and can be a great nuisance owing to their habit of sporing along the line where the cover touches the edge of the dish, thus shedding spores outside.

Rhizopus is readily distinguished from *Mucor* by the presence of stolons (runners), often several centimetres long, and of tufts of rhizoids (root-like hyphae) emerging from the points where the stolons touch the medium or surface of the glass. In addition, the apex of the sporangiophore is funnel-shaped (see Fig. 4.14). This type of structure is known as an **apophysis** (see also *Absidia*, p. 41). Zycha *et al.* (1969) accepted ten species. As species tend to merge separation is not as easy as first appears and such characters as relative growth at different temperatures are used. Inui *et al.* (1965) undertook a very intensive study of the genus. They investigated and compared the morphological, chemical and physiological characters of many isolates and accepted 14 species, arranged in three sections.

These fungi are of particular economic importance to the Japanese fermentation industry as they are used in food fermentations as well as in the Amylo process (Calmette, 1892 and later papers) for the production of alcohol. They are also important spoilage organisms. Only a few will be mentioned here.

R. stolonifer (Ehrenberg ex Fr.) Lindner (Lat. *stolo,* stolon; *fero,* to bear) (see Lunn, 1977h) (in early editions of this book cited as *R. nigricans*). A species of world-wide distribution and found on all kinds of mouldy material. It is frequently the first mould to appear on stale bread, and is often found on other foodstuffs. Colonies on most media spread very rapidly, completely filling tubes and dishes in a few days; stolons clearly differentiated, arising from and terminating in strong tufts of brown rhizoids (Fig. 4.15a); sporangiophores erect, arising in groups opposite the rhizoids, up to 2.5 mm long and about 20 μm diam.; sporangia globose, shining white at first, then turning black as the spores mature, up to 200 μm diam.; spores variously shaped, ellipsoid, polygonal or angular, striate, mostly 10–15 μm in long axis.

Although growing so luxuriously on most substrata, this species is unable to utilize nitrates and therefore, will not grow on Czapek's solution (see Chapter 13). It grows on Czapek agar but only very sparsely, presumably utilizing some impurity in the agar as a source of nitrogen.

R. oryzae Went & Prinsen-Geerligs (Lat. *oryzae,* of rice) (see Lunn, 1977f). Distinguished from *R. stolonifer* by its somewhat smaller spores (7–9 μm) and by its ability to grow at 37–40°C. During the second world war it was used by British prisoners in Java to make soya beans digestible, so as to eke out the meagre diet allowed.

R. arrhizus Fischer (Gr. *a-,* lacking; *rhiza,* root). Grows much less rampantly than the two preceding species; rhizoids short, pale and ragged; sporangiophores frequently arising from hyphal swellings not provided with rhizoids; sporangia globose, 100–200 μm diam.; spores 5–7 μm long. Grows well at 37°C.

R. sexualis (Smith) Callen (see Lunn, 1977g). The other species of this genus are mostly heterothallic; this species is homothallic and consistently produces zygospores in culture. No growth at 37°C.

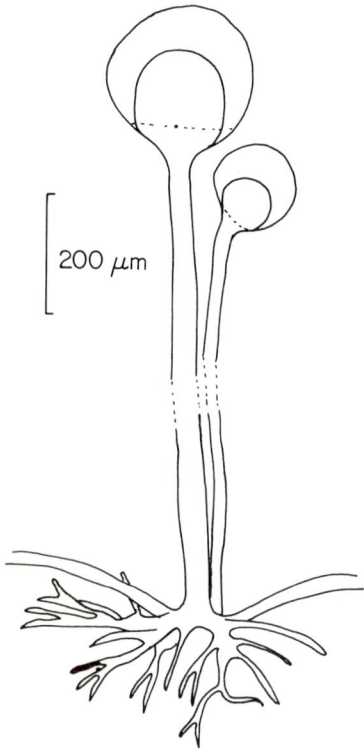

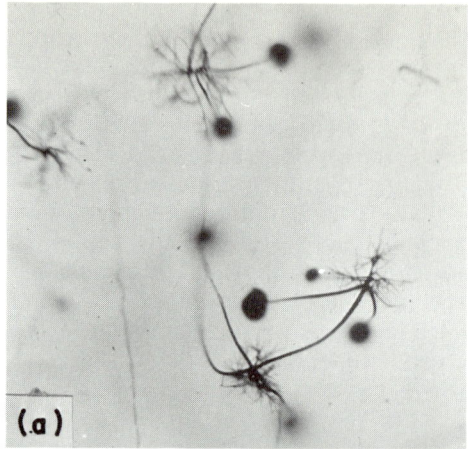

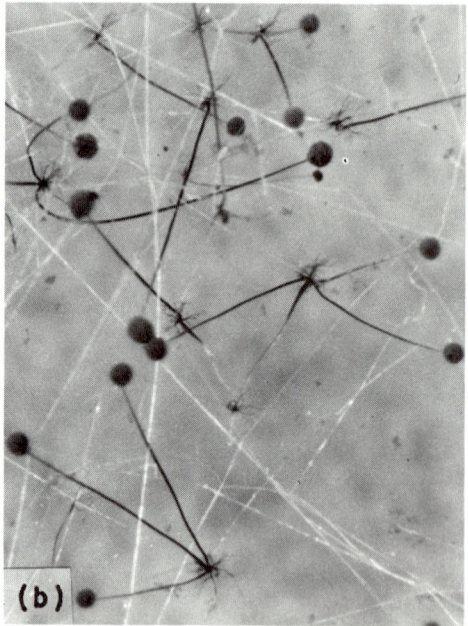

Fig. 4.14 (left)   *Rhizopus stolonifer*   Rhizoids and sporangiophore with funnel shaped apex.

Fig. 4.15 (right)   *Rhizopus stolonifer*   **(a)** Rhizoids and stolons (Petri dish culture); **(b)** tufts of sporangia (slide culture), ×16.

R. microsporus van Tieghem (see Lunn, 1977e). This species is characterized by its small, mainly limoniform spores 3–5 μm in length.

Rhizopus, Amylomyces rouxii Calmette and Fermentation Processes. In Indonesia and elsewhere in the orient fungi are often used in fermentations and to make fermented foods and drinks. For this purpose starters of concentrated fungal mycelium are prepared, referred to by various names such as Koji, Tempeh, Ragi and Chinese yeast. Hesseltine and his co-workers have made quite intensive studies of these fermentations, very full accounts being given by Hesseltine *et al.* (1976) and Batra and Millner (1976). Isolations made from the starters have shown that some of the species involved are members of the Mucorales. *Rhizopus oligosporus* is used in the production of Tempeh, an Indonesian soyabean food. *Rhizopus oryzae* is also used in a similar way. Ellis *et al.* (1976) discuss the production of Ragi and Chinese yeast which they found to be due to *Amylomyces rouxii* Calmette. This species is somewhat similar to *Rhizopus oryzae* (see Lunn, 1977f) but Ellis and his co-workers showed it to be distinct. As these fermentations have been mostly cottage industries (at least in the past) the organisms have tended to vary or be replaced, resulting in confusion of the identity of the organisms involved.

4.9 *Absidia* van Tieghem
(Gr. *apsis,* a loop; refers to the arched stolons) (Figs 4.16, 4.17)

This genus differs in several respects from *Rhizopus*. The rhizoids and stolons are not so clearly differentiated; the sporangiophores arise from the stolons and not from the points of attachment of the rhizoids; the sporangia are relatively small and pear-shaped; and, most characteristic of all, there is a well marked 'apophysis', i.e. a funnel shaped apex to the sporangiophores (Fig. 4.17a). The zygospores, of homothallic species, are surrounded by appendages originating from one or both suspensors (Fig. 4.17b). Zycha *et al.* (1969) described 21 species, whilst Naumov followed some earlier authors in splitting the genus into several smaller genera, with a total of 29 species and two varieties. Hesseltine and Ellis (1964, 1966) and Ellis and Hesseltine (1965, 1966) have published an extensive study of the genus.

   Some of the species are of common occurrence in soil and sometimes appear in mixed cultures from mouldy material, or as aerial contaminants. A number are thermophilic, with optimum temperature for growth close to 37°C, and some of these have been many times reported as pathogenic to various animals, including man. They can be a great nuisance in the bacteriological laboratory, for a single colony can almost cover a Petri dish in 24 hours at 37°C, and they thrive, better than most other moulds, on the usual bacteriological media.

A. corymbifera (Cohn) Saccardo & Trotter (Lat. *corymbus,* a cluster of flowers; *fero*, to bear) (see Lunn, 1977a). Synonyms: *A. lichtheimii* (Lucet & Costantin) Lendner and *A. ramosa* (Lindt) Lendner. Colonies at first white, then pale grey, up to 2 cm high; sporangiophores branched, often in whorls, rhizoids very sparingly produced; sporangia small and pyriform, columellae almost conical, with a well marked apophysis and often a short projection at the top; spores ellipsoid, 2–3.5 × 3–4.5 μm or variable; thermophilic and pathogenic.

20 μm

(a) (b)

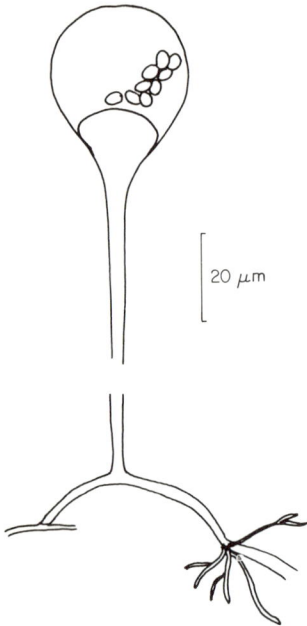

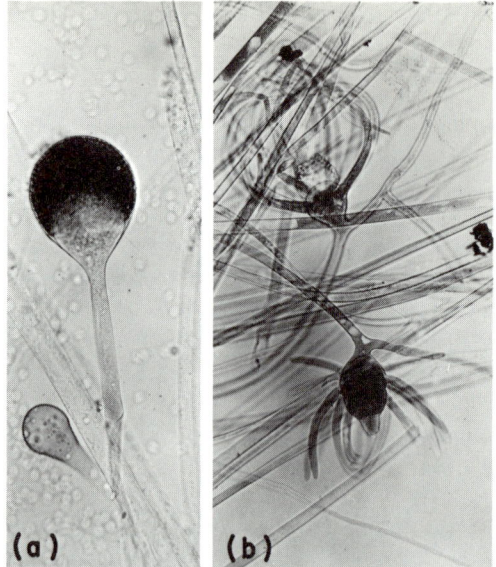

Fig. 4.16 (left) *Absidia corymbifera* Showing rhizoids, sporangiophore arising from stolon and funnel shaped apex to sporangiophore.

Fig. 4.17 (right) (a) *Absidia ramosa* Sporangium showing apophysis, ×250. (b) *Absidia spinosa* Formation of zygospores, showing outgrowths from one suspensor, ×250.

A. spinosa Lendner (Lat. *spinosus,* spiny). Colonies densely floccose. White then brownish grey; sporangiophores in clusters of 2–4, unbranched; columellae hemispherical, with one or more projections; a septum usually present in the sporangiophore, a little distance below the apophysis; spores cylindrical with slightly rounded ends, 4–5 μm long; zygospores abundant, 50–80 μm diam., with outgrowths from only one suspensor (Fig. 4.17b).

4.10 *Phycomyces* Kunze ex Fries
(Gr. *phykos,* seaweed; hence alga-like fungus) (see Benjamin and Hesseltine, 1959)

Species of this genus are readily recognized by their very characteristic sporangiophores, very stiff, long and with a metallic sheen, looking like oxidized steel wire. Several species are described but only two are of any importance.

P. nitens (Agardh) Kunze ex Fries (Lat. *nitens,* becoming shiny). Sporangiophores up to about 20 cm high; sporangia black when ripe, about 500 μm diam.; columellae cylindrical, constricted in the middle; spores ellipsoid, averaging 25 ×11 μm. Found on dung, in soil, and occasionally in empty oil barrels.

P. blakesleeanus Burgeff (A. F. Blakeslee, American mycologist). This species is distinguished from *P. nitens* by its larger sporangia, up to 1000 μm diam., and by its smaller spores, about 12 × 8 μm. It grows well on wort agar and some media made from plant extracts, but does not thrive on synthetic media, since it needs to be supplied with thiamin. Its sporangiophores are phototrophic.

4.11 *Thamnidium* Link ex Wallroth
(Gr. *thamnos*, a bush; *eidos*, like) (Fig. 4.18)

This differs from the preceding genera in that the main sporangiophores bear lateral clusters of sporangiola as well as large terminal sporangia. The sporangiola resemble miniature sporangia, containing from two to a dozen or more spores, and are found on highly branched outgrowths from near the base of the sporangiophore.

T. elegans Link ex Fries (Lat. *elegans,* elegant). This is by no means a common mould. It has been reported most frequently as infecting meat in cold storage and has been used to tenderize meat. It grows well on most culture media, cream coloured to greyish, granular. The sporing apparatus tends to vary according to temperature. At lower temperatures long sporangiophores are produced bearing terminal sporangia, at high temperatures the growth is low and dense clusters of sporangiola produced on dichotomously branched sporangiophores, the main sporangiophore often failing to produce a sporangium. At medium temperatures both sporangia and sporangiola are produced. Sporangia brown, 150–250 μm diam.; spores ellipsoid, 8–12 μm long. Hesseltine and Anderson (1956) have reported the formation of zygospores when certain isolates were mated (the species is heterothallic). The zygospores are produced only on particular media and only at low temperatures (6–7°C).

4.12 *Syncephalastrum* Schröter
(Gr. *syn*, completely; *kephale*, head; *aster*, a star) (see p. 31) (Figs 4.19 and 4.20)

When mounted and viewed at a low magnification the heads bear a striking resemblance to heads of *Aspergillus*. At a high magnification the chains of spores are seen to be enclosed in a merosporangial membrane and, if the development of the merosporangium is followed, it will be observed that the spores are formed in a similar manner to those of a typical sporangium. Figure 4.20 shows clearly that the spores are formed within, and distinct from, the merosporangial membrane.

Fig. 4.18 *Thamnidium elegans*   **(a)** Colony margin showing terminal sporangia and clusters of sporangiola, ×20; **(b)** cluster of sporangiola, ×200.

S. racemosum (Cohn) Schröter (Lat. *racemosus*, branched). A species of widespread distribution, but mainly of tropical or sub-tropical origin. It grows luxuriantly on laboratory media, the colonies resembling in texture those of *Rhizopus stolonifer*, with black heads produced on short branches from aerial hyphae. The merosporangia contain 5–10 spores, of somewhat irregular shape, 2.5–5 μm diam.

Fig. 4.19 *Syncephalastrum racemosum* Vesiculate sporangium and tubular sporangia.

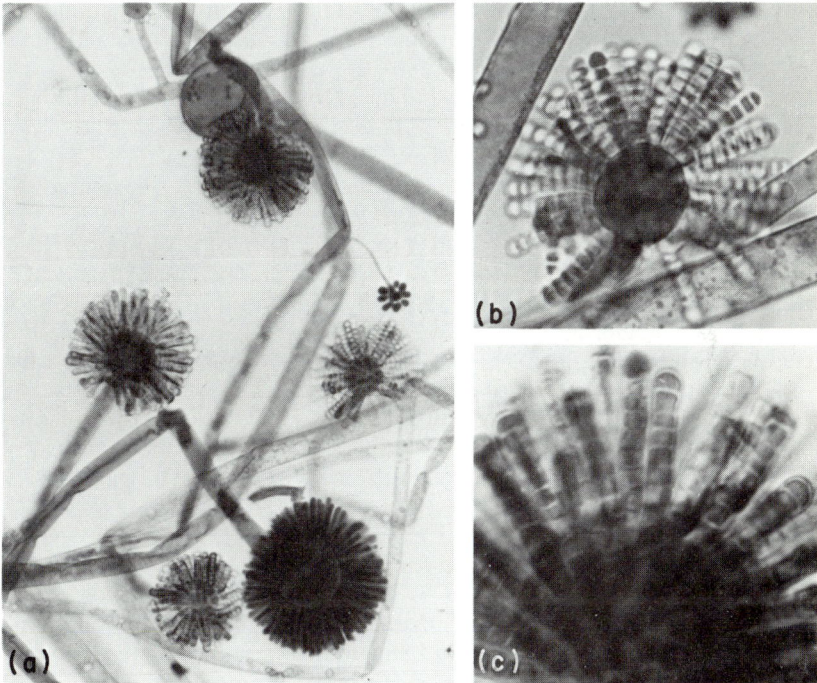

Fig. 4.20 *Syncephalastrum racemosum* **(a)** Spore heads, ×212, (note resemblance to *Aspergillus*); **(b)** enlarged heads to show tubular sporangia, ×470; **(c)** enlarged head, ×1100.

4.13 References

AINSWORTH, G. C. (1973a). Introduction and keys to higher taxa. In *The Fungi*, Vol. 4A. AINSWORTH, G. C., SPARROW, F. K. and SUSSMAN, A. S. (Eds). pp. 1–7. Academic Press, New York.

AINSWORTH, G. C. (1973b). Fungal nomenclature. *Rev. Plant Path.*, **52**, 59–68.

BATRA, L. R. and MILLNER, P. D. (1976). Asian fermented foods and beverages. *Developments in Industrial Microbiology*, **17**, 117–28.

BENJAMIN, C. R. and HESSELTINE, C. W. (1957). The genus *Actinomucor*. *Mycologia*, **49**, 240–9.

BENJAMIN, C. R. and HESSELTINE, C. W. (1959). Studies on the genus *Phycomyces*. *Mycologia*, **51**, 751–71.

BENJAMIN, R. K. (1959). The merosporangiferous Mucorales. *Aliso*, **4**, 321–433.

BENJAMIN, R. K. (1979). Zygomycetes and their spores. In *The Whole Fungus*, Vol. 2. KENDRICK, B. (Ed.) pp. 573–621. National Museums of Canada for the Kananaskis Foundation.

BENNY, C. L. and BENJAMIN, R. K. (1975). Observations on Thamnidiaceae (Mucorales). New taxa, new combinations, and notes on selected species. *Aliso*, **8**, 301–51.

BENNY, C. L. and BENJAMIN, R. K. (1976). Observations on Thamnidiaceae (Mucorales). II *Chaetocladium, Cokeromyces, Mycotypha* and *Phascolomyces*. *Aliso*, **8**, 391–424.

CALMETTE, A. (1892). Contribution à l'étude des ferments de l'amidon; la levûre chinoise. *Annls Inst. Pasteur, Paris*, **6**, 604–20.

CHIEN, C.-Y., KUHLMAN, E. G. and GAMS, W. (1974). Zygospores in two *Mortierella* species with 'stylospores'. *Mycologia*, **66**, 114–21.

ELLIS, J. J. and HESSELTINE, C. W. (1965). The genus *Absidia:* globose spored species. *Mycologia*, **57**, 222–35.

ELLIS, J. J. and HESSELTINE, C. W. (1966). Species of *Absidia* with ovoid sporangiospores. II. *Sabouraudia*, **5**, 59–77.

ELLIS, J. J. and HESSELTINE, C. W. (1974). Two families of Mucorales. *Mycologia*, **66**, 87–95.

ELLIS, J. J,. RHODES, L. J. and HESSELTINE, C. W. (1976). The genus *Amylomyces*. *Mycologia*, **68**, 131–43.

ELLIS, J. J., WANG, H. L. and HESSELTINE, C. W. (1974). *Rhizopus* and *Chlamydomucor* strains surveyed for milk clotting, amylolytic and antibiotic activities. *Mycologia*, **66**, 593–9.

EVANS, E. H. (1971). Studies on *Mortierella ramanniana*. I. Relationship between morphology and cultural behaviour of certain isolates. *Trans. Br. mycol. Soc.*, **56**, 201–16.

GAMS, W. (1969). Gliederungs-grinzipien in der Gattung *Mortierella*. *Nova Hedwigia*, **18**, 30–43.

GAMS, W. (1976). Some new or noteworthy species of *Mortierella*. *Persoonia*, **9**, 111–40.

GAMS, W. (1977). A key to the species of *Mortierella*. *Persoonia*, **9**, 381–91.

GAMS, W., CHIEN, C.-Y. and DOMSCH, K. H. (1972). Zygospore formation by the heterothallic *Mortierella elongata* and a related species, *M. epigamia* sp. nov. *Trans. Br. mycol. Soc.*, **58**, 5–13.

HAGEM, O. (1910). Neue Untersuchungen über norwegische Mucorineen. *Annls mycol.*, **8**, 265–86.

HANLIN, R. T. (1973). *Keys to the families, genera and species of the Mucorales.* Translated from the German. J. Cramer, Lehre.

HESSELTINE, C. W. (1952). A survey of the Mucorales. *Trans. N. Y. Acad. Sci.*, Ser. II, **14**, 210–14.

HESSELTINE, C. W. (1953). A revision of the Choanephoraceae. *Amer. Midl. Nat.*, **50**, 248–56.

HESSELTINE, C. W. (1954). The section Genevensis of the genus *Mucor*. *Mycologia*, **46**, 358–66.

HESSELTINE, C. W. (1955). Genera of Mucorales, with notes on their synonymy. *Mycologia*, **47**, 344–63.

HESSELTINE, C. W. (1957). The genus *Syzygites* (Mucoraceae). *Lloydia*, **20**, 228–37.

HESSELTINE, C. W. (1960). *Gilbertella* gen. nov. (Mucorales). *Bull. Torrey Bot. Club*, **87**, 21–30.

HESSELTINE, C. W. and ANDERSON P. (1956). The genus *Thamnidium* and a study of the formation of its zygospores. *Amer. J. Bot.*, **43**, 696–703.

HESSELTINE, C. W. and ANDERSON P. (1957). Two genera of moulds with low temperature growth requirements. *Bull. Torrey Bot. Club*, **84**, 31–45.

HESSELTINE, C. W. and BENJAMIN, C. R. (1957). Notes on the Choanephoraceae. *Mycologia*, **49**, 723–33.

HESSELTINE, C. W., BENJAMIN, C. R. and MEHROTRA, B. S. (1959). The genus *Zygorhynchus*. *Mycologia*, **51**, 173–94.

HESSELTINE, C.W. and ELLIS, J. J. (1961). Notes on Mucorales, especially *Absidia*. *Mycologia*, **53**, 406–26.

HESSELTINE, C. W. and ELLIS, J. J. (1964). The genus *Absidia: Gongronella* and cylindrical-spored species of *Absidia*. *Mycologia*, **56**, 568–601.

HESSELTINE, C. W. and ELLIS, J. J. (1966). Species of *Absidia* with ovoid sporangiospores. I. *Mycologia*, **58**, 761–85.

HESSELTINE, C. W. and ELLIS, J. J. (1973). Mucorales. In *The Fungi*, Vol. 4B. AINSWORTH, G. S., SPARROW, F. K. and SUSSMAN, A. S. (Eds) pp. 187–217. Academic Press, New York and London.

HESSELTINE, C. W. and FENNELL, D. I. (1955). The genus *Circinella*. *Mycologia*, **47**, 193–212.

HESSELTINE, C. W., SWAIN, E. W. and WANG, H. L. (1976). Production of fungal spores as inocula for oriental fermented foods. *Developments in Industrial Microbiology*, **17**, 101–15.

INGOLD, C. T. (1978). *The biology of* Mucor *and its allies.* Studies in Biology No. 88. Edward Arnold, London.

INGOLD, C. T. and ZOBERI, M. H. (1963). The asexual apparatus of Mucorales in relation to spore liberation. *Trans. Br. mycol. Soc.*, **46**, 115–34.

INUI, T., TAKEDA, Y. and IIZUKA, H. (1965). Taxonomical studies on genus *Rhizopus*. *J. Gen. Appl. Microbiol.*, Suppl. **11**, 1–181.

LENDNER, A. (1908). Les Mucorinées de la Suisse. *Materiaux pour la flore cryptogamique Suisse*, **3**, 1–180.

LINNEMANN, G. (1941). Die Mucorineen-Gattung *Mortierella* Coemans. *Pflanzenforschung* **23**. Berlin-Dahlem.

48   Zygomycetes

LUNN, J. A. (1977a). *Absidia corymbifera. CMI Descriptions of Pathogenic Fungi and Bacteria*, No. 521.
LUNN, J. A. (1977b). *Mucor miehei. CMI Descriptions of Pathogenic Fungi and Bacteria*, No. 528.
LUNN, J. A. (1977c). *Mucor pusillus. CMI Descriptions of Pathogenic Fungi and Bacteria*, No. 527.
LUNN, J. A. (1977d). *Mucor racemosus. CMI Descriptions of Pathogenic Fungi and Bacteria*, No. 529.
LUNN, J. A. (1977e). *Rhizopus microsporus. CMI Descriptions of Pathogenic Fungi and Bacteria*, No. 523.
LUNN, J. A. (1977f). *Rhizopus oryzae. CMI Descriptions of Pathogenic Fungi and Bacteria*, No. 525.
LUNN, J. A. (1977g). *Rhizopus sexualis. CMI Descriptions of Pathogenic Fungi and Bacteria*, No. 526.
LUNN, J. A. (1977h). *Rhizopus stolonifer. CMI Descriptions of Pathogenic Fungi and Bacteria*, No. 524.
MILKO, A. A. (1974). *Opredelitel Mukaral'nykh Gribov (Key to the Mucorales)*. Kiev.
NAUMOV, N. A. (1939). *Clés des Mucorinées (Mucorales)*. Translated from 2nd Russian edition, with additional notes by the author, by BUCHET, S. and MOURAVIEV, I. Paul Lechevalier, Paris.
NOTTEBROCK, H., SCHOLER, H. J. and WALL, M. (1974). Taxonomy and identification of Mucormycoses-causing fungi. I. Synonymity of *Absidia ramosa* with *A. corymbifera*. *Sabouraudia*, **12**, 64–74.
PIDOPLICKO, N. M. and MILKO, A. A. (1971). *Atlas mucoral'nykh gribov*. Kiev.
POVAH, A. H. W. (1917). A critical study of certain species of *Mucor*. *Bull. Torrey Bot. Club*, **44**, 241–59, 287–313, plates 17–20.
RICHARDSON, M. J. and WATLING, R. (1969). Keys to fungi on dung. *Bull Br. mycol. Soc.*, **3**, 86–8, 121–4.
SCHIPPER, M. A. A. (1969). Zygosporic stages in heterothallic *Mucor*. *Ant. van Leew.*, **35**, 189–208.
SCHIPPER, M. A. A. (1970). Two species of *Mucor* with oval and spherical spored strains. *Ant. van Leew.*, **36**, 475–88.
SCHIPPER, M. A. A. (1973). A study on variability of *Mucor hiemalis* and related species. *Studies in Mycology*, 4. CBS.
SCHIPPER, M. A. A. (1975). On *Mucor mucedo, Mucor flavus* and related species. *Studies in Mycology*, 10. CBS.
SCHIPPER, M. A. A. (1976). On *Mucor circinelloides, Mucor racemosus* and related species. *Studies in Mycology*, 12. CBS.
SCHIPPER, M. A. A. (1978). I. On certain species of *Mucor* with a key to all accepted species. II. On the genera *Rhizomucor* and *Parasitella*. *Studies in Mycology*, 17. CBS.
SCHIPPER, M. A. A., SAMSON, R. A. and STAPLERS, J. A. (1975). Zygospore ornamentation in the genera *Mucor* and *Zygorhynchus*. *Persoonia*, **8**, 321–8.
TALBOT, P. H. B. (1971). *Principles of Fungal Taxonomy*. Macmillan, London.
TURNER, M. (1963). Studies on the genus *Mortierella*. I. *Mortierella isabellina* and related species. *Trans. Br. mycol. Soc.*, **46**, 262–72.

WATERHOUSE, G. M. (1975). Key to the species *Entomophthora* Fres. *Bull. Br. mycol. Soc.*, **9**, 15–41.

WEBSTER, J. (1970). *Introduction to Fungi*. Cambridge University Press, Cambridge, London, New York, Melbourne.

ZYCHA, H. (1935). *Kryptogamenflora der Mark Brandenburg*. Band VIa, Pilze II, *Mucorineae*. Gebruder Borntraeger, Leipzig. (Reprinted 1963.)

ZYCHA, H., SIEPMANN, R. and LINNEMANN, G. (1969). *Mucorales. Eine Beschreibung aller Gattungen und Arten dieser Pilzgruppe*. J. Cramer, Lehre.

5

The Ascomycetes (Ascomycotina)

5.1 Introduction

The Ascomycetes constitute the largest class of fungi, the number of known species being approximately 32 000. As may be expected in such a large group there is considerable diversity of form and structure. At one end of the scale are the unicellular organisms commonly known as yeasts, and, at the other, species with extensive mycelium and large and elaborate fruiting structures, such as the truffles and morels.

The ascus, a structure which is peculiar to and which gives the name to the class, is a walled recepticle enclosing the spores and usually rupturing at maturity. The ascus is distinguished from the sporangium of the Phycomycetes (Mastigomycotina and Zygomycotina) in the origin and method of formation of the spores, details of which are to be found in any text-book of general mycology. A more obvious difference is that the sporangium contains an indefinite number of spores, whilst in the great majority of Ascomycetes the ascus invariably contains eight spores. In a few cases, which are of no importance here, the regular number is a multiple of eight, or more strictly, a higher multiple of two. In a few primitive members of the subdivision, notably in *Endomyces* and *Endomycopsis*, the number is four, whilst in the yeasts the number ranges from one to eight.

In the simpler Ascomycetes the asci are globose or ovate, with the spores packed tightly together. They may be formed singly, in irregular loose clusters, or arranged irregularly in a more or less definite fruitbody (the **ascoma** or **ascocarp**). The ascoma, when present, is usually more or less globose, with a fairly firm wall, or occasionally is nothing more than a loose web of hyphae around the asci. In the structurally complex members of the class the ascocarp is of more definite form, with club-shaped or cylindrical asci arranged in parallel series, often with elongated sterile cells (**paraphyses**) separating them. If the ascocarp is globose or flask-shaped and closed at maturity, except for a narrow passage (the **ostiole**), it is called a **perithecium**; if disc-like with the asci exposed to the air it is known as an **apothecium**. Sometimes the asci are produced in a hollow (**locule**) or in hollows in a tight cellular or hyphal mass (**stroma**).

5.2 Classification

At present there is no entirely satisfactory scheme for the classification of Ascomycetes available and those adopted in different text books vary considerably. Criteria used as the bases of classification in the 1950s are now being re-evaluated in the light of new autogenetic and ultrastructural information.

Papers and discussions have been produced by Miller (1949), Luttrell (1951, 1955) who related the differences between unitunicate and bitunicate asci to developmental types, von Arx and Müller (1975), Müller and von Arx (1973), Barr (1976) and von Arx (1979). Of these systems, only that of Barr (1976) is fully comprehensive but is not yet generally accepted.

For simplicity it is proposed to follow Ainsworth's (1966) system, which was enlarged in *The Fungi* (1973), with sections on the different groups by C. L. Kramer, D. I. Fennell, C. E. Yarwood, E. Müller and J. A. von Arx, E. S. Luttrell, R. K. Benjamin and R. P. Korf.

Only relatively few species are of interest to industry. Many Ascomycetes produce asexual spores as well as the sexual means of reproduction and it is often these species that effect industry.

5.3   Key to the classes of Ascomycotina

The key given here follows Ainsworth (1973).

Key to classes of Ascomycotina (following Ainsworth, 1973)

1.	Ascocarps and ascogenous hyphae lacking; thallus mycelial or yeast-like	*Hemiascomycetes*
1′.	Ascocarps and ascogenous hyphae present; thallus mycelial	2
2.	Asci bitunicate, ascocarp an ascostroma	*Loculoascomycetes*
2′.	Asci typically unitunicate, if bitunicate ascocarp an apothecium	3
3.	Asci evanescent, scattered within the astomous (without an ostiole) ascocarp which is typically a cleistothecium; ascospores without septa	*Plectomycetes*
3′.	Asci arranged regularly within the ascocarp, as a basal or peripheral layer	4
4.	Parasites of Arthropods	*Laboulbeniomycetes*
4′.	Not exoparasites of Arthropods	5
5.	Ascocarp a perithecium with an ostiole and asci inoperculate with an apical pore or slit	*Pyrenomycetes*
5′.	Ascocarp an apothecium or modified apothecium with both inoperculate and operculate asci	*Discomycetes*

5.4   The classes of Ascomycotina

Hemiascomycetes

In this class ascocarps and ascogenous hyphae are lacking. The asci are produced inside single cells. Most reproduction is asexual by budding or

division of single cells. Mycelium is usually fragmentary or lacking. The two
orders are the Endomycetales which include the ascosporogenous yeasts and
the Taphrinales which are parasites of vascular plants.

Loculoascomycetes

(See Luttrell, 1973.) The ascocarp is formed as a stroma within which the asci
are produced in hollows. The stroma may be thin and the hollow have an
apical opening so that it appears like a perithecium. However, the asci are
functionally bitunicate. At the time of spore liberation the outer wall layers
rupture and the ascospores are extruded within the inner wall layers from
which they are subsequently expelled.

Plectomycetes

(See Fennell, 1973.) The majority of the industrially interesting Ascomycetes
belong in this class. The asci are produced singly or in groups amongst
ascogenous mycelium or scattered within a non-ostiolate ascocarp. The
degree of wall development is often of specific importance. The ascospores
are aseptate. As well as a sexual (teleomorphic) means of reproduction,
asexual (anamorphic) vegetative spores (conidia) are usually produced. In
some genera these are more common than the ascospores and it may be easier
to identify them from the conidial state. In fact, similar conidial states may be
produced by more than one ascosporic genus, and it is easier to study these
forms together, for example the anamorphic genus *Aspergillus* has teleomorphs in *Eurotium, Sartorya, Eidemella* and several other genera, while
Penicillium anamorphs are produced by species of *Eupenicillium* and *Talaromyces. Aspergillus* (see Chapter 10) and *Penicillium* (see Chapter 11) are
dealt with as separate chapters.

Laboulbeniomycetes

These are very highly specialized minute exoparasites of Arthropods and thus
of no industrial interest.

Pyrenomycetes

The asci are produced in a perithecium, a flask-shaped body with an ostiole or
apical pore through which the ascospores are liberated. The asci are
unitunicate having a wall acting as a single layer during discharge. The
ascospores are usually liberated through a specialized pore at the apex of the
asci and this type of spore liberation is called inoperculate. Several species of
industrial interest belong in this group. As in the Plectomycetes many of these
have conidial states.

Discomycetes

The Discomycetes include about 4000 species, some of them are parasitic on plants, but many are saprophytic and often found growing on soil or dung. A number of larger forms are fleshy and a few, such as the morels, are edible. The ascocarp is an apothecium which is a flat dish-like open body on which the asci are produced in a regular even stand. The asci can be inoperculate or operculate (i.e. opening by an apical cap). They are often of a bright colour but are rarely of industrial importance.

5.5　Genera of Ascomycotina

Byssochlamys Westling
(Gr. *byssos*, cotton; *chlamys*, a cloak) (Fig. 5.1)
This is a genus of interest to the systematic mycologist, as it produces clusters of eight spored asci without any surrounding wall (**peridium**), and thus forms a link between the Endomycetales and the Plectomycetes with their more or less definite ascocarps. The relationship to the genus *Paecilomyces* was discussed by Brown and Smith (1957). Ram (1968) made a study of the genus and three species have been described. Stolk and Samson (1971) in their studies on *Talaromyces* and related genera gave concise descriptions and indicated the means of ascus formation. The conidial state was further studied by Samson when he monographed *Paecilomyces* (1974), and he described another new species in 1975 (Samson and Tansey, 1975).

B. fulva Olliver & G. Smith (1933) (Lat. *fulvus*, tawny). This species is of considerable importance, since it is the cause of serious spoilage of canned and bottled fruits. The mature ascospores can withstand a temperature which is lethal to the spores of most fungi, and sometimes survive the commercial sterilizing process. The fungus is able to grow in an atmosphere containing very little oxygen, or completely submerged in fluid, where the only oxygen available is the small amount dissolved. Spoilt fruit frequently show no visible colonies of the mould and the cans are not blown as when the spoilage is due to anaerobic bacteria, the only evidence of the presence of the fungus being a general softening, or sometimes complete disintegration of the fruit, such as would occur if cooking had been unduly prolonged. Research still continues as the recent papers by Bayne and Michener (1979), King *et al.* (1979), Beuchat and Toledo (1977) and Rice and Beuchat (1978) show.

The mould grows well on all ordinary media, with colonies of loose cottony texture, becoming fucous (tan to sandy brown) as spores develop. The conidial state is of the *Paecilomyces* type (see Chapter 11), with long chains of ovate spores, 4–9×2.3–$2.5\ \mu$m (Fig. 5.1a). The asci develop rapidly in compact clusters, without any trace of peridium (Fig. 5.1b), each ascus containing 8 ovate spores, 6–6.5×4.3–$4.5\ \mu$m.

This species is common in some soils, particularly in strawberry beds, with the result that spores are carried into the canning factories along with the early soft fruits.

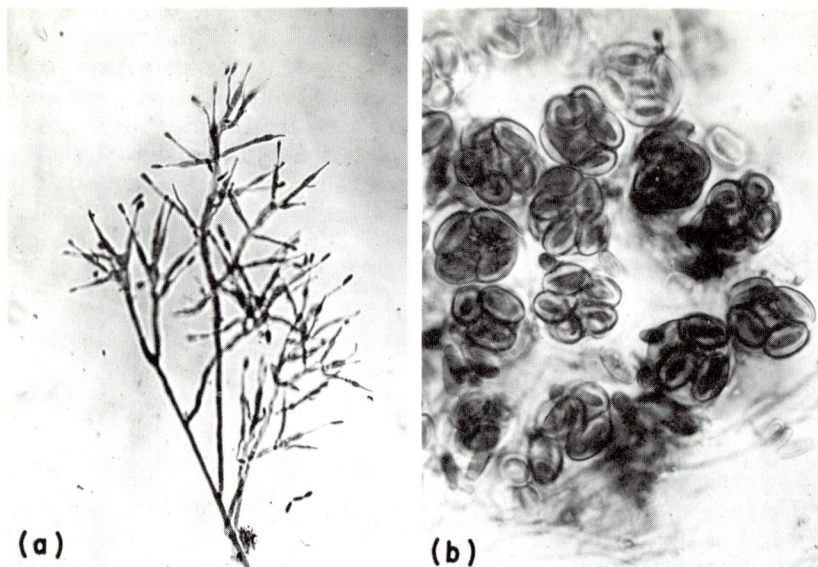

Fig. 5.1 *Byssochlamys fulva*   **(a)** Conidiophores (from slide culture), ×250; **(b)** cluster of asci, ×1000.

B. nivea Westling (Lat. *niveus*, snow white) is the type species of the genus. Colonies are persistently white or very slightly brownish in age and the ascospores are somewhat smaller than those of *B. fulva*. It also differs from the latter in producing more bacillar conidia much less abundantly and in forming very numerous macrospores (small terminal chlamydospores). In previous editions this was reported as not common, but recently it has repeatedly been received for identification particularly from mouldy hay and feed stuffs and would at present appear to be more common than *B. fulva*. The characteristic chlamydospores are usually smooth, but occasionally are slightly rough which suggests that separation from its variety *B. nivea* var. *languculariae* may not be very distinct.

Eurotium, Sartorya, Eupenicillium and *Talaromyces*
These and other common genera of the Plectomycetes are often better known by their conidial (anamorphic) states. Many of them are of industrial interest and are frequently associated with moulding problems.

The best known species of the genus *Eurotium* are more commonly known as the *Aspergillus glaucus* series. The conidial and ascosporic states were long thought to be separate and distinct species, until De Bary in 1854 showed their real relationship. The connections of the conidial and ascosporic states of many other species have now been worked out. It is proposed to describe these common moulds in the chapters on the Fungi Imperfecti, *Aspergillus* and *Penicillium*, as it is most often the conidial state that is evident in the industrial situation. However, attention will be drawn to the ascosporic states in these descriptions.

Monascus van Tieghem
(Gr. *monos*, single; hence forming a single ascus) (Fig. 5.2)
When the genus was first described it was thought that the cleistothecium consisted of a single, many spored ascus. It is now known that normal asci are produced, but that these break up as the spores ripen, leaving the somewhat thin walled cleistothecium full of spores. Although a member of the Plectomycetes the ascosporic state tends to be dominant.

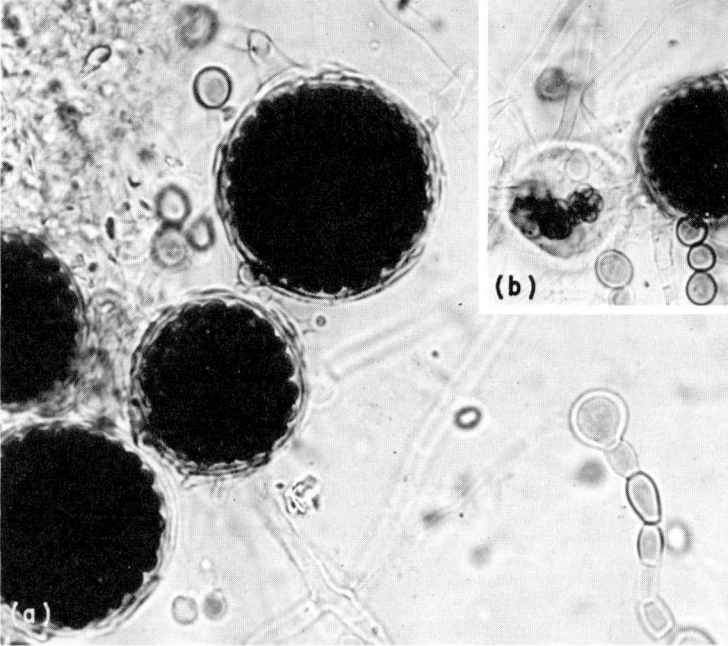

Fig. 5.2 *Monascus ruber* **(a)** Mature cleistothecia and conidia, ×630; **(b)** young cleistothecia with 8-spored asci and conidia, ×450.

M. ruber van Tieghem (= *M. purpureus* Went) (Lat. *purpureus*, purple). The species is not uncommon, particularly in dairy products. It has been isolated a number of times from commercial lactose and from life rafts. It is also the organism which gives the characteristic colour to Chinese red rice. Colonies form a thin, spreading growth, with mycelium turning reddish or purple, becoming greyish as conidia and cleistothecia develop, and with reverse dull purplish red. Cleistothecia are produced singly on stalks, whilst the conidia are formed in short chains (Fig. 5.2). Ascospores are ovoid, smooth, colourless, 5.5–6 × 3.5–4 μm. Conidia are brown, ovate to barrel-shaped, with a truncate base, 9–10.5 × 7–9 μm. Cole and Kendrick (1968) investigated the method of conidial spore formation and showed the spores to be annellospores.

M. bisporus (Fraser) von Arx. This very xerophilic fungus was described by Fraser (1953) as *Xeromyces bisporus* and it was transferred to *Monascus* by von Arx (1970). It is often regarded as rare, but this may be due to its inability to grow on commonly used isolation media. It grows well in the presence of 60 per cent sugar, was originally found in licorice (Fraser, 1953) and is seen from time to time from this substrate (Pitt, 1975). It has also been reported from prunes (Pitt and Christian, 1968), tobacco, currants, chocolate sauce and other foodstuffs (Dallyn and Everton, 1969) and has been received for identification at the Commonwealth Mycological Institute from food-stuffs, including packaged cakes and biscuits.

The growth is slow and white, and slightly granular due to the cleistothecia. The smooth D shaped ascospores are born two in the ascus. The conidial state consists of white truncate annellospores produced singly or in short chains.

Chaetomium Kunze ex Stundel
(Gr. *Chaitome*, a plume of hair) (Fig. 5.3)

This is the most common of the genera of Pyrenomycetes encountered in industrial work and its species are of interest not only as a cause of spoilage and mycotoxin formation (Udagawa *et al.*, 1979) but also for the possible recycling of cellulose waste and protein production (Chahal and Hawksworth, 1976; Chahal and Wang, 1978).

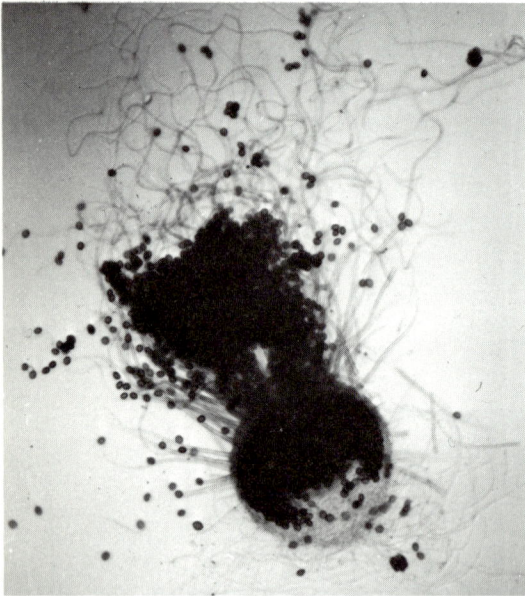

Fig. 5.3 *Chaetomium globosum* Perithecium exuding ascospores and showing sinuous hairs (bleached to render transparent), ×100.

The genus includes a fairly large number of species, all characterized by forming dark perithecia with short or no distinct necks and beset with dark coloured hairs, which are variously straight, branched, or curled. They are found chiefly on cellulosic materials and thrive particularly on paper. When kept in culture a strip of filter paper only partially immersed in the culture medium or laid on top of the medium usually ensures satisfactory production of perithecia.

Perhaps still the most useful papers for the identification of common species are two papers by Skolko and Groves (1948, 1953); the last including a complete key to all the species of *Chaetomium* known at the time. The paper by Millner (1975) is also useful in including a key to many widespread species.

Since 1953 numerous additional species have been described and several further major studies and monographs of the genus made. These include Udagawa (1960), Ames (1963), Seth (1972), Hawksworth and Wells (1973) which includes a stereoscan microscope study of hair types, Hawksworth (1975) with a key to the *Chaetomia* with *Botryotrichum* anamorphs, Dreyfuss (1976), Millner (1977), on growth and Millner *et al.* (1977) on ascospore types.

Separation in the genus usually rests on the type of terminal hairs, shape and size of the ascospores, and the shape of the asci.

Two very common species in the British Isles are *C. globosum*, with sinuous unbranched hairs, and *C. elatum* with dichotomously branched hairs. Both species have lemon shaped ascospores.

Neurospora Shear & B. Dodge
(Gr. *neuron*, nerve, vein; refers to the striate spores) (Fig. 5.4)
The red bread mould *Monilia sitophila* (Montagne) Saccardo and related species are the conidial state of some *Neurospora* species. Until the 1960s

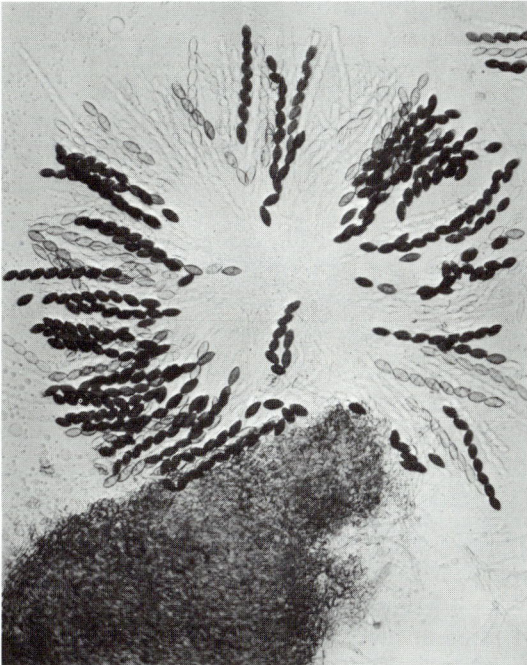

Fig. 5.4 *Neurospora sitophila*   Crushed perithecium, ×100.

only species producing the conidial state had been seen, and most isolates had been heterothallic (Shear and Dodge, 1927). A key to the species is given by Frederick *et al.* (1969). The ascospores are very resistant to heat and are frequently able to survive in the interior of a loaf during the baking process. In addition, it is difficult to secure germination of the spores unless they are first heated to a fairly high temperature. This probably explains some of the outbreaks of 'red mould disease' in sliced and wrapped bread.

Isolates deficient in enzymes are often produced and these are used in biological assay and studies on genetics. Genetic recombination can be studied by mating isolates with pale and dark ascospores and producing asci with different combinations of pale and dark spores.

Sordaria Cesati & de Notaris
(Lat. *sordes*, dirt, filth)

Species of this genus are common on dung; hence the name. As described in older mycological works the genus included a miscellany of Ascomycetes with dark, one-celled spores, provided with appendages or surrounded by mucus. Moreau (1953) has adopted a more limited conception of the genus, restricting it to species having spores without appendages, but with a colourless outer membrane which swells in water. The species with appendages at one or both ends are regarded as belonging to different genera. A detailed discussion is given by Lundquist (1972).

S. fimicola (Roberge) Cesati & de Notaris (Lat. *fimus*, dung, excrement; *colo*, to inhabit) is a fairly common mould, found on many substrates other than dung.

It spreads extremely rapidly on wort agar or other vegetable media, producing abundant perithecia, but grows very poorly or not at all on synthetic media unless thiamin and biotin are added. The perithecia are brown and opaque, almost black in mass, and have short necks through which the spores are extruded or shot forcibly towards light at maturity. The spores are dark brown, ovate to biconvex, $17–24 \times 11–13\ \mu$m produced in single rows in long cylindrical asci. The spores have no tails. Figure 5.5 shows a crushed perithecium.

Microascus Zukal
(Gr. *micro*, small; *ascus*, a sack)

In a recent monograph von Arx (1975) gives a key to the genus *Microascus*, species of which are found to have conidial states in *Doratomyces*, *Scopulariopsis* and *Wardomyces*. The monograph of Barron *et al.* (1961) is still most useful.

Microascus forms dark ostiolate perithecia which may have a long neck and small asymmetrical one-celled, smooth, yellow ascospores with a single germpore at the base. The most common species encountered is *Microascus cinereus*.

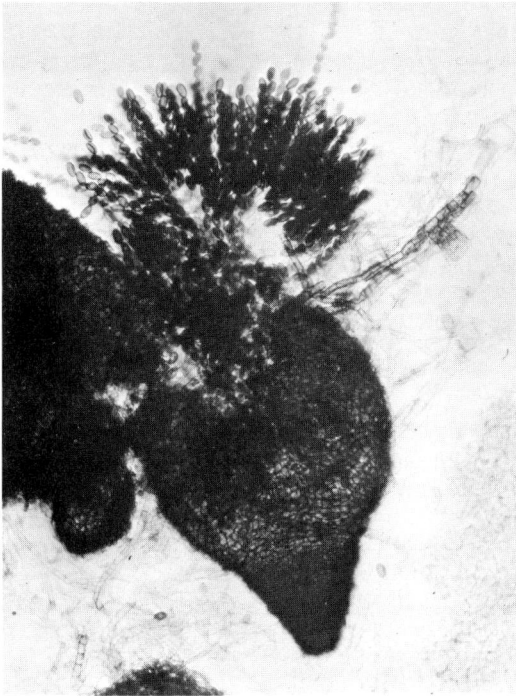

Fig. 5.5 *Sordaria fimicola*
Crushed perithecium, ×100.

Ceratocystis Ellis and Hallsted (including *Ophiostoma*)
Species of the genus *Ceratocystis* are the cause of some serious plant diseases. Although not of industrial significance it seems hardly possible to discuss Ascomycetes without mentioning *C. ulmi*, the cause of 'Dutch Elm Disease' which has caused the destruction of so many trees in recent years in the United Kingdom (see Booth and Gibson, 1973).

The perithecia have a swollen base with a very long neck through which the ascospores are exuded to form a sticky mass at the apex. Conidia are also produced in a wet spore mass at the apex of coremia, consisting of parallel bundles of hyphae (see Fig. 5.6). The disease is spread by bark boring beetles, especially *Scolytus multistriatus*, which tunnel into the wood and bark to feed and lay eggs. The sticky spores adhere to their bodies and are carried to other parts of the tree or to other trees.

Eremascus Eidam
This is often classified with the Endomycetacae but as the asci are produced singly on hyphae in the hyphal mass they will be mentioned here instead of with the yeasts.

There are eight spores in the ascus and they are somewhat asymmetrical in shape (Fig. 5.7). They are of interest as spoilage organisms as they grow in very dry conditions such as dry mustard, on prunes and produce good growth on media containing up to 60% sugar.

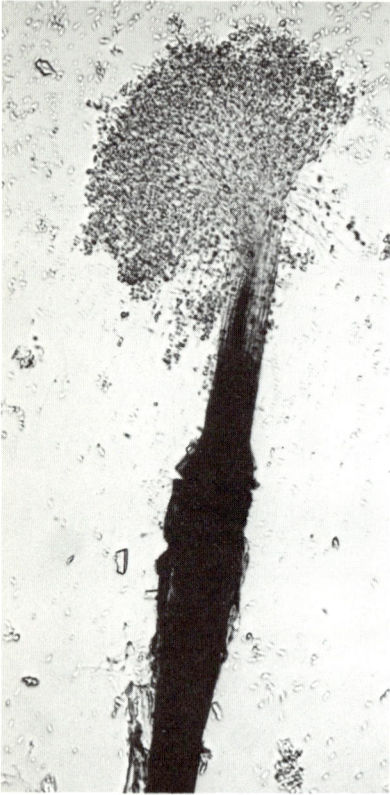

Fig. 5.6 *Ceratocystis ulmi* Dark synnema bearing conidia, ×250.

There are only two species, *E. albus* and *E. firtilis*. These are fully discussed and described by Harrold (1950) and more recently by Paugh and Gray (1969). A recent study of ultra structure of the hyphae and ascospores was made by Kreger van Rij, Veenuhuis and Leemberg van der Graaf (1974). *E. albus* Eidam is pure white, has coiled ascus initials and flattened spherical ascospores, 5–5.5 μm diam. In *E. fertilis* Stoppel the growth is restricted, the colony more brownish and the smaller ascospores are elongate to almost pointed, approximately 5 × 3 μm. The ascospore initials are lumpy.

Pleospora Rabenhorst (Gr. *pleos*, full; refers to the many celled spores) (Fig. 5.8)
Many species have been described but only one is of importance in the present connection. Wehmeyer (1961) monographed the genus.

P. herbarum (Pers. ex Fries) Rabenh. (Lat. *herbarum*, of plants) (see Booth and Pirozynski, 1967) is found frequently on fruit of various kinds in storage. It forms black spots on the surface and may, on occasions, spread to the interior. The fungus grows rapidly on most culture media, forming a dense floccose mat of greyish mycelium, soon producing conidia. The conidial (anamorphic) state is *Stemphylium botryosum* Wallroth. It is of the Loculo-

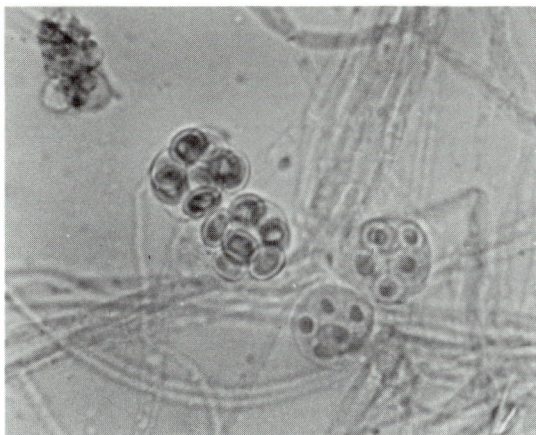

Fig. 5.7 (above) *Eremascus albus* Cluster of asci, ×800.

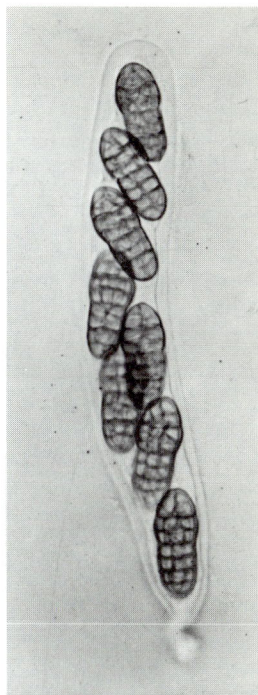

Fig. 5.8 (right) *Pleospora herbarum* Ascus showing typical septate ascospores, and apparently double wall, ×500.

ascomycetes and forms spore-producing bodies which look like perithecia, and mature after several weeks. The 8-spored asci are bitunicate with a double wall and contain the spores either in a single row or more or less in two rows. The golden-brown ascospores are very characteristic, having seven cross septa and numerous longitudinal septa. Conidia are produced fairly freely when the fungus is first isolated, but are seldom to be found in subsequent cultures.

Other Pyrenomycetes are occasionally isolated in the industrial laboratory. Some of these have pale coloured walls through which the spores, and sometimes the asci can be seen. Dark coloured perithecia with walls which are opaque under the microscope, can be distinguished from pycnidia of similar gross appearance by gently crushing under a coverglass, or, if too hard, by squashing them with a needle before putting on the coverglass. If the perithecia are not completely mature it is nearly always possible in this way to find immature asci containing the normal eight spores.

5.6 References

AINSWORTH, G. C. (1966). A general purpose classification for fungi. *Bibl. Syst. Mycol.*, **1**, 1–4.

AINSWORTH, G. C. (1973). Introduction and keys to higher taxa. In *The Fungi*, Vol. 4A. AINSWORTH, G. C., SPARROW, E. K. and SUSSMAN, A. S. (Eds) pp. 1–7. Academic Press, London.

AMES, L. M. (1963). A monograph of the Chaetomeaceae. *U.S. Army Res. & Devel. Series*, No. 2.

ARX, J. A. VON (1970). *The genera of fungi sporulating in pure culture*. 1st edition. J. Cramer, Lehre.

ARX, J. A. VON (1974). *The genera of fungi sporulating in pure culture*. 2nd edition. J. Cramer, Vadug.

ARX, J. A. VON (1975). Revision of *Microascus* with description of a new species. *Persoonia*, **8**, 191–7.

ARX, J. A. VON (1979). 13. Ascomycetes as Fungi Imperfect. In *The Whole Fungus*. KENDRICK, B. (Ed.) pp. 201–13. National Museums of Canada for the Kananaskis Foundation.

ARX, J. A. VON and MÜLLER, E. (1975). A re-evaluation of the bitunicate Ascomycetes with keys to families and genera. *Studies in Mycology*, **9**, 1–159.

BARR, M. E. (1976). Perspectives in the Ascomycotina. *Memoirs of the New York Botanical Garden*, **28**, 1–8.

BARRON, G. L., CAIN, R. F. and GILMAN, J. C. (1961). The genus *Microascus*. *Can. J. Botany*, **39**, 1609–31, 6 plates.

BARY, A. DE (1854). Uber die entwickelung und den zusammenhang von *Aspergillus glaucus* und *Eurotium*. *Botan. Ztg.*, **12**, 425–34, 441–51, 465–71.

BATRA, L. R. (1975). Ascomycetes in Pakistan: Plectomycetes. *Biologia*, **21**, 1–37.

BAYNE, H. G. and MICHENER, H. D. (1979). Heat resistance of *Byssochlamys* ascocarps. *Appl. & Environ. Microbiol.*, **37**, 449–53.

BEUCHAT, L. R. and TOLEDO, R. T. (1977). Behaviour of *Byssochlamys nivea* ascospores in fruit syrups. *Trans. Br. mycol. Soc.*, **68**, 65–71.

BOOTH, C. and GIBSON, I. A. S. (1973). *Ceratocystis ulmi*. *CMI Descriptions of Pathogenic Fungi and Bacteria*, No. 361.

BOOTH, C. and PIROZYNSKI, K. A. (1967). *Pleospora herbarum*. *CMI Descriptions of Pathogenic Fungi and Bacteria*, No. 150.

BROWN, A. H. S. and SMITH, G. (1957). The genus *Paecilomyces* and its perfect stage *Byssochlamys*. *Trans. Br. mycol. Soc.*, **40**, 17–89.

CHAHAL, D. S. and HAWKSWORTH, D. L. (1976). *Chaetomium cellulolyticum*, a new thermotolerant and cellulolytic *Chaetomium*. I. Isolation, description and growth rate. *Mycologia*, **68**, 600–10.

CHAHAL, D. S. and WANG, D. I. C. (1978). *Chaetomium cellulolyticum*, growth behaviour on cellulose and protein production. *Mycologia*, **70**, 160–70.

COLE, G. T. and KENDRICK, W. B. (1968). Conidium ontogeny in Hyphomycetes. The imperfect state of *Monascus ruber* and its meristem arthrospores. *Canad. J. Bot.*, **46**, 987–92.

DALLYN, H. and EVERTON, J. R. (1969). The xerophilic mould, *Xeromyces bisporus*, as a spoilage organism. *J. Fd Technol.*, **4**, 399–403.

DREYFUSS, M. (1975 (1976)). Taxonomische Untersuchungen der Gattung *Chaetomium* Kunze. *Sydowia*, **28**, 50–133.

FENNELL, D. I. (1973). Plectomycetes; Eurotiales. In *The Fungi*, Vol. 4A. AINSWORTH, G. C., SPARROW, E. K. and SUSSMAN, A. S. (Eds) pp. 45–68. Academic Press, London.

FRASER, L. (1953). A new genus of the Plectascales. *Proc. Linn. Soc. N.S.W.*, **78**, 241–6.

FREDERICK, L., UECKER, F. A. and BENJAMIN, C. R. (1969). A new species of *Neurospora* from soil of West Pakistan. *Mycologia*, **61**, 1077–84.

HARROLD, C. E. (1950). Studies in the genus *Eremascus*. *Ann. Bot.*, **14**, 127–48.

HAWKSWORTH, D. L. (1975). *Farrowia*, a new genus in the Chaetomiaceae. *Persoonia*, **8**, 167–85.

HAWKSWORTH, D. L. and WELLS, H. (1973). Ornamentation on the terminal hairs in *Chaetomium* Kunze ex Fr. and some allied genera. *Mycol. Pap.*, **134**, 1–24.

KING, A. D., BAYNE, H. G. and ALDERTON, G. (1979). Non logarithmic death rate calculations for *Byssochlamys fulva* and other microorganisms. *App. & Envir. Microbiol.*, **37**, 596–600.

KREGER VAN RIJ, N. J. W., VEENHUIS, M. and LEEMBURG VAN DER GRAAF, C. A. (1974). Ultrastructure of hyphae and ascospores in the genus *Eremascus* Eidam. *Antonie van Leeuwenhoek*, **40**, 533–42.

LUNDQUIST, N. (1972). Nordie Sordariaceae s. lat. *Symb. Bot. Upsal.*, **20**, 1–374.

LUTTRELL, E. S. (1951). Taxonomy of the Pyrenomycetes. *Univ. Missouri Studies*, **24 (3)**, 1–120.

LUTTRELL, E. S. (1955). The ascostromic Ascomycetes. *Mycologia*, **47**, 511–32.

LUTTRELL, E. S. (1973). Loculoascomycetes. In *The Fungi*, Vol. 4A. AINSWORTH, G. C., SPARROW, F. K. and SUSSMAN, A. S. (Eds) pp. 135–219. Academic Press, London.

MILLER, J. H. (1949). A revision of the classification of the Ascomycetes with special emphasis on the Pyrenomycetes. *Mycologia*, **41**, 99–127.

MILLNER, P. D. (1975). Ascomycetes of Pakistan: *Chaetomium*. *Biologia*, **21**, 39–73.

MILLNER, P. D. (1977). Radial growth responses to temperature by 58 *Chaetomium* species, and some taxonomic relationships. *Mycologia*, **69**, 492–502.

MILLNER, P. D., MOTTA, J. J. and LENTZ, P. L. (1977). Ascospores, germ pores, ultrastructure, and thermophilism of *Chaetomium*. *Mycologia*, **69**, 720–33.

MOREAU, C. (1953). Les genres *Sordaria* et *Pleurage*. Leur affinities systematiques. *Encycl. mycol.*, **25**. Lechevalier, Paris.

MÜLLER, E. and ARX, J. A. VON (1973). Pyrenomycetes, Meliolales, Coronophorales, Sphaeriales. In *The Fungi*, Vol. 4A. AINSWORTH, G. C., SPARROW, E. K. and SUSSMAN, A. S. (Eds) pp. 87–132. Academic Press, London.

OLLIVER, M. and SMITH, G. (1933). *Byssochlamys fulva* sp. nov. *J. Bot., Lond.*, **7**, 196–7.

PAUGH, R. L. and GRAY, W. D. (1969). Studies on the growth of the osmophilic fungus *Eremascus albus*. *Mycologia*, **61 (2)**, 281–8.

PITT, J. I. (1975). Xerophilic fungi and the spoilage of foods of plant origin. In *Water Relations of Foods*. DUCKWORTH, R. B. (Ed.) pp. 273–307. Academic Press, London.

PITT, J. I. and CHRISTIAN, J. H. B. (1968). Water relations of xerophilic fungi isolated from prunes. *Appl. Microbiol.*, **16**, 1853–8.

RAM, C. (1968). Timber-attacking fungi from the state of Maranhão, Brazil. Some new species of *Paecilomyces* and its perfect stage *Byssochlamys* Westl. VIII. *Nova Hedwigia*, **16**, 305–14.

RICE, L. R. and BEUCHAT, L. R. (1978). Polygalacturonase, biomass, and ascospore production by *Byssochlamys fulva*. I. Effects of acids found in fruits. *Mycopathologia*, **63**, 29–34.

SAMSON, R. A. (1974). *Paecilomyces* and some allied Hyphomycetes. *Studies in Mycology*, **6**, 1–119.

SAMSON, R. A. and TANSEY, M. R. (1975). *Byssochlamys verrucosa* sp. nov. *Trans. Br. mycol. Soc.*, **65**, 512–14.

SETH, H. K. (1972). A monograph of the genus *Chaetomium* Beih. *Nova Hedwigia*, **37**, 1–134.

SHEAR, C. L. and DODGE, B. O. (1927). Life histories and heterothallism of the red bread mould fungi of the *Monilia sitophila* group. *J. agric. Res.*, **34**, 1019–42.

SKOLKO, A. J. and GROVES, J. W. (1948). Notes on seed borne fungi. V. *Chaetomium* species with dichotomously branched hairs. *Canad. J. Res.*, Sect. C, **26**, 269–80.

SKOLKO, A. J. and GROVES, J. W. (1953). Notes on seed borne fungi. VII. *Chaetomium*. *Canad. J. Bot.*, **31**, 779–809.

STOLK, A. C. and SAMSON, R. A. (1971). Studies on *Talaromyces* and related genera. I. *Hamigera*, gen. nov. and *Byssochlamys*. *Persoonia*, **6**, 341–57.

UDAGAWA, S. (1960). A taxonomic study on the Japanese species of *Chaetomium*. *J. Gen. Appl. Microbiol.*, **6**, 223–51.

UDAGAWA, S., MUROI, T., KURATA, H., SEKITA, S., YOSHIHIRA, K., NATORI, S. and UMEDA, M. (1979). The production of chaetoglobosins, sterigmatocystin, O-methyl sterigmatocystin, and chaetocin by *Chaetomium* spp. and related fungi. *Can. J. Microbiol.*, **25**, 170–7.

WEBSTER, J. (1970). *Introduction to Fungi*. 1st edition. University Press, Cambridge. (Paperback edition, 1977.)

WEHMEYER, L. E. (1961). *A World Monograph of the Genus* Pleospora and its Segregates. Univ. Michigan Press, Ann Arbor.

6

Yeasts and Yeast-like Organisms

by R. R. Davenport
　　University of Bristol, Long Ashton Research Station,
　　Long Ashton, Bristol 9AF BS18, U.K.

6.1　Introduction

Yeasts are microfungi in which a unicellular form is usually the predominant cell type. However, this statement is not strictly true since some unicellular structures can be induced in certain filamentous fungi (e.g. *Mucor rouxii*). This change from filamentous state to yeast is dependent on various environmental parameters such as temperature and substrate. Some yeasts only form individual cells and sometimes short chains of similar cells while others produce a range of cell forms, including filaments of various types more in keeping with other filamentous microfungi. Morphological, physiological and biochemical characteristics of yeasts are variable; these aspects will be described later. Thus yeasts are not a uniform group of microorganisms which can be assigned to one particular group within the fungal kingdom; indeed only diffuse boundaries are recognizable between yeasts and other organisms. Also there is some controversy among specialists concerning methods, criteria and nomenclature. This situation can be rather confusing for investigators familiar with other microbiological disciplines. Therefore the purpose of this chapter is to assist those who wish to begin yeast studies. Emphasis is placed on basic knowledge for those with the minimum of biological training. The methods described are particularly suited to industrial laboratories and students about to embark on research.

　　In this discourse the term 'yeast' is taken to mean both yeasts and yeast-like organisms.

6.2　Distribution

It is common knowledge that yeasts are of great economic importance. Selected strains are used for industries such as baking, brewing, distilling and production of wine. These strains belong to the species, *Saccharomyces cerevisiae* and sometimes other closely related yeasts are used for beverages and foods (e.g. *S. bayanus* for wine-making). Industrial raw materials, products and commodities are liable to yeast spoilage problems which can mean great economic losses. Additionally, certain yeasts can be biodeteriogenic agents while other species are animal pathogens, including man. Table 6.1 gives some examples of species which can be undesirable inhabitants in many situations.

Table 6.1 Examples of yeasts causing spoilage problems

Yeasts (with some synonyms *)	Examples of products spoilt
Saccharomyces cerevisiae (S. ellipsoideus, S. sake, S. willianus, S. carlsbergensis, S. uvarum, S. aceti, S. diastaticus)	Fruit products (juices, pulps and wines)
Zygosaccharomyces bailii (S. bailii, S. acidefaciens, S. elegans)	High sugar products, wines, acid foods
Kluveromyces lactis (S. lactis, Z. lactis, Z. casei, S. sociasii)	Dairy products
Hypopichia burtonii (Endomycopsis burtonii, E. chodatii, Trichosporon variabile, Candida chodatii, Pichia burtonii)	Bakery products
Candida valida (C. mycoderma, Mycoderma valida)	Beers, wines, bread
Rhodotorula glutinis (Torula glutinis, Torulopsis glutinis)	Frozen vegetables, dairy products
Saccharomycodes ludwigii (Saënkia bispora)	Alcoholic beverages
Saccharomycopsis lipolytica (C. lipolytica)	Butter, mineral oils

* Considering past, present and future nomenclature

It is vital to understand that, in general, yeast names are not indicative of their potential significance either as beneficial or disastrous organisms. This is important since names may change but the activities of the organisms are always paramount. Thus it is better first to consider the significance of a yeast and then its probable identity. The identification of yeasts is discussed later. Meanwhile it is pertinent to consider the presence of yeasts in various environments.

True inhabitants are those which reproduce in a habitat while others may just survive and others eventually die. The status of a yeast in an environment is determined by the interaction of the prevailing chemical, physical and biological features, i.e. pH, a_w (water activity, see p. 330), water content of habitats, oxygen present, carbon, nitrogen and vitamin sources available, and the influence of other living organisms, for example animal and plant surfaces (Davenport, 1980). It is axiomatic that yeast characteristics may equally be important, so that the combined yeast properties plus the environmental parameters determine whether a species can live successfully or merely tolerate its environment (Davenport, 1973, 1975). This principle was further expanded to integrate both ecological and taxonomic studies where yeasts were isolated by adjusting cultural conditions to match the environment, followed by a selection of identification tests (Davenport, 1974a), thus simultaneously establishing the role or significance and most probable identity. Davenport (1980) formed ecological groups of yeasts within environments according to their distribution patterns: (1) **general** – genera which have species usually found in various habitats (e.g. the genus *Torulopsis* Berlese) and (2) **restricted** – general and/or species confined to one or very few habitats (e.g. the genus *Schwanniomyces* Klöcker occurring only in selected soils, and the species *Pichia chambardii* found only in tanning fluids).

Yeasts can be found in most habitats, for example animals, plants, all plant surfaces (genus *Nematospora* Peglion, the only true yeast plant pathogens), fresh water, marine sources, Antarctic seas, snow, beverages and foods, soils tanning fluids and mineral oils. Significance and detection is a consequence of the laboratory methods used; therefore it is absolutely essential that the right objective is defined at the outset of any yeast investigation. Otherwise two difficulties will be encountered. First, other microorganisms, rather than yeasts, may be determined more readily; second, more than one type of yeast may be present. One should also bear in mind the relevance of any organism interactions and/or succession of microorganisms. (For further yeast details for isolation and other techniques see Beech and Davenport, 1969, 1971 and Davenport, 1973.)

The foregoing approach to yeast studies can be illustrated by the following examples. Pink (carotenoid) yeast colonies can be obtained on a wide selection of microbiological media due to their ubiquitous distribution and broad nutritional activities. The generic classification of these isolates and the separation from non-carotenoid (pink) yeasts can be differentiated (see Table 6.2). The genus *Brettanomyces* Kufferath et van Lear is strongly characterized from all other yeasts by the copious production of acetic acid, cultures which are generally short-lived, formation of a characteristic aroma and its only established habitat being certain beverages and their associated environs. Therefore any yeast showing one or more of the above characteristics is likely to be a *Brettanomyces* species. The genus *Dekkera* van der Walt has two ascogenous species which are identical to two ascosporogenous species within the genus *Brettanomyces*, i.e. perfect (*Dekkera*) and imperfect (*Brettanomyces*) states. Davenport (1979) made a biotaxonomic study of these yeasts in which he used a 'grouping/recognition' as well as classical identification systems. Table 6.3 sets out a suitable simple identification method for distinguishing *Brettanomyces* species.

Mechanisms of yeast distribution, i.e. dissemination (particularly by animals, air and water) are not discussed here. This information may be found in various publications (e.g. Carmo-Sousa, 1969 and Davenport, 1976).

6.3 Morphology

The diversity of morphological features exhibited by yeasts ranges from having a single vegetative cell typical of some imperfect species (e.g. some *Torulopsis** species) to organisms which have single cells as well as multicellular and/or modified forms which are part of the life-cycle. Yeasts with life-cycles belong either to the ascosporogenous genera (e.g. genus *Metschnikowia* Kamienski) or to the basidiosporogenous yeasts (e.g. genus *Rhodosporidium* Banno). It is rare and often difficult to observe many life-cycle components since their presence are dependent on environmental and mating of cells. Therefore for routine investigations it is easier in the first place to use the well known names and simple structures. Thus one should record any usual features and bear in mind the possibility of a more complicated yeast. Table 6.4 gives the perfect and imperfect states of some common yeasts.

* Future classification may reject the genus *Torulopsis* – many of these strains are now in perfect genera or transferred to the genus *Candida* (Kreger van Rij, 1979, pers. comm.).

Table 6.2 A simplified method for the separation of carotenoid yeast genera from *Pulcherrimin* species

Pink colonies
|
Pigment type
— Carotenoid
— Pulcherrimin

Pulcherrimin:
Pigment soluble in methanolic potash solution,* also slightly in water, indicated by a pink-grey halo on some media (e.g. potato dextrose agars)

Species	Principal habitats
Metchnikowia pulcherrima (*Candida pulcherrima*)‡	Flowers and fruits
Kluyveromyces dobzhanskii, *K. drosophilarum*, *K. fragilis* (*C. pseudotropicalis*)+, *K. lactis* (*Torulopsis sphaerica*) and *K. marxianus* (*C. macedoniensis*)	Warm blooded animal gut contents and animal products (e.g. yoghurt, milk and cheese)

Carotenoid:
Pigment soluble in organic solvents
— Budding cells
— Other morphological features

Rhodotorula, *Cryptococcus*, *Phaffia* (−)

Ballistospores (+):
— (+) *Sporobolomyces*, *Sporidiobolus*
— (−) *Rhodosporidium*

Further separation**

	I	F	N	M	SS	Principal habitats
Phaffia	−	+	−	−	−	Tree exudates and cactus plants (R)
Cryptococcus	+	−	±	−	−	Various (C)
Rhodotorula	−	−	±	+	−	Ubiquitous (C)
Rhodosporidium	−	−	±	+	+	Antarctic sea water/plant surfaces (R)
Sporobolomyces	±	±	±	−	+	Plant surfaces/air (C)
Sporidiobolus	±	−	+	−	+	Plant surfaces (R)

Key * Pigment intensified on agar media supplemented by biotin (0.002%) and ferric ammonium citrate (0.05%), thence a dark pink-red pigmentation is obtained for these yeasts (except *T. sphaerica*). In addition most *Metchnikowia* (*Candida*) *pulcherrima* strains become dark-red to purple with a metallic sheen (Beech and Davenport, 1971).

** I = assimilation of inositol; F = fermentation of glucose; N = nitrate assimilated; M = mycellium and/or pseudomycelium formed; SS = sexual states; ‡ = common imperfect state; ± = positive and negative strains; − = negative (all strains); + = positive (all strains); R = rare (usually need special cultural conditions); C = common.

Table 6.3 Simple identification of *Brettanomyces/Dekkera* species (from Davenport, 1979)

Common characteristics

Resistant to >100 ppm Acitidione (cycloheximide) and strong acid production – use WL (Difco) or Acitidione agar (Oxoid). Mixture of cells, including lemon, ogive (Gothic arch), oval-circular or cylindrical and various mycelial forms; multipolar budding. High vitamin (especially thiamin) level essential for growth. Characteristic aromas produced – 'Brettanomyces' smell (acidic, fruity, butyric), 'mouse' smell in some wines and beers.

Differential characteristics

	L	EMB	VRB agar	Nitrate	Nitrite	Fermentation Galactose	Fermentation Lactose	Fermentation Trehalose	Cyclohexane	Principal habitats
Group 1										
B. abstinens	−	−	−	+	−	−	−	W	−	Mineral waters and
B. naardenensis	±	−	+	+	−	−	−	−	−	non alcoholic beverages
Group 2										
B. anomalus*	−	±	−	±	±	+	±	+	−	Lambic+ and other
B. claussenii	−	±	−	+	−	+	+	+	−	old (spoilt) beers,
B. custersii	+	±	+	+	−	W	−	+	−	ciders and wines
B. intermedius*/										
D. intermedia	−	±	−			−	±	±	±	
B. bruxellensis*/										
D. bruxellensis	−	±	−	+	±	−	−	+	−	
B. lambicus*	±	±	±	+	−	±	−	+	−	
B. custerianus	−	±	−			−	+	W	−	

Key L = Littman oxgall crystal voilet agar
EMB = Eosin methylene blue agar
VRB = Violet red bile agar
Nitrate and nitrite = nitrogen assimilation tests
Cyclohexane = carbon assimilation test
± = positive and negative strains
− = negative for all strains
+ = positive for all strains
+ *Brettanomyces* species are always present in lambic beers
* Commonest species found

Colonies formed by the more common yeasts can usually be readily distinguished from bacteria and common fungi. Fortunately there is a range of yeast colony types, whose descriptions allow the investigator a means of partial organism recognition before carrying out a full identification programme. Indeed colonial and cellular appearance, plus some ecological data, is usually enough to place yeasts into morphological groups and genera from which one can programme for subsequent selective identification procedures (Davenport, 1974a). Briefly this means that certain tests and possible yeast identities are either accepted or rejected as observations are recorded without the full compliment of classical tests (e.g. those given by Lodder, 1970). The important features of yeast colonies and streak appearance are colour (range,

Table 6.4   Perfect/imperfect states of common yeasts

Ascosporogenous genera	Basidiosporogenous genera
Metschnikowia/Candida	*Filobasidiella/Cryptococcus*
	Filobasidium/Leucosporidium and *Candida*
	Rhodospiridium/Rhodotorula
	Sporidiobolus/?Sporobolomyces
	?Rhodotorula

i.e. black to brown, pink-red, orange-cream-white), cross section (i.e. flat, raised, convex, pulvinate, umbonate), edge features (entire, undulate, lobate, filamentous) and surface topography (smooth, rough, shiny, matt, dull, mucoid, leathery, hairy). It is strongly recommended that thumb-nail sketches should be made of each different colony accompanied by drawings of cells. Some essential points must be always considered with yeast morphological characteristics.

1. There are several factors which can influence both macroscopic and microscopic features of yeasts; for example nutrient status, pH of media, residues of sample on the agar surface, cultural methods (surface plating or pour plate techniques) and incubation conditions such as temperature and oxygen concentration.

2. The age of colonies can sometimes show different characters such as some butyrous colonies become 'frosty' or 'hairy' due to the development of chains of normal or elongated cells (pseudomycelium) or cells resembling threads (hyphae or mycelium).

3. Yeast cells in various habitats (i.e. before culturing) often look very different from the same cells grown under laboratory conditions. These differences are usually seen with cell contents and cell size and shape, but the mode of vegetative reproduction (i.e. budding, bud-fission or fission) remains constant.

Yeast cell shapes are round or spherical; cylindrical; egg-shaped (examples of these shapes can be seen with *Saccharomyces cerevisiae* strains in Fig. 6.1a and b); lemon-shaped (e.g. *Kloeckera apiculata*, Fig. 6.2c; elongated ellipse; triangular; bottle; ogive or gothic arch (e.g. *Brettanomyces bruxellensis*, Figs 6.2a and 6.5) and elongated threads (*Saccharomycopsis* and *Geotrichum*, Figs 6.3a, c and 6.4). The principal form of yeast reproduction is a vegetative process where at certain points the cell surface swells to form a bud which eventually becomes detached to form another individual (e.g. *B. bruxellensis*, Fig. 6.2a). At the detachment points a scar is left (e.g. *Kluyveromyces fragilis*, Fig. 6.1c). The latter yeast is an example of a multipolar budding yeast where numerous sites are situated over the whole cell surface. In complete contrast, members of the genus *Schizosaccharomyces* do not bud but divide by a fission process (Fig. 6.2b) and some genera reproduce by combined budding and fission (called bud-fission). With this mode a bud forms either at one end (monopolar budding) or both ends (bipolar budding) and, when mature, separates by fission. This happens only

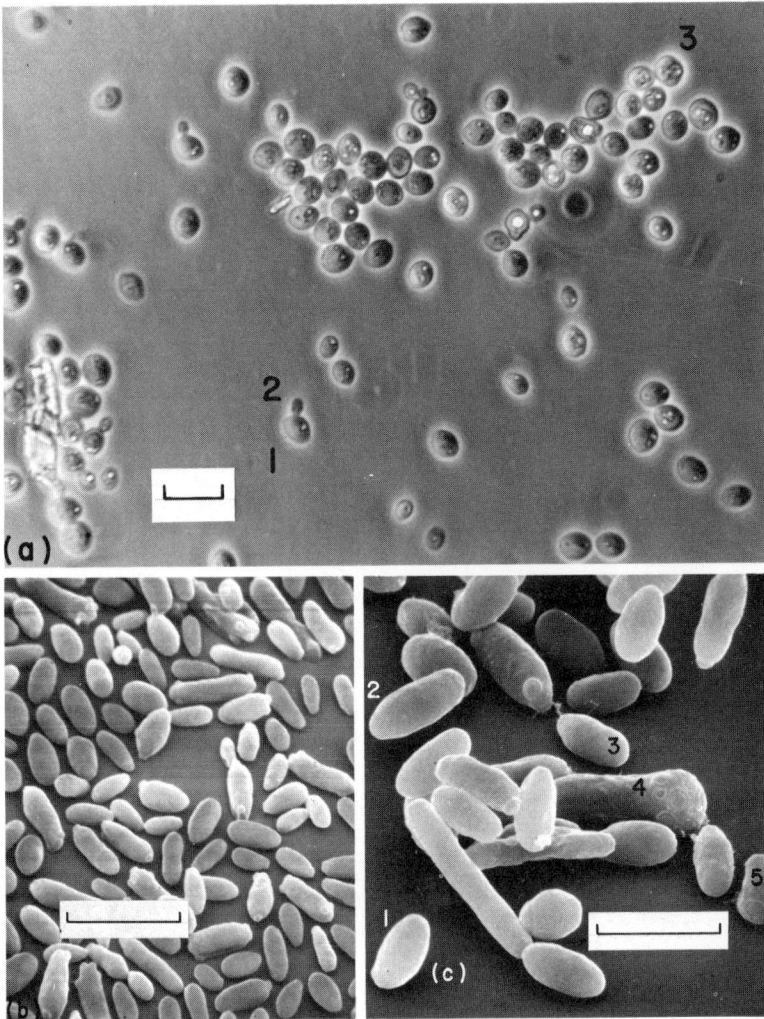

Fig. 6.1 **(a)** *Saccharomyces cerevisiae* (brewing strain) An example of a multipolar budding yeast. **1,** mother cell; **2,** daughter cell: **3,** ascus with four ascospores. Scale = 10 μm. **(b)** *Saccharomyces cerevisiae* (wine strain) Note the scars indicating multipolar budding. Scale = 2.2 μm. **(c)** *Kluyveromyces fragilis* (from a culture collection) Multipolar budding yeast with different cell shapes. **1,** oval; **2,** elongated; **3,** cylindrical; **4,** cylindrical, showing multiple scars, i.e. multipolar budding; **5,** oval with bud scars. Scale = 5 μm.

on the polar axes, never on the lateral surfaces. In addition, the process can be repeated many times over, leaving several scar annulations (Figs. 6.2 and 6.4) rather than open scars (Fig. 6.1). Some multipolar budding genera, within the basidiomycetous yeasts, have special cells bearing external pegs (sterigmata) which produce on the peg tip a secondary cell (a conidium). This conidium is ejected and is therefore a form of ballistospore. These structures are confined to the genera *Aessosporon* van der Walt, *Bullera* Derx,

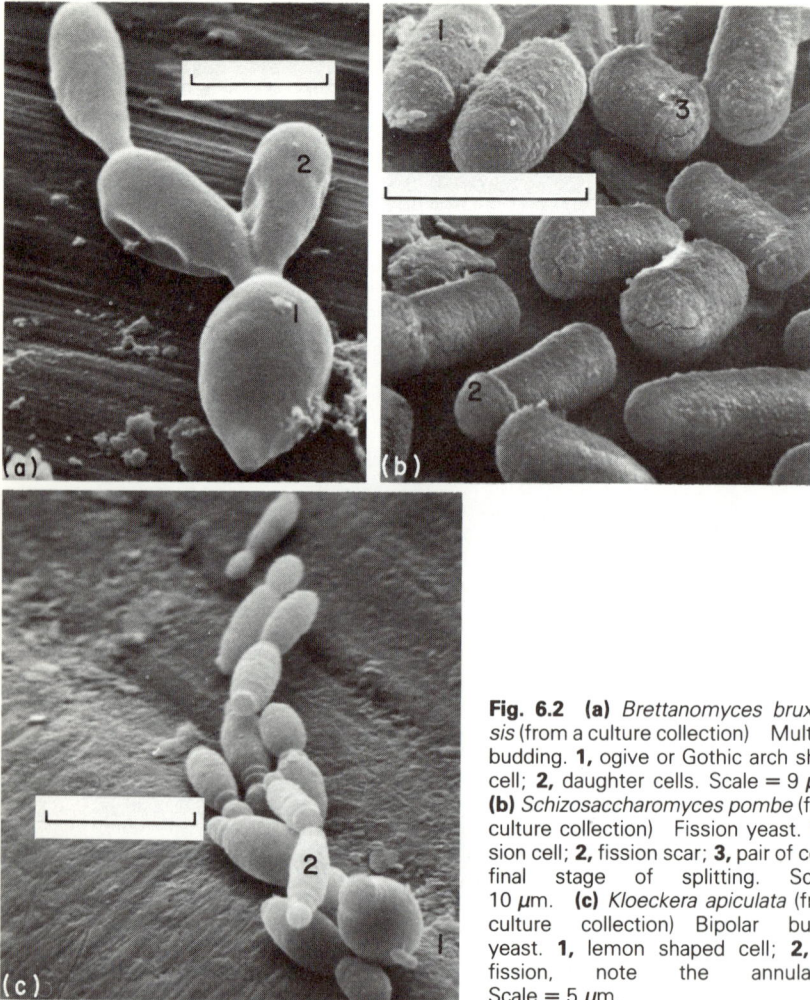

Fig. 6.2 (a) *Brettanomyces bruxellensis* (from a culture collection)   Multipolar budding. **1,** ogive or Gothic arch shaped cell; **2,** daughter cells. Scale = 9 μm.
(b) *Schizosaccharomyces pombe* (from a culture collection)   Fission yeast. **1,** fission cell; **2,** fission scar; **3,** pair of cells at final stage of splitting.   Scale = 10 μm.   **(c)** *Kloeckera apiculata* (from a culture collection) Bipolar budding yeast. **1,** lemon shaped cell; **2,** bud-fission, note the annulations. Scale = 5 μm.

Sporidiobolus Nyland and *Sporobolomyces* Kluyver and van Niel. Many yeasts have either ascomycetous or basidiomycetous mechanisms for sexual reproduction which form the basic diagnostic characters for current classical taxonomy. This part of yeast studies can be difficult to demonstrate as well as interpret, hence only a simple outline will be given of selected ascospore forming yeasts since many of the latter and most of the basidiomycetous types are rarely of industrial importance. Yeast cells containing ascospores (asci) can be distinguished from vegetative cells by the presence of firm bodies of definite shapes – including spherical (e.g. *Saccharomycodes ludwigii*, Fig. 6.4b); elliptical; oval; walnut; kidney; hat (e.g. *Pichia membranaefaciens*, Fig. 6.4c); saturn; helmet; and needle (e.g. *Metschnikowia pulcherrima*, Fig. 6.4d). In addition ascospore walls may be smooth, rough, warty, thick walled, double walled, starchy or contain oil drops. The number of spores per

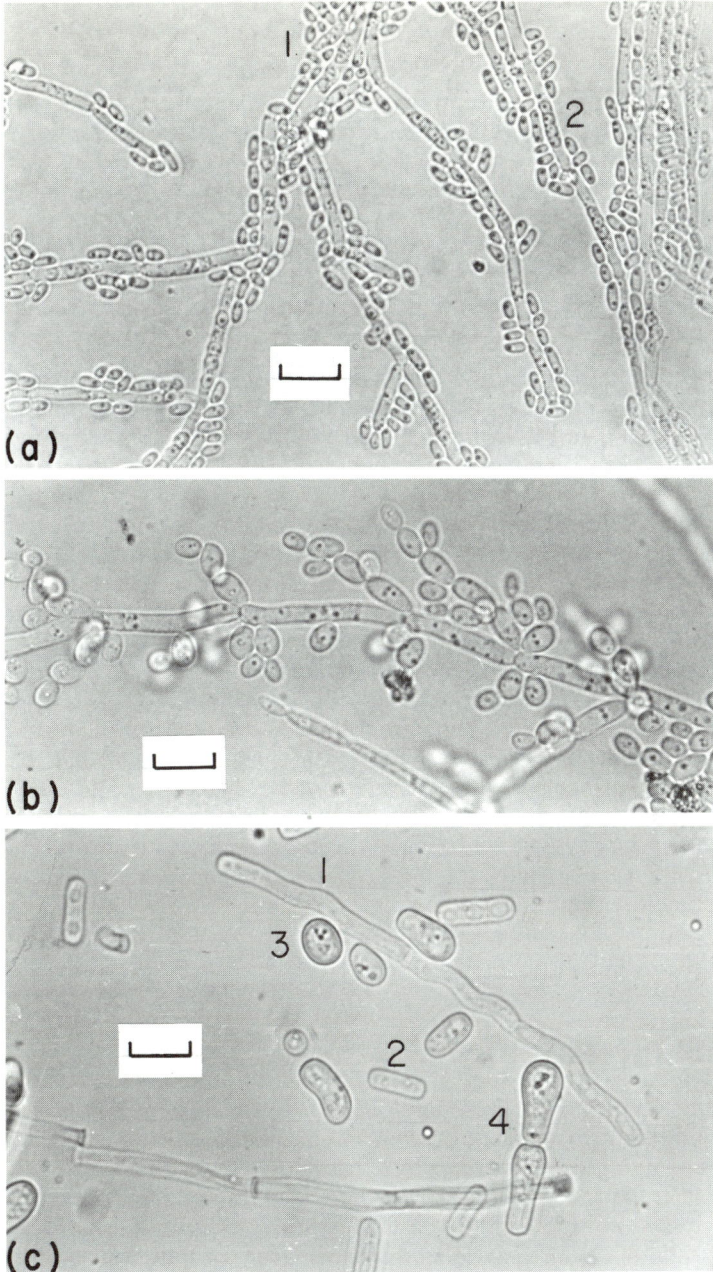

Fig. 6.3 **(a)** *Saccharomycopsis vini* (from insects) **1,** budding cells; **2,** mycelium with ascospores. Scale = 10 μm. **(b)** *Pichia membranaefaciens* (from wine) Pseudomycelium formation. Scale = 10 μm. **(c)** *Geotrichum lactis* (from a culture collection) **1,** mycelium; **2,** arthrospores (recently formed); **3,** arthrospores (rounded, i.e. older cells); **4,** pair of arthrospores not separated. Scale = 10 μm.

ascus is usually between 1–4, rarely 8–16 and only one species (*Kluyveromyces polysporus*) produces an exceptional unknown number of ascospores. Yeast ascospores arise in one of three ways.

1. **Isogamic conjugation**   Conjugation of two equal cells, for example *Zygosaccharomyces bailii*. (Heterogamic conjugation (see below) can also occur in this species.)
2. **Heterogamic conjugation**   Conjugation of unequal cells, for example *Zygosaccharomyces florentinus*.
3. **No trace of conjugation**   In this case sexuality has been exhibited prior to ascus formation, for example fusion of a pair of ascospores to give a cell (zygote) which then buds to give other generations (e.g. *Saccharomycodes ludwigii*, Fig. 6.4b).

It is very important to remember that not all yeasts produce ascospores or ballistospores. Conditions may not be favourable for the production of the perfect state. On repeated sub-culturing many yeasts fail to sporulate and some yeasts have as yet never shown a sexual state (e.g. *Schizoblastosporon ciferri*). In this case such yeasts are placed in the imperfect fungi. Nevertheless it is wise to interpret tests as for perfect yeasts even if sexual states are not observed, since current taxonomic practise assigns all yeasts to either the ascomycetous or basidiomycetous genera. Thus yeasts often have two sets of names – one for the sexual state and one for the asexual state. Additionally, many yeasts acquire synonyms with advance and revision of knowledge.

Sometimes many yeasts form chains of cells which fail to separate; this trait ranges from a link of a few normal sized cells which may (e.g. *Pichia membranaefaciens*, Fig. 6.3b) or may not be branched to fully formed mycelium (e.g. *Saccharomycopsis vini*, Fig. 6.3a). This mycelium can be non-septate or septate; some yeasts have one form or the other while a few genera have both pseudomycelium and mycelium. Usually these mycelial yeasts form arthrospores by fragmentation of the mycelium (e.g. *Hyphopichia burtonii*, Fig. 6.4a). Occasionally thick-walled cells, chlamydospores, capsules and other cellular modifications (e.g. teliospores and sporidia) are encountered in some genera. These characters are all used in classical taxonomy (Lodder, 1970).

6.4   Physiology and biochemistry

Yeasts, like other fungi, do not have any photosynthesis mechanism; therefore carbon requirements have to be drawn from a suitable substrate. A wide range of carbon compounds are utilized by yeasts; some compounds are used by all yeasts (e.g. fructose and glucose) but there is great variability in other carbon sources used throughout the yeast domain. This forms the basis of classical separation of yeast species and strains. The commonest carbon sources selectively used are hexose sugars and their derivatives, disaccharides, trisaccharides, some polysaccharides, pentose sugars and their derivatives, alcohols, certain organic acids (e.g. DL-lactic acid), and sometimes hydrocarbons such as alkanes and alkenes. Carbohydrate utilization by yeasts is considered in two ways in taxonomy; this concept can equally be applied

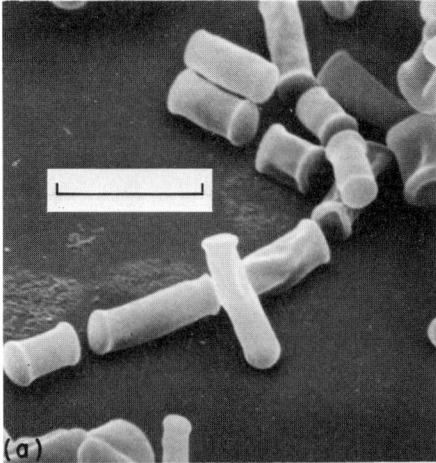

Fig. 6.4 **(a)** *Hyphopichia burtonii* (from a culture collection) Arthrospore formation. Scale = 10 *μ*m. **(b)** *Saccharomycodes ludwigii* (from cider) **1,** lemon shaped cell showing annulation scars of bipolar (bud-fission) budding; **2,** conjugating ascospores (this usually takes place within the ascus, ascospores probably released during preparation for S.E.M.). Scale = 3.6 *μ*m. **(c)** *Pichia membranaefaciens* (from wine) Hat shaped ascospores. Scale = 9 *μ*m. **(d)** *Metschnikowia pulcherrima* (from insects) Needle shaped ascus. Scale = 10 *μ*m.

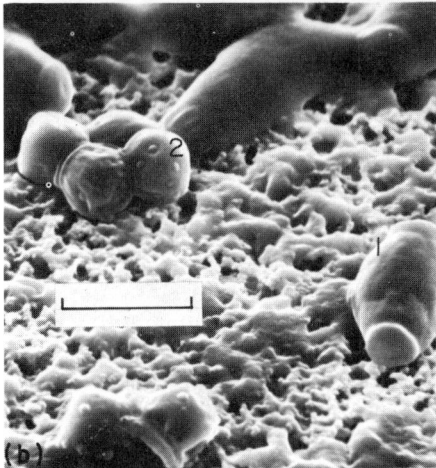

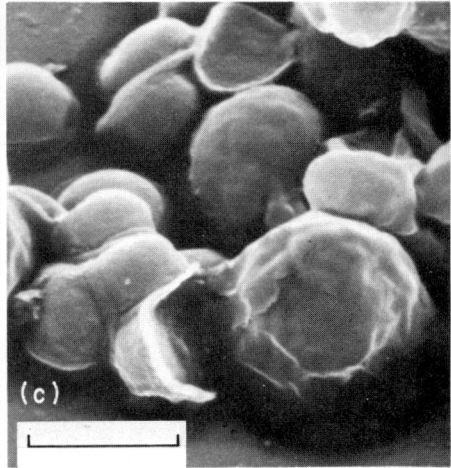

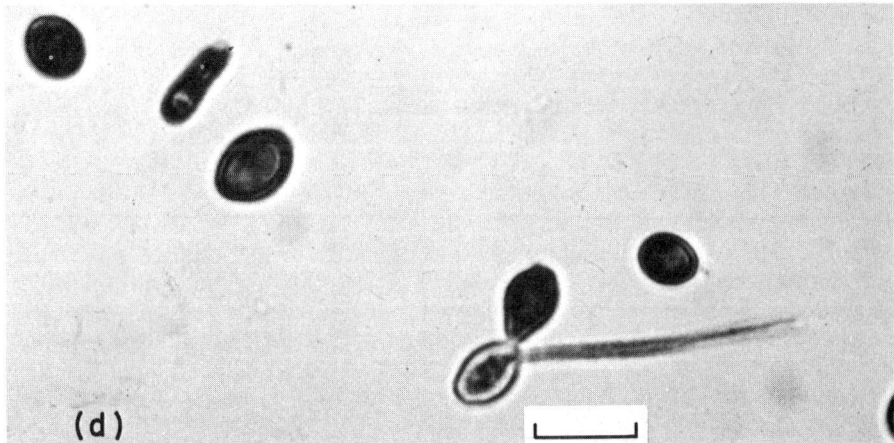

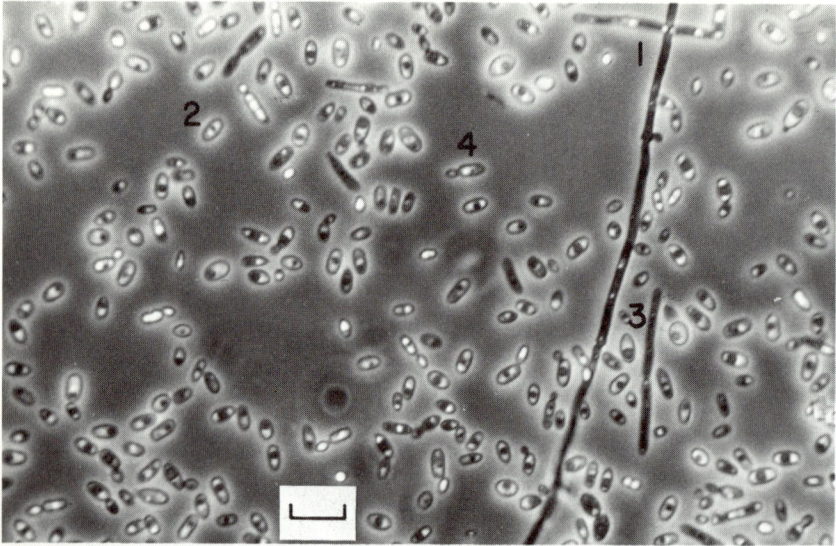

Fig. 6.5 *Brettanomyces bruxellensis* (from lambic beer)   Mixture of cells. **1,** mycelium; **2,** oval cell; **3,** cylindrical cell; **4,** ogive cell (see Fig. 6.2(a) for comparison). Scale = 10 μm.

here. First, **fermentation** is an anaerobic process where carbohydrates are metabolized to give ethanol and carbon dioxide (usually this activity is depressed by molecular oxygen, i.e. the pasteur effect).* Second, **assimilation** is an aerobic process where the yeast uses the compound by respiration. In both cases there is uptake by the cell so the terms fermentation and assimilation can sometimes be misleading – it is simplest to remember that yeasts which can ferment a compound can also assimilate it but the converse is not true, and that fermentation = the production of 'gas + acid'.

Many nitrogen sources can be used by most yeasts providing there is an adequate amount of essential growth factors (e.g. vitamins) present and the concentration of certain nitrogen compounds (e.g. urea) is not a toxic level. All yeasts can use peptone, asparagine, ammonium sulphate and urea but only selected species and/or strains can use potassium nitrate, sodium nitrite, aliphatic amines and specific amino acids – these substances are often employed in yeast taxonomy (Lodder, 1970; Barnett, Payne and Yarrow, 1979).

Among the other yeast physiological and biochemical activities are growth in low or high temperature environments (i.e. <10°C and >37°C); formation of acid, esters, taints and odours, and extracellular compounds (e.g. starch); fat splitting; resistance to beverage and food preservatives, and anti-microbial compounds (e.g. cycloheximide); growth in high osmotic pressure habitats; and synthesis of its own vitamin requirements. Not all these characteristics

---

* Some yeasts exhibit an opposite effect (e.g. *Brettanomyces* species). This is described as the negative pasteur or custers effect (Wiken and Scheffers, 1961).

are common to all yeasts, hence it is possible to use these features for both identification and classification procedures.

6.5   Taxonomy

Yeast taxonomy can be difficult for everyone. The present state of this subject is not helpful to specialist or non-specialist alike because of the many advances in yeast studies over the past decade or so. These advances have not been directed toward the routine use required by many laboratories. Therefore, the following is an outline restricted to the important yeast genera which are most likely to be encountered. Sufficient characterization and methodology have been described to cope with the selected genera (Table 6.5). A full list of genera is given in section 6.8 and a key to the separation of the principal industrial yeast species in section 6.7. The methodology used is based on modifications (Table 6.6) of classical systems (Lodder and Kreger van Rij, 1952; Beech, Davenport, Goswell and Burnett, 1968; Davenport, 1973, 1974a, 1978, 1979 and 1980; Barnett and Pankhurst, 1974; Barnett, Payne and Yarrow, 1979). This approach has a dual advantage since it is simple and economic, and is part of standard identification practice within yeast culture collections.

6.6   Exposition of identification methods

Before attempting to start identification tests, it is essential that full consideration should be given to the principles described earlier, i.e. history, macro- and microscopic examinations followed by selected tests. Surface plating techniques are preferable for yeasts since pour plates tend to inhibit some aerobic (non-fermenting) species and do not allow any satisfactory count or examination. A preliminary screening into two groups may be made applying a colony staining method described by van der Walt and Hopsu-Havu (1976) and later, further developed by incorporating into a simplified yeast identification system (Davenport, 1978, 1980).

The Dalmau technique is used for observations of pseudomycelium and mycelium on either corn-meal or potato dextrose agar for both aerobic and anaerobic inoculations (Fig. 6.6). Observations are best made daily up to five days. Drawings and notes should be made of the cellular features. Sometimes ascospores, chlamydospores and arthrospores as well as other features may be observed with this technique. Pseudomycelium and mycelium may sometimes be evident along the edges of colonies on isolation plates; such an observation can often dispense with the need for a separate test.

Yeast ascospores can be more readily seen in fresh isolates and sometimes in spoiled products. Wet mounts are essential to determine spore shape, and staining by the malachite green/safrannin technique of Schaeffer and Fulton for bacterial spores only for confirmation. In using this technique a smear is prepared, fixed by flaming three times, flooded with malachite green (5% aqueous solution allowed to stand for half an hour and filtered) and heated to steaming three or four times within 30 seconds. Excess stain is washed off

Table 6.5 Characteristics of selected yeast genera

1. Multipolar budding yeasts and yeast-like genera

	DBB test (A or B)*	Ballistospores	Ascospores	Mycelium	Pseudomycelium	Circular	Oval–cylindrical	Mixture	Arthrospores	Fermentation	Nitrate	Comments
(a) Generally rough or hairy colonies – filamentous growth produced												
Aureobasidium	A	–	+	+	+	+	+	+	–	–	+	Extremely variable genus
Geotrichum	A	–	±	+	–	–	+	–	+	–	±	Perfect state, rare
Hyphopichia	A	–	+	+	+	–	+	+	+	+	–	Many synonyms (see Table 6.1)
Saccharomycopsis	A	–	+	+	+	+	+	+	+	+	–	Name in Lodder (1970) no longer valid
Trichosporon	A/B	–	±	+	+	±	±	±	+	–	±	Budding cells of various sizes and shapes
(b) Carotenoid pigment												
Cryptococcus	B	–	–	–	–	++	–	–	–	–	±	Inositol assimilated; starch formation ±
Rhodotorula	B	–	–	–	–	++	++	–	–	–	±	Yeast cells only
Sporobolomyces	B	+	–	±	±	–	±	–	–	–	±	Common species yeast cells and ballistospore formation only
(c) Pulcherrimin pigment												
Metschnikowia	A	–	+	–	+	+	+	+	–	+	–	Single species, *M. pulcherrima* (see Table 6.2)
Kluyveromyces	A	–	+	±	+	+	+	+	–	+	–	Selected species only (see Table 6.2)
(d) Mucoid – cream/white colonies												
Cryptococcus	A & B	+	++	–	–	++	–	–	–	–	±	Inositol assimilated; starch formation ±
Torulopsis	A	–	+	–	–	++	–	–	–	±	±	Often imperfect states of other genera

(e) Smooth or rough colonies

Genus	DBB*								
Brettanomyces	A	−	+	+	+	+	+	±	Very strong acid formation
Candida	A & B	+	+	+	+	+	+	±	Often imperfect states of other genera
Cryptococcus	A & B	+	−	−	−	−	−	±	Inositol assimilated; starch formation ±
Dekkera	A	+	+	+	+	+	−	+	Perfect states of two *Brettanomyces* species
Hansenula	A	+	±	+	+	+	+	+	Esters often produced
Kluyveromyces	A	+	±	+	+	+	+	−	Hydrocarbons assimilated
Lodderomyces	A	+	−	−	−	+	+	−	Perfect state of *C. parapsilosis*
Metschnikowia	A	+	−	−	−	+	+	−	Characteristic 'needle' ascospores
Pichia	A	+	+	+	+	+	±	−	Negative or very weak fermentation
Saccharomyces	A	+	−	±	±	−	+	−	Vigorous fermentation, one or more common sugars
Torulaspora	A	+	−	−	±	±	+	−	Species formerly in genus *Saccharomyces*
Torulopsis	A	−	−	±	±	−	±	±	Often imperfect states of other genera
Zygosaccharomyces	A	+	+	±	+	+	+	−	Ascospores with conjugation process

2. Bipolar budding yeasts and yeast-like genera

Genus		Description
Kloeckera/ Hanseniaspora	No spores Ascospores	Acid and/or acetaldehyde production resistant to 100 ppm cycloheximide
Saccharomycodes	Ascospores	Large lemon shaped cells; not resistant to cycloheximide (i.e. <10 ppm); highly resistant to free SO$_2$ (i.e. >200 ppm)

3. No budding cells – fission only

Schizosaccharomyces

* DBB Test: Diazonium Blue B salt (1 mg ml^{-1} in 0.1 M Tris-HCl buffer (at 4°C), one drop on the yeast colony (van der Walt and Hopsu-Havu; Davenport, 1978 and 1980)
A = No reaction = Ascomycetous genus
B = Red colouration = Basidiomycetous genus
± = positive and negative species within the genus

Table 6.6 Characteristics to be examined and methods used in yeast examination

Characteristics to be examined	Media/methods and comments
Pigment, macro- and micromorphology	Direct examination of samples Isolation and purification media Dalmau plates for pseudomycelium and mycelium
Ascospores and ballistospores	Isolation and purification media* Selection of media-appendix
Cell sizes, pellicle and odour formation	Malt wort liquid: glucose, peptone, yeast-extract broth Direct examination of samples
Fermentation and assimilation	Combined test or separate methods N.B. Fermentation may be observed on examination of the sample
Assimilation only (carbon and nitrogen sources)	Solid and liquid methods
Acid formation	Chalk agar
Cycloheximide resistance	WLD** (Difco) or Acitidione agar (Oxoid) N.B. Acid formation also indicated on these media

* Old contaminated agar plates and membrane filters often reveal sporing yeasts
** WLD Agar = WL Differential agar

under the tap for about 30 seconds, 0.5% aqueous safrannin solution applied for 30 seconds and then the smear is washed, blotted, dried and examined with oil immersion lens. Immersion oil is placed directly on the dried stained smear. A medium power (e.g. ×50) oil immersion lens is preferable to a high power one (i.e. ×100). Spores stain green, vegetative cells pink-red.

Ballistospore formation is usually easy to detect since mirror-image shadow-like colonies are formed on the petri dish lid (from plates incubated bottom uppermost) or large colonies surrounded by similar smaller colonies – this gives the appearance of a planet with satellites. To confirm this observation, two Petri dishes are poured with a suitable medium (e.g. malt extract agar) and the lids discarded. One dish is inoculated with the test yeast – this forms the top part, and the other the lower part. Both parts are then held together with sellotape. After about 5–7 days' incubation, any ballisto-spores shot off from the parent colonies will have grown to produce a mirror-image growth pattern on the lower dish medium.

Fermentation and assimilation for common carbohydrates can be conveniently carried out simultaneously (Davenport, 1974b, 1978 and 1975) by using small test-tubes or screw-cap bottles with Durham tube inserts. Each test-tube or bottle contains a basal medium (yeast nitrogen base – Difco) and 2% (w/v) of a single carbon source. After inoculation and incubation the results indicate fermentation by gas collecting in the Durham tube or assimilation by turbidity only. This method can be standardized by using a known inoculum and both incubation and interpretation conditions (Davenport, 1975). Fermentation tests can also be carried out by a broth containing 0.5% yeast extract (van der Walt, 1970) and an indicator, Bromothymol Blue (Wickerham, 1951). Van der Walt (1970) gives further details of this test and

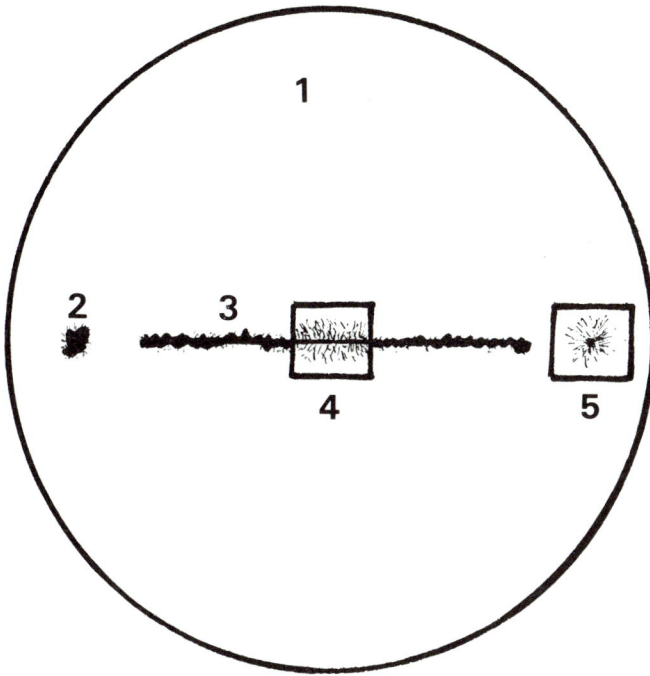

Fig. 6.6 The Dalmau plate method for pseudomycelium and mycelium for yeasts and yeast-like organisms. Inoculations made on corn-meal agar are to produce growth from aerobic point **2** and streak **3** as well as anerobic streak **4** and point **5** sites. The anaerobic sites are areas covered with alcohol flamed cover glasses. A hand lens or low-power microscope can be used to give an overall picture of yeast development. The medium-power microscopy can be used to examine the edges of the main inoculum areas where filamentous growth is most likely to occur, as shown on the diagram.

the responses for the majority of known yeasts, and more recently other yeast responses have been reported (Barnett *et al.*, 1979).

Assimilation tests can be performed either on solid or in liquid media. For carbon tests, yeast nitrogen base (Difco) is used and for nitrogen tests, yeast carbon base (Difco) is used. These media are chemically defined, containing all necessary growth substances except that a single source of either carbon or nitrogen must be added to the appropriate test medium. For liquid carbon assimilation tests the yeast must be depleted of any reserve stored carbon source by growth in a minimal medium (yeast nitrogen base plus 0.05% glucose) for 3–4 days. The culture is then centrifuged, washed in fresh yeast nitrogen base solution followed by preparing a standard inoculum ($\simeq 10^6$ cells ml^{-1}), of which 0.1 ml is added to tubes containing 5 ml test medium. Thus each tube contains the same basal medium but a different carbon source at the same concentration as for the glucose tubes. Carbon sources used are selected from those given for the fermentation tests and others as suggested by van der Walt (1970) and Barnett *et al.* (1979). Incubation should be for at least 14 days and the growth read daily – yeast growth is evident by turbidity and there may be other signs such as pigment formation and pellicle formed. For nitrogen assimilation the most important compound to be tested is potassium nitrate $(0.78\,g\,l^{-1})$ in yeast carbon base solution. Cultures are standardized as for the carbon tests, inoculated and then incubated for one week. Tubes not showing turbidity (i.e. no growth) are negative and are discarded. Turbid tubes are re-inoculated by taking one drop, from the tube, for inoculating into a fresh tube which is incubated for one week. Any tubes showing signs of growth are deemed positive. If required, the results and inoculum can be standardized by a simple card method suggested by Wickerham (1951) and modified by Davenport (1980). Assimilation tests with solid media are quicker than the liquid method but standardization of the test compound concentration is not possible unless impregnated discs are used. These are not always satisfactory and can be expensive. However, for a simple test take 4% washed agar (18 ml) (van der Walt, 1970) or noble agar (Difco) plus test yeast cells suspended in 2 ml of either yeast carbon base (for nitrogen sources) or yeast nitrogen base (for carbon sources) (Davenport, 1968). Mixing must be at 50°C and the total volume is enough for one 9 cm Petri dish. A minute amount of control compound (i.e. glucose for carbon tests and peptone for nitrogen tests) is placed at one quadrant and the remaining three each receive a similar amount of test compound. This issues into the basal medium and a positive growth response is shown by a diffuse halo. Plates must be examined daily (up to 7 days) as there can be rapid growth with some species.

Acid formation can be demonstrated very well by inoculating on chalk agar (yeast extract, 1 g; glucose, 5 g; calcium carbonate, 0.5 g; agar, 2 g; distilled water, 100 ml) or acidity can be observed using WL and WLD agars in which acid is indicated by the medium turning from blue-green to yellow.

Cycloheximide resistance can best be tested during 'detection/isolation' procedures by using a commercial media containing this compound, or a stock solution (0.5% aqueous cycloheximide) can be used to add specific amounts to others such as malt or wort agar.

Further tests and other identification details are described by Lodder and Kreger van Rij (1952), Beech et al. (1968) and Davenport (1973, 1974a, 1980).

6.7 Guide for the separation of the principal yeasts and yeast-like organisms of industrial significance

METHOD   First make a short list of the tentative identities of the organism or organisms. This is carried out by considering the habitat and by direct microscopic examination of a sample followed by growth on a solid medium (e.g. wort agar). Separation can then be achieved by regarding the short list and subsequent selection of the observations and tests. The identification process can be started or stopped at any point depending on the initial observations and amount of detail required (i.e. groups, genera or species).

PRIMARY SEPARATION   Based on colony characteristics
See Table 6.7 (p. 84).

SECONDARY SEPARATION   Based on habitats and cellular characteristics
See Table 6.8 (p. 85).

TERTIARY SEPARATION
See Table 6.9 (pp. 86–7).

6.8   Genera of yeasts and yeast-like organisms

This list (Table 6.10 (pp. 88–90)) is intended as a guide for those who wish to proceed with other yeast studies. During the last two decades there have been many new genera defined; these have either been newly discovered organisms or known organisms which have been 're-classified'. There are many reasons for this 're-classification'; details may be found in the appropriate literature.

Table 6.7 Primary separation, based on colony characteristics (see Tables 6.2 and 6.5 for pink and red yeasts)

GROUP 1   Young colonies often butyrous, later becoming filamentous, rough/tough, or hairy and usually dark pigmented (e.g. black, grey, brown-orange, green-grey).
Genera: *Aureobasidium, Geotrichum* and *Trichosporon*

GROUP 2   Colonies usually butyrous, various colour shades of white-grey, cream-light brown, but colours and texture never becoming as Group 1. Older colonies tops or whole surface may become darker than younger colonies.
Genera: *Hanseniaspora/Kloeckera, Saccharomycodes, Saccharomyces, Zygosaccharomyces, Kluyveromyces* and *Schizosaccharomyces*

GROUP 3   Colonies butyrous later rough, or rough only. Various colour shades of white-grey, cream to light brown, occasionally pinkish* but colours and textures never as Group 1. Older colonies may become slightly darker.
Genera: *Candida, Brettanomyces/Dekkera, Hansenula, Pichia, Torulopsis* and *Debaryomyces*.

* Not carotenoid pigment

Table 6.8 Secondary separation, based on habitats and cellular characteristics

	Habitats	Pinkish colonies*	Slow growth**	Odour	Mixture	Oval-circular	Oval-cylindrical	Ogive	Lemon	Mycelium	Pseudomycelium	Arthrospores	Vegetative reproduction	Ascospores
Aureobasidium	U	–	–	–	+	+	+	–	–	+	+	R	M	–
Geotrichum	An:S:Fp	–	–	–	–	–	+	–	–	+	–	+	F	–
Trichosporon	An:S:Fp	–	–	AC	+	+	+	–	–	+	+	C	MPB/F	–
Hanseniaspora/Kloeckera	B:Fp	–	–	–	–	+	–	–	+	–	–	–	BPB	+
Saccharomycodes	B	–	–	–	>	+	+	–	+	–	P	–	BPB	+
Saccharomyces	U	–	–	–	V	+	+	–	–	–	P	–	MPB	+
Zygosaccharomyces	B:Fp	–	–	–	–	+	+	–	–	–	P	–	MPB	+
Kluyveromyces	B:Fp	+	–	–	+	+	+	–	–	>	+	–	MPB	+
Schizosaccharomyces	Fp:B	–	–	–	+	+	+	–	–	***	+	–	F	–
Candida	Fp:B	–	>	–	+	+	+	–	–	>	+	–	MPB	–
Brettanomyces/Dekkera	B	>	>	C	+	+	+	+	+	>	+	–	MPB	+
Hansenula	Fp:B	–	–	E	+	+	+	–	–	>	+	–	MPB	+
Pichia	B:Fp	–	–	–	–	+	+	–	–	–	+	–	MPB	+
Debaryomyces	Fp:B	–	–	–	–	+	+	–	–	–	+	–	MPB	+
Torulopsis	Fp:B	–	–	–	–	+	–	–	–	–	–	–MPB	+	+

* Not carotenoid pigment: unknown pigment for *Brettanomyces/Dekkera* and Pulcherrim pigment for some *Kluyveromyces* species (see Table 6.2).

** Slow growth, i.e. >7 days up to 8 weeks depending on strain differences.

*** Mycelium present only for *Schizosaccharomyces japonicus*.

U = ubiquitous; An = animal sources; B = beverages; Fp = food products; S = soil; – = negative all strains; + = positive all strains; AC = acetaldehyde; C = characteristic mouse and/or acid-rancid smell of *Brettanomyces*; E = estery; V = variable, i.e. some strains +, others –; R = rare; P = primitive, i.e. short chains of normal or elongated cells; M = multilateral budding; F = fission; MPB = multipolar budding; BPB = bipolar budding.

Table 6.9 Tertiary separation

	Habitats	Acid production*	Cycloheximide resistance**	Ethylamine HCl***	Fermentation	Nitrate	Ascospores****	Utilization of carbon compounds+										High sugar medium++	High sugar medium+++	Nitrite assimilation
								GA	SU	MA	LA	RA	CE	TR	SOR	ME	LAC			
VEGETATIVE REPRODUCTION BY MULTIPOLAR BUDDING																				
Brettanomyces anomalus	B	+	+	+	+	+	−	F	F	F	−									
Brettanomyces/Dekkera bruxellensis	B	+	+	+	+	+	R	F					F	−						
B. intermedius/D. intermedia	B	+	+	+	+	+	R						F	F						
B. claussenii	B	+	+	+	+	+	−	F					F	F						
B. custersii	B	+	+	+	+	+	−	F			F		F	F						
B. lambicus	B	+	+	+	+	+	−	F			A		A	F						
B. abstinens	B	+	+	+	−	+	−	−			−		A	−						
B. naardenensis	B	−	+	+	+	+	−	A			−		A	−						
Hansenula anomala	U	−	−	+	+	+	C	F	F	F								V	−	−
H. subpelliculosa	Fp	−	−	+	+	+	C	A	A	A								+	−	−
Candida curiosa	FF	−	−	+	+	−	−	F	F	−			A	F	A			−	−	−
C. parapsilosis	Fp:B	−	−	+	+	−	−	F	A	A			−	A	A			−	−	−
C. tropicalis	Fp	−	−	+	W	−	−	F	F	F			A	F	A			−	−	−
C. zeylanoides	B	−	−	+	W	−	−	A	−	−			A	−	A			−	−	−
C. valida/Pichia membranaefaciens	U	−	−	+	W	−	+	−	−				−	−	−					
P. fermentans	Fp	−	−	+	+	−	+	−	−				−	−	−					
Torulopsis famata/Debaryomyces hansenii	U	−	−	+	W	−	+	W	W	W			A	A	A			V	−	+
Kluyveromyces bulgaricus	B	−	+	+	+	−	+	W										−	−	−
K. fragilis	Fp	−	+	+	+	−	R											−	−	−
K. lactis	Fp	−	+	+	+	−	R													
K. maxianus	Fp	−	+	+	+	−	R											V	+	+

Species	Reproduction								Fermentation / Assimilation			
Zygosaccharomyces cidri	B	–	+	+	–	+	+	+	F	F	F	F · A
Z. florentinus	B	–	+	+	–	+	+	+	F	F	F	–
Z. microellipsodes	B	–	+	+	–	+	+	+	F	F	–	A
Saccharomyces exiguus	B	–	+	+	–	R	–	R	F	F	–	A
S. unisporus	B:Fp	–	+	R	+	–	F	–	F	–	–	–
Z. bailii	B:Fp	+	>	>	+	+	C	A	A	V	A	+ · –
Z. bisporus	Fp:B	–	>	>	+	+	R	A	A	–	F	+ · +
Z. rouxii	Fp	–	>	>	+	+	R	A	A	V	V	+ · +
Torulaspora delbrueckii	B:Fp	–	>	>	+	–	+	–	A	V	V	A · –
S. cerevisiae⊕	U	–	–	–	+	C	V	F	F	F	–	–

VEGETATIVE REPRODUCTION BY BUD-FISSION (i.e. bipolar budding)

Species	Reproduction										
Saccharomycodes ludwigii	B	–	–	+	+	–	C	F	–	–	
Hanseniaspora osmophila	Fp	+	+	+	+	–	C	–	A	–	
H'spora uvarum/Kloeckera apiculata	B:Fp	+	+	+	+	–	C	–	–	–	
H'spora valbyensis/Kloeckera apiculata	B:Fp	+	+	+	+	–	C	–	–	–	

VEGETATIVE REPRODUCTION BY FISSION ONLY

Species	Reproduction									
Schizosaccharomyces japonicus	B:Fp	+	+	+	+	+	F	–	F	+ · –
S. malidevorans	B	+	+	+	+	–	–	–	+ · –
S. octosporus	Fp	+	+	+	–	–	F	F	+ · –
S. pombe	Fp:B	+	+	+	+	C	F	F	+ · –

* Detected by Custer's medium (5 g glucose, 0.5 g yeast extract, 0.5 g calcium carbonate, 3 g agar and 100 ml distilled water). Acid formation indicated by clear zones.

** Growth on either actidione agar (Oxoid) or WL differential agar (Difco or BBL). Acid formation also indicated by medium colour change from blue-green to yellow.

*** 1.17 g Difco yeast nitrogen base and 0.064 g ethylamine HCl in 100 ml distilled water – filter sterilize.

**** Check samples and all inoculated media from isolation and identification tests, especially old agar plates – not essential to demonstrate since other features are enough to separate these yeasts.

B = beverages; U = ubiquitous; FF = frozen foods (incubation <20°C); R = rare; C = common; F = fermentation (gas formation + growth); A = assimilation (no gas + growth); W = weak, growth and/or fermentation; V = variable (i.e. strain differences, some +, some –); – = all strains negative; + = all strains positive; Fp = food products.

GA = galactose; SU = sucrose; MA = maltose; LA = lactose; RA = raffinose; CE = cellobiose; TR = trehalose; SOR = sorbose; ME = melibiose; LAC = lactic acid.

NB. ⊕ Current classification includes many former *Saccharomyces* species (e.g. *S. carlsbergensis, S. uvarum, S. bayanus, S. heterogenicus* and others).

Table 6.10 List of genera of yeasts and yeast-like organisms

	Major classical works					Comments
	Lodder and Kreger van Rij (1952)	Lodder (1970)	von Arx (1974)	Barrett et al. (1974)	Barrett et al. (1979)	+From previously either described yeasts within genera or single species
Ascomycetous genera						
Ambrosiozyma van der Walt			*	*	*	Pichia and Endomycopsis
Ascoidea Brefeld			*			
Arthroascus von Arx			*	*	*	Endomycopsis javanensis
Botryoascus von Arx			*			
Cephaloascus Hanawa			*			
Citeromyces Santa Maria		*	*	*	*	++Torulopsis
Cyniclomyces van der Walt			*			Saccharomycopsis
Debaryomyces Klöcker	*	*	*	*	*	
Dekkera van der Walt		*	*	*	*	++Brettanomyces (2 species)
Dipodascus Lagerh			*			++Cryptococcus albidus
Endomyces Reess	*		*			++Geotrichum
Eremascus Eidam in Cohn			*			
Eremothecium Borzi			*			
Guilliermondella Nadson and Krassilnikov			*	*	*	E. selenospora
Hanseniaspora Zikes	*	*	*	*	*	++Klockera
Hansenula H. & P. Sydon	*	*	*	*	*	
Hormoascus von Arx			*	*	*	E. platypodis
Hypopichia von Arx and van der Walt			*		*	E. burtonii
Kluyveromyces van der Walt	*	*	*	*	*	Saccharomyces
Lipomyces Lodder & Kreger van Rij		*	*	*	*	
Loddermyces van der Walt		*	*	*	*	++C. parapsilosis
Metschnikowia Kamienski		*	*	*	*	++Candida
Nadsonia Sydow	*	*	*	*	*	
Nematospora Peglion	*	*	*	*	*	
Pachysolen Boidin & Adzet		*	*	*	*	

Genus				Related species
Pachytichospora van der Walt	*	*	*	
Pichia Hansen	*	*	*	*S. transvaalensis*
Saccharomyces Meyen emend. Reess	*	*	*	
Saccharomycodes Hansen		*	*	
Saccharomycopsis Schiönning	*	*	*	*Endomycopsis*
Schizosaccharomyces Linder	*	*	*	
Schwanniomyces Klöcker	*	*	*	++*Candida*
Stephanoascus Smith, van der Walt and Johannsen				
Torulaspora Linder		*	*	*Saccharomyces*
Wickerhamia Soneda		*	*	
Wickerhamiella van der Walt and Liebenberg			*	*Torulopsis domercii*
Wingea van der Walt		*	*	
Zendera Redhead and Malloch			*	*Pichia robertsii*
Zygosaccharomyces Barker			*	*E. ovetensis* *Saccharomyces*

Basidiomycetous-like genera

Genus				Related species
Aessosporon van der Walt	*		*	
Bullera Derx		*	*	++*Sporobolomyces salmonicolor*
Filobasidiella		*	*	
Filobasidium Olive		*	*	
Itersonia Derx			*	
Leucosporidium Fell, Statzell, Hunter and Phaff	*	*	*	++*Candida*
Rhodosporidium Banno		*	*	++*Rhodotorula*
Sporidiobolus Nyland			*	
Sporobolomyces Kluyver and van Niel	*	*	*	++*C. albicans, C. claussenii* and *C. stellatoidea*
Syringospora (Robin) Berkhout				

Imperfect genera

Genus			
Telletiopsis Derx		*	
Tilletiaria Bandoni and Johri		*	

Sometimes these genera are states of either the Ascomycetous (A) or the (B) Basidiomycetous-like genera? = unknown

continued overleaf

Table 6.10 *continued*

	Major classical works					Comments
	Lodder and Kreger van Rij (1952)	Lodder (1970)	von Arx (1974)	Barnett et al. (1974)	Barnett et al. (1979)	+From previously either described yeasts within genera or single species
Imperfect genera, cont.						
Aciculoconidium King & Jong					*	*Trichosporon aculeatum*
Aureobasidium (de Bary) Arnaud			*		*	(A)
Brettanomyces Kufferath and van Laer						(A) *Dekkera* (2 species)
Candida Berkhout	*	*	*	*	*	(A) and (B)
Cryptococcus Kützing	*	*	*	*	*	(A) and (B)
Geotrichum Butler and Petersen			*		*	(A) *Endomyces* or *Diploascus*
Kloeckera Janke	*	*	*	*	*	(A) *Hanseniaspora*
Oosporidium Stautz				*	*	?
Phaffia Miller, Yoneama and Soneda					*	?
Pityrosporum Sabouraud			*			(A)
Protendomycopsis Windish						
Rhodotorula Harrison						(B)
Sarcinosporon King & Jong						(B) *Trichosporon inkin*
Schizoblastosporon Ciferri						?
Selentila Yarrow			*	*		?
Selenozyma Yarrow					*	?
Torulopsis Berlese	*	*	*	*	*	(A) and (B)
Trichosporon Brenrend	*	*	*	*	*	? (A) and ! (B)
Trigonopsis Schachnea	*	*	*	*	*	? (A)

Yeast-like organisms which give yeast cells or slimy mycelium but no other characteristics on laboratory media – thus identification is only possible by examination of structures on the living host plant (von Arx, 1974).

Exobasidium Woronin, *Microstroma* Neissl, *Protomyces* Unger, *Taphrina* Fr., *Kabatiella* Bubak, *Ustilago* (Pers.) Roussel, *Tilletia* Til., and *Urocystis* Rabenh.

*Included in major classical works   +Some generic representatives now placed within a separate genus   ++Imperfect state

6.9 References

BARNETT, J. A., PAYNE, R. W. and YARROW, D. (1979). *A Guide to Identifying and Classifying Yeasts*. Cambridge University Press.

BARNETT, J. A. and PANKHURST, R. J. (1974). *A New Key to the Yeasts*. North-Holland Publishing Company, Amsterdam.

BEECH, F. W., DAVENPORT, R. R., GOSWELL, R. W. and BURNETT, J. K. (1968). Two simplified schemes for identifying yeast cultures. In *Identification Methods for Microbiologists*. GIBBS, B. M. and SHAPTON, D. A. (Eds). The Society for Applied Bacteriology, Technical Series, No. 2, Part B, pp. 150–75. Academic Press, London and New York.

BEECH, F. W. and DAVENPORT, R. R. (1969). The isolation of non-pathogenic yeasts. In *Isolation Methods for Microbiologists*. SIMPSON, D. A. and GOULD, G. W. (Eds). Society for Applied Bacteriology, Technical Series, No. 3. Academic Press, London and New York.

BEECH, F. W. and DAVENPORT, R. R. (1971). Isolation, purification and maintenance of yeasts. In *Methods in Microbiology*, Vol. 4. NORRIS, J. R., RIBBONS, D. W. and BOOTH, C. (Eds). Academic Press, London and New York.

CARMO-SOUSA, L. D. (1969). Distribution of yeasts in Nature. In *The Yeasts*, Vol. 1. ROSE, A. H. and HARRISON, J. S. (Eds). Academic Press, London.

DAVENPORT, R. R. (1968). *The Origin of Cider Yeasts*. Ph.D. Thesis. Institute of Biology, London.

DAVENPORT, R. R. (1973). Vineyard yeasts – an environmental study. In *Sampling Microbiological Monitoring of Environments*. BOARD, R. G. and LOVELOCK, D. H. (Eds). The Society for Applied Bacteriology Technical Series, No. 7, pp. 143–74. Academic Press, London and New York.

DAVENPORT, R. R. (1974a). A simple method using Stripdex equipment for the assessment of yeast taxonomic data and identification keys. *J. of Appl. Bact.*, **37**, 269–71.

DAVENPORT, R. R. (1974b). *Mycology and Taxonomy of Fungi in Fruit Juices*. Lecture and Laboratory Course Manual, pp. 1–250. Published jointly by the Organization of the American States, Washington D.C. and Instituto de Botanica, Sao Paulo, Brazil.

DAVENPORT, R. R. (1975). *The Distribution of Yeasts and Yeast-like Organisms in an English Vineyard*, pp. 1–480. Ph.D. dissertation, University of Bristol.

DAVENPORT, R. R. (1976). Distribution of yeasts and yeast-like organisms from aerial surfaces of developing apples and grapes. In *Microbiology of Aerial Plant Surfaces*. DICKINSON, C. H. and PREECE, T. F. (Eds), pp. 325–59. Academic Press, London.

DAVENPORT, R. R. (1978). *Yeasts and Yeast-like Organisms in Beverages and Foods – with Special Application to their Significance and Simple Identification*. Course manual – lectures. 120 pp. Course manual – laboratory exercises and identification keys. 98 pp. (Course sponsored by Ministry of Commerce, Conafruit Organization, University of Mexico City and the National Fruit Institute (Mexico) on secondment from the Agricultural Research Council to the Ministry of Overseas Development and British

Council (with help from the Tropical Products Institute, London, England.)

DAVENPORT, R. R. (1979). *Biotaxonomic and Ecological Studies on 'Lambic' Yeasts*. Report, unpublished.

DAVENPORT, R. R. (1980). An introduction to yeasts and yeast-like organisms. In *Biology and Activities of Yeasts*. SKINNER, F. A., PASSMORE, S. M. and DAVENPORT, R. R. (Eds). Academic Press, London and New York.

LODDER, J. and KREGER VAN RIJ, N. J. W. (1952). *The Yeasts, a Taxonomic Study*. North-Holland Publishing Company, Amsterdam.

LODDER, J. (1970). *The Yeasts*. North-Holland Publishing Company, Amsterdam.

VAN DER WALT, J. P. (1970). Criteria and methods used in classification. In *The Yeasts, a Taxonomic Study*. LODDER, J. (Ed.). pp. 34–107. North-Holland Publishing Company, Amsterdam, London.

VAN DER WALT, J. P. and HOPSU-HAVU, V. K. (1976). A colour reaction for the differentiation of ascomycetous and hemibasidiomycetous yeasts. *Antonie van Leeuwenhoek*, **42**, 157–63.

WICKERHAM, L. J. (1951). *Taxonomy of Yeasts*. Technical Bulletin, No. 1029. United States Department of Agriculture, Washington, D.C.

WIKEN, T. and SCHEFFERS, W. A. (1961). On the existence of a negative Pasteur effect in yeasts classified in the genus *Brettanomyces*. *Antonie van Leeuwenhoek*, **27**, 401–33.

7
The Basidiomycotina (Basidiomycetes)

7.1 Introduction

Fructifications of members of the Basidiomycetes are familiarly known as toadstools, mushrooms or bracket fungi. Frequently their massed mycelium is evident to the naked eye amongst woodland litter. The Basidiomycetes are a large group showing much diversity of habit and habitat, from microscopic plant pathogens to decayers of worked timber producing conspicuous fructifications. However, all the Basidiomycetes form microscopic **basidia** (sing. basidium), sexual reproductive cells which typically give rise to four **basidiospores,** each borne externally on a short protuberance, the **sterigma** (plur. sterigmata) arising from the top of the basidium. This contrasts with the Ascomycetes, where there are typically eight ascospores held within an ascus. Many Basidiomycetes produce their basidia on elaborate macroscopic fructifications known as basidiocarps; these are formed, as in the larger Ascomycetes, by a proliferation of interwoven and compressed hyphae.

7.2 General classification

The Basidiomycetes may be divided into two major groups. If the basidium arises from a thick-walled cell the fungi are Hemibasidiomycetes (Teliomycetes), subdivided into the Uredinales and the Ustilaginales. The fungi of both these families are plant parasites and are therefore of limited industrial significance. The second sub-group contains all the other Basidiomycetes – namely those where basidia do not arise from thick-walled cells. Unlike the Uredinales and Ustilaginales these fungi all produce macroscopic basidiocarps and their classification depends on the form of the basidiocarp. They may be difficult to grow in the laboratory. In these fungi the microscopic basidia are usually massed together to form a **hymenium** (plur. hymenia) and it is the arrangement of the hymenium which is important in classification (Webster, 1970) (see key to the Basidiomycotina).

The Hymenomycetes are a large group containing the fleshy toadstools and mushrooms all of which belong to the Agaricales. The Aphyllophorales contain the bracket fungi and similar leathery fungi, whilst the jelly fungi are included in the Tulasnellales. It is amongst the groups the Agaricales and Aphyllophorales that many of the edible and wood destroying fungi are found. The basidiospores in the Hymenomycetes are actively projected at maturity from the exposed hymenium.

General key to the groups of the Basidiomycotina (following the classification of Ainsworth, 1966 and Webster, 1970)

1.	Basidia arising from thick-walled cells	*Hemibasidiomycetes*
1'.	Basidia not arising from thick-walled cells	2
2.	Hymenium freely exposed at maturity	*Hymenomycetes* 3
2'.	Hymenium enclosed	*Gasteromycetes*
3.	Basidia not segmented or forked	4
3'.	Basidia segmented or forked	*Tulasnellales*
4.	Fruit body fleshy	*Agaricales*
4'.	Fruit body usually leathery, corky or woody	*Aphyllophorales*

The Gastromycetes are distinguished by the hymenium not being exposed on the basidiocarp at maturity and by the basidiospores not being actively projected at maturity. Their nutrition has not been well investigated. The group contains the Lycoperdales (puff-balls and earth stars), Sclerodermatales (earth-balls), Phallales (stinkhorns) and Nidulariales (birds'-nest fungi). They are not commercially important at present.

The asexual reproductive structures of Basidiomycetes are not nearly as varied as are those of the Ascomycetes, many Basidiomycetes producing unicellular oidia directly on a vegetative mycelium, and sometimes sclerotia. Most species exist with at least two different mating strains which segregate at basidiospore production. The basidiospores germinate to form monokaryotic mycelium which must fuse with another strain to produce dikaryotic mycelium before basidiocarps can be formed. The dikaryotic mycelium, which is usually the dominant vegetative phase of basidiomycetes, may exhibit clamp connections (see Rogers, 1936) which are unique to basidiomycetes.

7.3   Distribution

The distribution of this group is world-wide, as with the other major groups of fungi. However, the range of habitats suitable for Basidiomycetes is more restricted. Many are inhabitants of soil and plant debris, some are dung fungi whilst a number of species are extremely important destroyers of both standing and worked timber. Indeed, it is amongst the Basidiomycetes that the major utilizers of the very recalcitrant and ubiquitous substance, lignin, are found. Although the ascocarps of some Ascomycetes such as the truffles are prized as human food, it is the basidiocarp of the common mushroom and certain of its allies, which are important foodstuffs throughout the world.

7.4   Laboratory culture and isolation

It is not easy to isolate Basidiomycetes into culture, principally because of their slow growth compared with many mould fungi. Several selective media have been suggested which usually rely on the preferential inhibition of non-Basidiomycete fungi by sub-lethal concentrations of fungicides. Most work has been done on the isolation of wood destroying basidiomycetes and such selective agar techniques have recently been reviewed by Hale and Savory (1976). Useful papers on the techniques of obtaining inocula and subsequent culturing have been published by Grant and Savory (1969) and by Carey (1975).

7.5   Economic importance

7.5.1   Cellulose decomposers

Many Hymenomycetes are cellulose decomposers, and a significant number can also utilize lignin. When growing in timber those that break down only cellulose produce the friable, so-called 'brown rot', whilst those capable of decomposing lignin, all of which are also cellulose decomposers, produce the fibrous stringy 'white rot'. The superficial decay of wood known as 'soft rot' is caused by Ascomycetes and other microfungi.

7.5.2   Basidiomycetes and industry

Industrially the Basidiomycetes are likely to be of concern either as wood deteriogens or for food.

Food production

At present the Basidiomycetes which are grown for food, and thus could be considered for industrial handling, are *Agaricus bisporus* (the cultivated mushroom), *Lentinus elodes* (the Shiitake) and *Volvariella volvacea* (padi straw mushroom). *Tricholoma nudum* (the Blewitt) is also receiving interest. It is best to consult specialist taxonomic works for identification (Jordan, 1975; Soothill and Fairhurst, 1978). Basidiomycetes have been considered industrially for the production of biomass using a number of waste substrates, including crop waste and sawdust. The commercial growth of mushrooms is discussed in Chapter 15.

Wood-destroying fungi

Although wood has not entered into industrial constructions to any extent recently, as renewable resources become more important so the damage caused by basidiomycetes is likely to be of more concern. As wood destroying fungi frequently do not produce fructifications until the latter stages of decay have occurred, it is important to recognize the type of damage as early as possible. Decay is divided into two types – dry rot and wet rot.

Dry rot
This well-known decay of softwood timbers is brought about by the basi-

diomycete *Serpula lacrimans*, the sole causative agent of dry rot in the United Kingdom. The name of the decay is rather misleading, for as with all fungal decay, moist conditions are required to initiate the growth (a moisture content of 20% or more). Drier timbers are rendered susceptible to attack only by the ability of this fungus to transmit water and nutrients over considerable distances using specialized hyphal strands (rhizomorphs). Conditions of 'static dampness' are particularly favourable to *S. lacrimans,* whereas wet rot fungi are able to tolerate more fluctuating conditions (Butler 1957). In badly ventilated areas of high humidity the hyphae in the wood may manifest themselves on the surface, giving rise to a white felty covering which later develops into a flat rust coloured sporophore with a white margin. The colour of the sporophore is caused by the spores, and the red spore dust is often an indication of a deep seated and hidden infestation. Timber damaged by this fungus often has a typical cubical cracking. As the fungus is able to penetrate structures to a great depth, using its rhizomorphs to transport moisture and nutrients from the original infected wood, remedial measures often have to be extensive and include brickwork (travelled over and sometimes penetrated by the fungus) and apparently sound timbers. (See Figs 7.1 and 7.2) Dry rot is difficult and costly to eradicate (Butler, 1957).

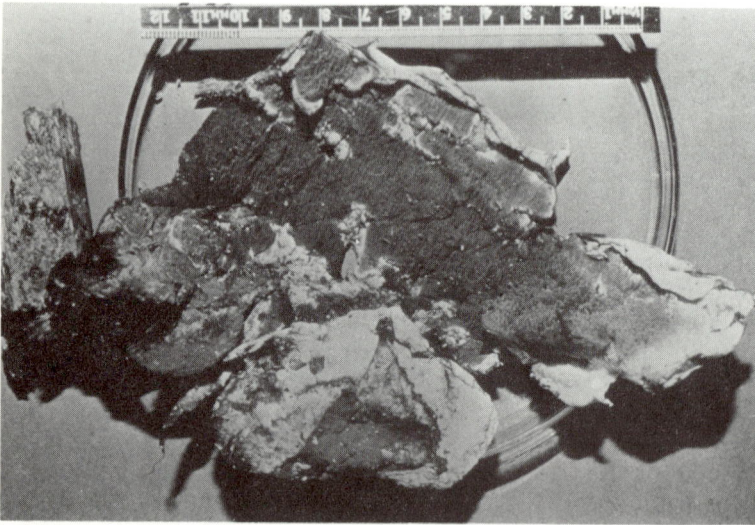

Fig. 7.1 *Serpula lacrimans*   Fruiting body. (Photograph by courtesy of Professor G. J. F. Pugh.)

Wet rot

Wet rot is caused by a number of basidiomycetes, of which the commonest are *Coniophora puteana*, *Fibroporia* spp., especially *F. vaillantii*, and *Phellinus contiguus* on external joinery. High humidity is required and a moisture content of over 50%. The use of impervious floor covering may tend to aggravate local dampness and under such conditions the fungus first appears as yellow mycelial felt. In conditions of high humidity, strands of

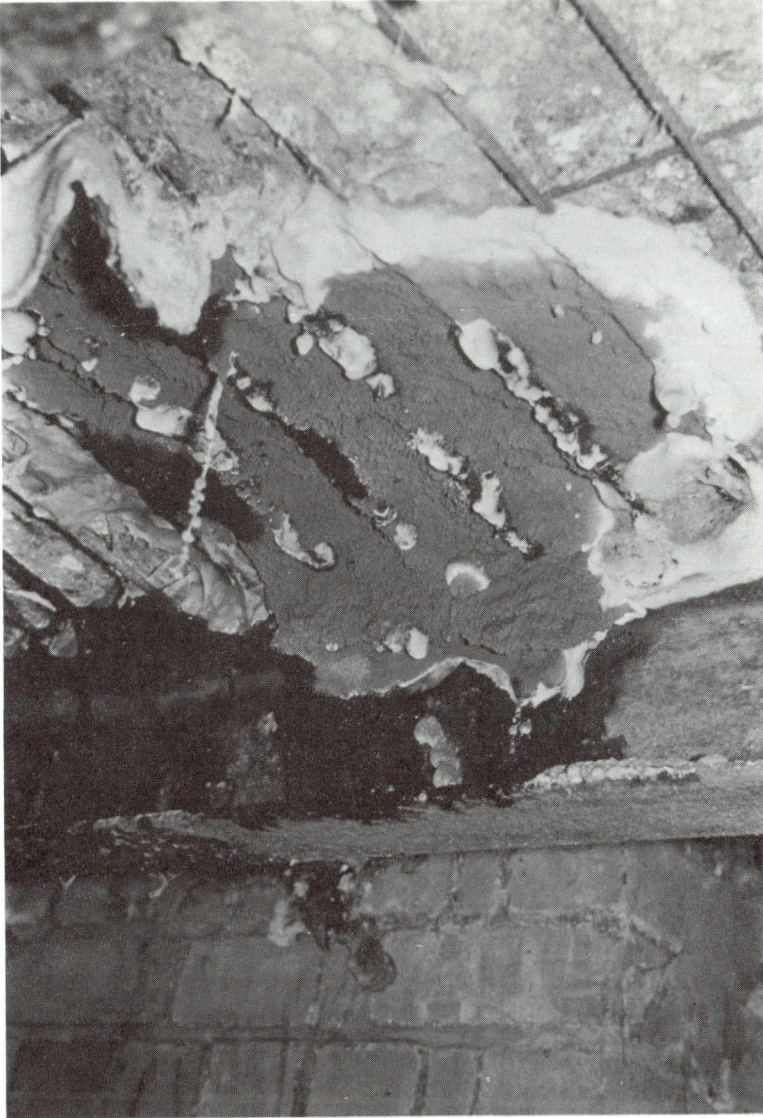

Fig. 7.2 *Serpula lacrimans* Growth from a cellar ceiling. (Photograph by courtesy of Professor G. J. F. Pugh.)

mycelium may occur which darken with age, but never attain the thickness of the rhizomorphs of dry rot. The fruiting body is a dull green-brown in colour with a warty surface. The fungus causes considerable shrinkage of wood, and cracking may occur, not unlike that caused by dry rot. Treatment is simpler than for dry rot as the growth is checked once the dampness is dealt with. A good account of wet rot is to be found in Cartwright and Findlay (1958).

7.6 References

BUTLER, G. M. (1957). The development and behaviour of mycelial strands in *Merulius lacrymans*. *Ann. Bot.*, **21**, 523–37.

CAREY, J. K. (1975). Notes on the isolation and characterization of wood inhabiting fungi. *Building Research Establishment Current Paper,* CP 93/75.

CARTWRIGHT, K. ST. G. and FINDLAY, W. P. K. (1958). *Decay of Timber and its Prevention.* HMSO, London.

GRANT, C. and SAVORY, J. G. (1969). Methods for isolation and identification of fungi on wood. *Int. Biodet. Bull.*, **5**, 77–94.

HALE, M. D. C. and SAVORY, J. G. (1976) Selective agar media for the isolation of Basidiomycetes from wood – A review. *Int. Biodet. Bull.*, **12**, 112–5.

JORDAN, M. (1975). *A Guide to Mushrooms.* Millington, London.

ROGERS, D. P. (1936). Basidial proliferation through clamp formation in a new *Sebacina. Mycologia,* **28**, 347.

SOOTHILL, E. and FAIRHURST, A. (1978). *A New Field Guide to Fungi.* Michael Joseph, London.

WEBSTER, J. (1970). *Introduction to Fungi.* Cambridge University Press.

8

Deuteromycotina

8.1 Introduction

The subdivision Deuteromycotina was previously referred to by the rather appropriate name Fungi Imperfecti. As has already been stated in Chapter 3, by definition the Deuteromycotina strictly contain organisms which fail to produce either ascospores or basidiospores and reproduce by conidia alone. All Zygomycetes are excluded from the subdivision although they may fail to produce zygospores.

The fact that these fungi do not produce a perfect state (teleomorph) may be due to many reasons, the main ones being lack of the correct conditions or because they have lost the ability to do so. In many cases mycologists have not yet been able to correlate teleomorphs and anamorphs. The known perfect states (teleomorphs) of many fungi classified in the Deuteromycotina are, however, steadily increasing. As such correlations are not always readily obtained and often are infrequently seen it is still sometimes convenient to use the imperfect state (anamorph) names, even though in a strict nomenclatural sense the name of the perfect state has priority.

It is sometimes found that similar imperfect states (anamorphs) are produced by fungi with dissimilar perfect states and conversely that similar perfect states can produce widely different imperfect states. This apparent conflict may be due to the fact that not enough knowledge is available about one of the other states. Von Arx (1979) gives many anamorphs and teleomorphs and shows they are often related. However, any name applied in the Deuteromycotina does not necessarily indicate a true relationship and is merely a convenience to indicate a grouping of similar imperfect structures. Thus terms used in classifying the Deuteromycotina are often terms for identification and do not have quite the same significance they have in the Ascomycotina and Basidiomycotina. Because of this it is sometimes useful to distinguish 'form genera' and 'form species' for members of the Deuteromycotina, as opposed to 'genus' and 'species' in Ascomycotina and Basidiomycotina. However, in common practice the 'form' is often omitted as being rather pedantic. A number of species (form species) will be dealt with under the Deuteromycotina in this book even though the perfect states are well known. This is because the anamorphs (imperfect states) are much more likely to be encountered in the industrial situation.

8.2 Systems of classification

The number of species and genera of the Deuteromycotina is very large and varied and hence the construction of a satisfactory scheme of classification has proved difficult. The first complete scheme was that of Saccardo (1880, 1884, 1886) and included all genera known at the time. However, there have been drawbacks in its use. Much of Saccardo's original thought is still useful and present classification is largely based on his works, Ainsworth's modification (1966, 1973) being widely accepted. (See key to the classes of Deuteromycotina.) The Blastomycetes are discussed in the chapter on yeasts.

Key to the classes of Deuteromycotina (following Ainsworth, 1973)

1.   Budding (yeast or yeast-like) cells with or without pseudomycelium characteristic, true mycelium lacking or not well-developed .... *Blastomycetes*
1'. Mycelium well developed, assimilative budding cells absent 2

2.   Mycelium sterile or bearing spores directly on special branches (conidiophores) which may be variously aggregated but not in pycnidia, pycnothria, acervuli or stromata   *Hyphomycetes*
2'. Spores in pycnidia, acervuli, pycnothyria or stromata ....   *Coelomycetes*

8.3 Coelomycetes

This group is taken (following Grove, 1935; see Sutton, 1973) to include those imperfect fungi with a sporing apparatus which is produced within a cavity of the substrate in which the fungus grows or within a cavity of fungal tissue. The conidia are produced within pycnidia, pycnothyria or stromata formed of fungal material or in a cavity (**acervulus**) in the host tissue.

Since Hughes' (1953) work on the importance of conidiogenesis in the classification of the Hyphomycetes, an increasing interest has been taken in the methods of spore formation in the Coelomycetes and these have been shown to be similar to those of the Hyphomycetes (see Sutton, 1971). This suggests that the relationship of these two groups must be very close.

In view of the relatively few genera likely to be met with in the industrial field these will be described and keyed with the Hyphomycetes. There are no pycnothyrial or stromatic Coelomycetes of industrial significance.

Key to the Coelomycetes

1.   Pycnidial ....   *Sphaeropsidales*

2.   Acervular ....   *Melanconales*

Sphaeropsidales
(Gr. *opsis,* like; hence like the Sphaeriales, an order of Pyrenomycetes)

The pycnidium, which is the fruiting structure characteristic of the order is a more or less flask-shaped body, superficially resembling a perithecium. However, instead of containing asci, the inside is lined with conidiophores or conidiogenous cells. When the pycnidium is squashed the spores are liberated in irregular masses.

Most of the species are parasites of plants, or saprophytes found on decaying plant material, but a few are found fairly frequently on industrial products and are of some practical importance. Almost all species in this category belong to the one genus *Phoma* (p. 149).

Melanconiales
(Gr. *melano-,* black; *konis,* dust)

As in the case of the previous order, most of the species are either parasitic or occur as saprophytes on fallen parts of plants. In their natural habitats they are at first produced within the host tissue but later often form pustular growths, consisting either of a saucer-shaped receptacle, somewhat resembling a pycnidium opened out flat, or of a dense cluster of short conidiophores breaking through the epidermis of the host, without any definite basal mycelial layer, and thus resembling a sporodochium. In some schemes of classification the Melanconiales are grouped together with the Tuberculariaceae (see below), on the grounds that the distinctions made by Saccardo are impractical. It has often been found that one and the same species can produce fructifications of either type, depending on host or cultural conditions.

The genus most likely to be isolated in the laboratory (*Pestalotiella*) produces scattered or more or less confluent, black pustules, consisting of bundles of very short conidiophores bearing slimy masses of spores.

8.4 Hyphomycetes
(Consisting of hyphae, i.e. not forming definite receptacles)

This class includes most of the common moulds. Saccardo divides the class into families as shown in the following key (p. 102).

He then subdivided each family into sections according to the septation and structure of the conidia. In addition, Saccardo often regarded the host on which the fungus was growing to be of specific importance. Although many fungi are still regarded as host specific it has been shown that most common species occur on many hosts and varying substrata, and can be identified in the absence of the host.

The limitations of Saccardo's scheme, which is simply an identification key with no attempt at a rational classification, became increasingly obvious and a succession of authors tried to produce alternative schemes which reflected natural fundamental relationships.

Key to the Hyphomycetes (following Saccardo)

1.	Mycelium sterile	*Agonomycetes* or *Mycelia sterilia*
1'.	Mycelium bearing spores	2
2.	Conidiophores separate	*Moniliales* 3
2'.	Conidiophores in synnemata	*Stilbellaceae*
2''.	Conidiophores in sporodochia	*Tubercular- iaceae*
3.	Mycelium and conidia hyaline	*Moniliaceae*
3'.	Mycelium and/or conidia dark coloured	*Dematiaceae*

Whereas Saccardo stressed the importance of spore structure, Mason (1933, 1937), who surveyed the work of previous authors, considered that the spore function was of more importance and proposed division of the Hyphomycetes into wet spored species (distributed by water) and dried spored species (distributed by air). This scheme was followed by Wakefield and Bisby (1941) in their list of British Hyphomycetes. Ingold (1942) added a third group, the truly aquatic species.

The methods of development of conidia have always been of interest and were stressed in particular by Vuillemin (1910, 1912) and Mason (1933). However, it was when Hughes (1953) brought out a scheme for classification of Fungi Imperfecti based on conidiophore and conidial development that its real significance became apparent. Hughes divided the Fungi Imperfecti into eight main sections according to the method of spore production. This work was rapidly followed with further studies by Tubaki (1958), Barron (1968), Subramanian (1963), von Arx (1970) and others in which more sections were added and varying interpretations made of the different organisms. As each author modified terminology or introduced his own terms to describe conidia and their development much confusion arose when interpreting different descriptions. To bring some order to this confusion and thereby allow accurate descriptions of fungi to be prepared, a conference of specialists was held at Kananaskis, Canada in 1969 at which an attempt was made to standardize the classificatory system and the terminology while at the same time recommend the rejection of confused or obsolete terms.

A report of this work was published by Kendrick (1971) and the system was adopted by Ellis in his monograph on *Dematiaceous Hyphomycetes* (1971). Work still continues on conidium ontogeny and as more sophisticated techniques become generally available (e.g. electron microscopy, scanning electron microscopy, time lapse photomicrography, etc.) more refinements to the scheme are suggested.

A study on conidial development with numerous drawings and photographs has been published by Cole and Samson (1979) which summarizes and elaborates much of this work. The whole pattern is not yet complete, but progress continues and a more complete picture is gradually emerging. As tends to happen in biology, with increasing knowledge it is becoming evident that the distinctions between what were thought to be distinct entities or sections are not as sharp as first appeared. In addition, some of the processes are far more complex than was first thought. It is hoped that systems for classification and identification – systems that are not too complicated or difficult to apply for ordinary workers to use with ordinary equipment – will be developed. It is proposed in this book to mention only characters that can be studied with normal laboratory equipment, and give only an outline of the present system of classification, keeping the descriptions of the terms as simple as possible.

The present systems of classification are based on morphological characters, even though some of them deal with fine structures. It seems inevitable that ultimately, as in bacteriology, the use of biochemical, enzymic and similar characters, by means of the modern techniques available to present day biochemists, will become of increasing importance as knowledge and information builds in this field. When this happens it will be interesting to see how well the two approaches to classification correspond.

Kendrick and Carmichael (1973), in their survey of the Hyphomycetes for the book *The Fungi*, did not wholly rely on conidium ontogeny as a means of separation and classification. In fact, they reverted to Saccardo's spore forms as convenient for primary means of separation, quoting for each genus the Saccardonian spore group, the general arrangement of the conidia with respect to each other, and the colour as well as the method of conidium production. This suggests that at present even the specialists find it convenient to use simple characters in identification.

The Saccardonian conidial spore forms according to Ainsworth (1971) are shown in Table 8.1 and Fig. 8.1.

Table **8.1**   The Saccardonian conidial spore forms according to Ainsworth (1971)

Spore type	Name
Spores 1-celled	Amerosporae
Spores 2-celled	Didymosporae
Spores with 2 or more cross septa	Phragmosporae
Spores with both longitudinal and cross septa	Dictyosporae
Spores filiform or thread-like	Scolecosporae
Spores forked or star-shaped	Staurosporae
Spores spirally coiled	Helicosporae

Note: The suffix 'phaeo' (dark) or 'hyalo' (pale) is sometimes added to the basic name to indicate the colour of the spore, for example phaeodictyosporae, hyalophragmosporae.

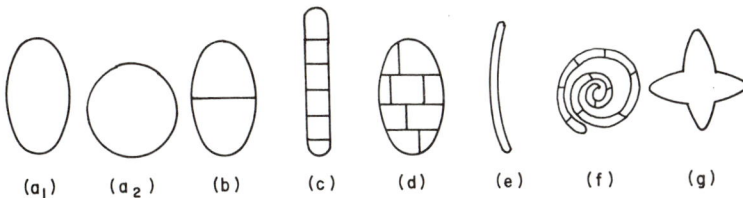

(a₁)   (a₂)   (b)   (c)   (d)   (e)   (f)   (g)

Fig. 8.1   Diagrammatic representation of Saccardo's spore forms: (**a₁**) and (**a₂**) amerospores; (**b**) didymospore; (**c**) phragmospore; (**d**) dictyospore; (**e**) scolecospore; (**f**) helicospore; (**g**) staurospore.

8.5   Standard terms and basic scheme for classification of Hyphomycetes

It was recommended in the standardization of terms (Kendrick, 1971) that descriptions should be basically adjectival using as few nouns as possible.

Mycelium

A fungus is made up of a mass (**mycelium**) of filamentous strands (**hyphae**) with cross septa (walls). These may be aggregated together to form a **sclerotium** or hard hyphal mass.

Conidiophores are any hyphal structure supporting a **conidigenous cell,** the cell actively involved in producing a conidium. The point on the cell from which the conidium is produced is the **conidiogenous locus.** The cell or part of the cell which will become a conidium is the **conidium initial.** When a conidigenous cell is not different from the rest of the mycelium it is **micronematous.** If it is specialized and different from the rest of the mycelium it is **macronematous.** The conidiogenous cell may be part of the main conidiophore and then it is called **integrated.** If, however, it is of distinctive shape and separate from the rest of the conidiophore it is **discrete.**

Types of spore production

There are two basic methods of spore production – thallic and blastic.
Thallic (Fig. 8.2)   The conidium is produced from the whole conidigenous cell and there is no enlargement of the conidial initial before the septum or septa are laid down differentiating the conidium. This occurs when a hypha breaks up into sections to form individual cells (conidia) as in *Geotrichum,* or when one cell, either terminal or intercalary, is cut off by septa, swells up and develops a thick wall to form a resting spore or **chlamydospore.** Relatively few genera develop conidia in this way.

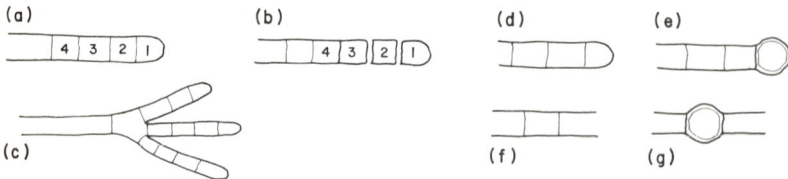

Fig. 8.2   Thallic spore development: **(a)** and **(b)** without swelling; **(c)** branched; **(d)** and **(e)** swelling to produce terminal chlamydospore; **(f)** and **(g)** swelling to produce intercalary chlamydospore.

Blastic (Fig. 8.3)   The conidium is produced by the enlargement of a part of the conidiogenous cell and enlargement of the conidium initial occurs before a septum is laid down. The majority of Fungi Imperfecti are produced in this way.

Blastic conidia

The various forms of blastic conidia depend on the relationship of the new conidia to the conidiogenous cell wall.

Holoblastic   Both the inner and outer walls of the conidiogenous cell swell out to form a new conidium, and form part of it (Fig. 8.4).

Enteroblastic   The conidium is produced from within the conidiogenous cell either including the inner part of the cell wall or no part of the cell wall (Fig. 8.5).

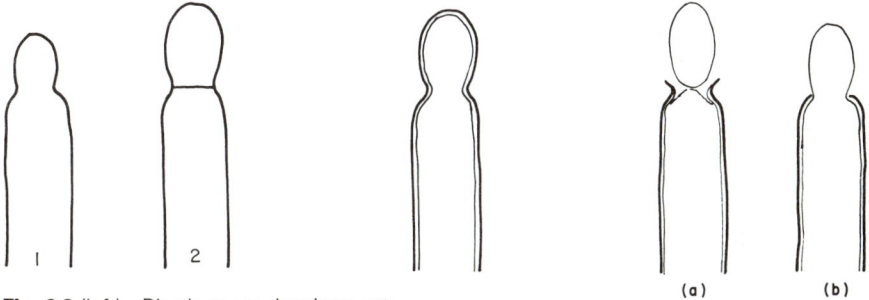

Fig. 8.3 (left)   Blastic spore development.

Fig. 8.4 (centre)   Holoblastic conidium.

Fig. 8.5 (right)   Enteroblastic conidia: **(a)** phialidic; **(b)** tretic.

These two basic types of blastic conidiogenesis can be further divided according to the details of development.

Holoblastic conidial forms

Holoblastic conidiogenesis is perhaps one of the simplest forms to understand, but probably produces the most diverse modifications of the basic concept.

Monoblastic conidia are produced at only one point on the conidiogenous cell. When only one conidium is produced at the apex, and the conidiogenous cell ceases growth it is **determinate**. In some cases the first conidium produces a second conidium which develops at its own apex or as a branch and the second produces a third until a chain is produced with the youngest conidium at the apex. This is called an **acropetal** chain (Fig. 8.6a).

Annellides (Fig. 8.6c)   In other cases after the first conidium is formed the hyphal wall grows out through the scar left by the first conidium and produces a second conidium at a higher level; a third may be produced in a similar way and so on. Again a chain of conidia is formed, though the conidia are not connected in the same way as in an acropetal chain. The youngest conidium is at the base and the conidiogenous cell progressively elongates and is left with a series of scars or annellations at the apex. This form of conidiogenous cell is called **annellidic**. The distance between the scars varies considerably from one species to another as does the size and number of the scars. A good example

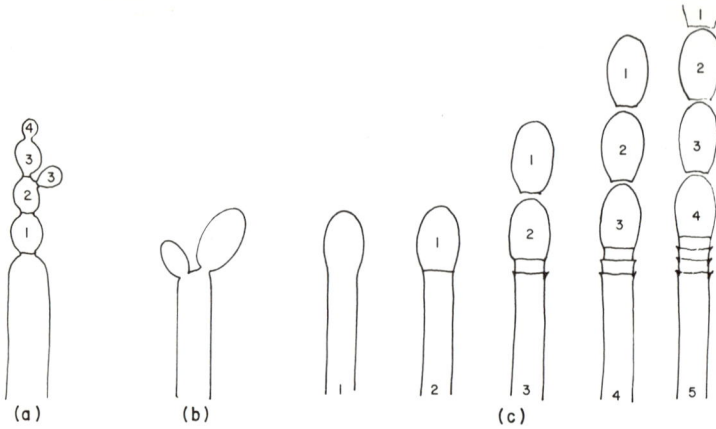

Fig. 8.6 Holoblastic conidia: **(a)** acropetal chain; **(b)** polyblastic, produced at several points on the conidiogenous cell; **(c)** annellidic, basipetal chain, youngest conidium at top.

of this is seen in *Scopulariopsis*. A chain with the youngest conidium at the base is called **basipetal**.

Polyblastic conidia are produced at several points on the conidiogenous cell (Fig. 8.6b).

Enteroblastic conidial forms

There are two main types of enteroblastic conidia – tretic and phialidic.

Tretic (Fig. 8.7a, b)   The wall of the conidia is an extension of the inner wall of the conidiogenous cell and conidia are produced through a pre-determined opening in the outer wall of the conidiogenous cell.

Phialides (Fig. 8.7c, d)   The first conidium is produced from within the conidiogenous cell by rupture or dissolution of all layers of the cell wall, subsequent conidia are produced from a conidiogenous locus through this aperture, the walls of the conidia being newly formed and not part of the original conidiogenous cell wall.

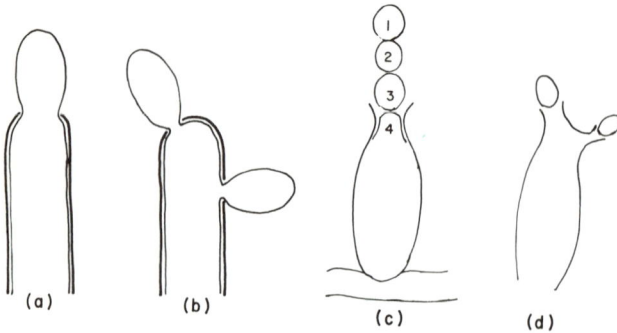

Fig. 8.7 Enteroblastic conidia: **(a)** monotretic; **(b)** polytretic; **(c)** phialidic, basipetal chain, youngest conidium at the apex; **(d)** polyphialidic.

As in the holoblastic series, enteroblastic conidia can be produced from a single locus or from more than one locus on the conidiogenous cell thus giving the terms **monotretic, polytretic, monophialidic** and **polyphialidic.**

Other variations of conidial production

Sympodial growth (Fig. 8.8)   When a conidium has been produced at the

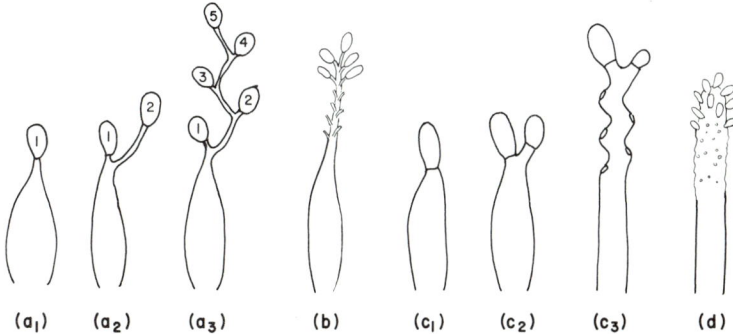

Fig. 8.8 Sympodial growth: **(a)** long narrow elongations; **(b)** short narrow elongations; **(c)** short and broad elongations; **(d)** swollen conidiogenous cell bearing conidia irregularly.

apex of a conidiogenous cell growth may again develop from the side of the apical scar resulting in the production of another conidium. This may then be repeated on the other side, producing a zig-zag line of growth with the youngest conidium at the apex. This is called **sympodial growth.** The hyphal elongations vary from long and narrow to short and broad. In some cases the differentiation is difficult to see and a swollen conidiogenous cell bearing many conidia irregularly over its surface is all that can be seen, for example *Nodulisporium*. There are variations on simple sympodial growth and in many genera the loci are neither regular nor strictly sequentially produced.
Synchronous growth (Fig. 8.9)   Occasionally the conidiogenous cell is swollen and bears the conidia synchronously all over its surface as in *Botrytis* sp.

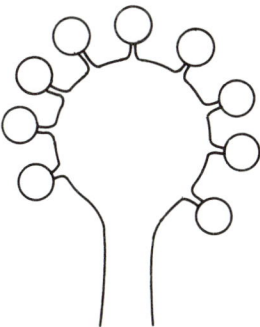

Fig. 8.9 Synchronous growth.

Figure 8.10 summarizes the main methods of conidial production.

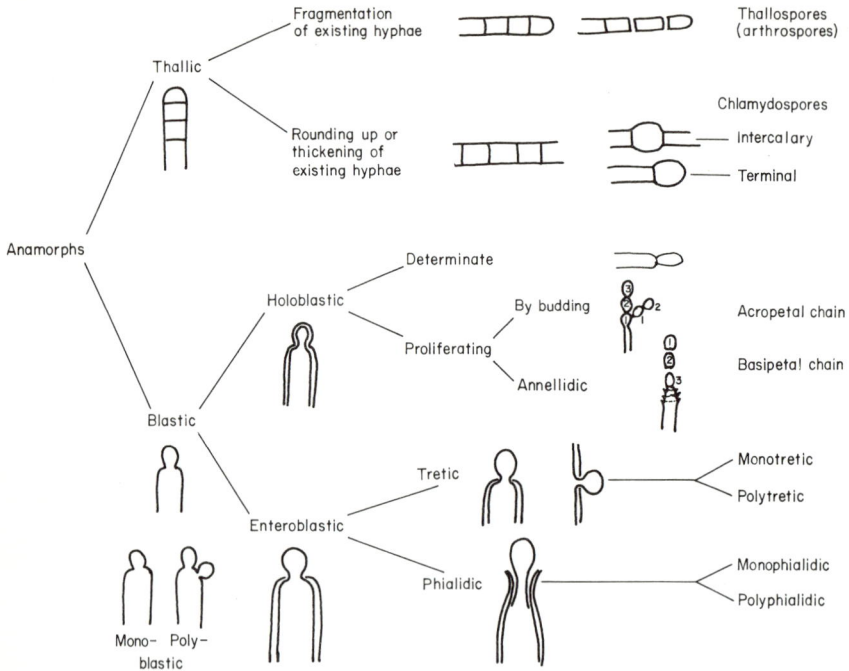

Fig. 8.10   Main methods of conidial production (anamorph production).

8.6   Agonomycetes or Mycelia sterilia

In addition to the many fungi producing conidia there are a number of fungi which are, so far as is known, permanently sterile. They produce mycelium and in many cases sclerotia, but no true reproductive structures or conidia. These have no place in any of the families or groups founded on spore characteristics and are usually classed apart as the *Mycelia sterilia* or, as Ainsworth called them, *Agonomycetes*.

Purely mycelial forms occur fairly often in cultures but it must not be hastily assumed that these belong to the *Mycelia sterilia*. Many of them are parasitic fungi, Ascomycetes and Basidiomycetes, which do not form spores on the usual culture media. In addition some fungi, after a period in cultivation, cease to spore. Before assuming a culture to belong in the *Mycelia sterilia* considerable efforts should be made to induce it to spore, such as incubation at several temperatures on several media and exposure to black light. Some of these cultures are so characteristic that experts can identify them to genera and species by their vegetative characteristics. However, they are not of significance in the industrial field and it is not proposed to mention them further in this book.

8.7 References

AINSWORTH, G. C. (1966). A general purpose classification for fungi. *Bilbiography of Systematic Mycology,* **1,** 1–4.

AINSWORTH, G. C. (1971). *Dictionary of the Fungi,* 6th edition. Commonwealth Mycological Institute, Kew, Surrey, England.

AINSWORTH, G. C. (1973). Introduction and keys to higher taxa. In *The Fungi,* Vol. 4A. AINSWORTH, G. C., SPARROW, E. K. and SUSSMANN, A. S. (Eds). Academic Press, London.

ARX, J. A. VON (1970, 2nd edition, 1974). *The genera of fungi sporulating in pure culture.* J. Cramer, Lehre.

ARX, J. A. VON (1979). Ascomycetes as Fungi Imperfecti. In *The Whole Fungus,* Vol. 1. KENDRICK, B. (Ed.). pp. 201–14. National Museums of Canada for the Kananaskis Foundation.

BARRON, G. L. (1968). *The genera of Hyphomycetes from soil.* Williams & Wilkins, Baltimore.

COLE, T. C. and SAMSON, R. A. (1979). *Patterns of Development in Conidial Fungi.* Pitman, London, San Francisco and Melbourne.

ELLIS, M. B. (1971). *Dematiaceous Hyphomycetes.* Commonwealth Mycological Institute, Kew, Surrey, England.

GROVE, W. B. (1935). *British Stem and Leaf Fungi (Coelomycetes). 1.* Cambridge University Press, London and New York.

HUGHES, S. J. (1953). Conidiophores, conidia and classification. *Canad. J. of Bot.* **31,** 577–659.

INGOLD, C. T. (1942). Aquatic hyphomycetes of decaying alder leaves. *Trans. Br. mycol. Soc.,* **25,** 339–417.

KENDRICK, B. (Ed.) (1971). *Taxonomy of Fungi Imperfecti.* University of Toronto Press, Toronto and Buffalo.

KENDRICK, B. (Ed.) (1979). *The Whole Fungus.* Vols 1 and 2. National Museums of Canada for the Kananaskis Foundation.

KENDRICK, W. B. and CARMICHAEL, J. W. (1973). Hyphomycetes. In *The Fungi,* Vol. 4A. AINSWORTH, G. C., SPARROW, F. K. and SUSSMAN, A. S. (Eds). pp. 323–509. Academic Press, New York and London.

MASON, E. W. (1933). Annotated account of fungi received at the Imperial Mycological Institute. List II (Fascicles 2). *Mycological Papers,* No. 3.

MASON, E. W. (1937). Annotated account of fungi received at the Imperial Mycological Institute. List II (Fascicle 3). *Mycological Papers,* No. 4.

SACCARDO, P. A. (1884). *Sylloge fungorum omnium hucusque cognitorum.* Vol. 3. Pavia, Italy.

SACCARDO, P. A. (1886). *Sylloge fungorum omnium hucusque cognitorum.* Vol. 4. Pavia, Italy.

SACCARDO, P. A. (1880). Conspectus generum fungorum italiae inferiorum nempe ad Sphaeropideas, Melanconiaes et Nyphomycetas pertinentium, systemate sporologico dispositorum. *Michelia,* **2,** 1–38.

SUBRAMANIAN, C. V. (1963) ('1962'). Classification of the Hyphomycetes. *Bull. Bot. Surv. India,* **4,** 249–59.

SUBRAMANIAN, C. V. (Ed.) (In press). *Proceedings of the International Symposium on Taxonomy of Fungi.* University of Madras.

SUTTON, B. C. (1971). Conidium ontogeny in pycnidial and acervular fungi. In

Taxonomy of Fungi Imperfecti. KENDRICK, B. (Ed.). University of Toronto Press, Toronto and Buffalo.

SUTTON, B. C. (1973). Coelomycetes. In *The Fungi,* Vol. 4A. AINSWORTH, G. C., SPARROW, E. K. and SUSSMANN, A. S. (Eds). pp. 513–82. Academic Press, London.

TUBAKI, K. (1958). Studies on the Japanese hyphomycetes. V. Leaf and stem group, with a discussion of the classification of Hyphomycetes and their perfect stages. *J. Hattori bot. Lab.,* **20,** 142–244.

TUBAKI, K. (1963). Taxonomic study of Hyphomycetes. *Ann. Rep. Inst. Ferm. Osaka I (1961–62),* 25–54.

VUILLEMIN, P. (1910). Matériaux pour une classification rationelle des Fungi Imperfecti. *C.R. Acad. Sci., Paris,* **150,** 882–4.

VUILLEMIN, P. (1912). *Les champignons. Essai de classification.* O Doin et Fils, Paris.

WAKEFIELD, E. M. and BISBY, G. R. (1941). List of Hyphomycetes recorded for Britain. *Trans. Br. mycol. Soc.,* **25,** 49–126.

9
Hyphomycetes

You will find a fungus and determine its characteristics. You turn to the books and decide on its genus. Then you look for the species. And you look and you look, and after a while you find it! In another genus!

J. J. Davis, fide Gilman, *Mycologia*, 1953.

9.1 Introduction

As indicated in Chapter 7, the class Hyphomycetes includes all the Fungi Imperfecti (Deuteromycotina) which do not form pycnidia or stromata but instead produce conidiophores openly on any part of the mycelium. The great majority of moulds which are of industrial importance belong in this group.

There is a large number of generic names to be found in the literature and many may be synonyms. Some genera have been founded on insufficient data obtained from the study of single specimens in their natural habitats. In many cases the descriptions have been hopelessly inadequate and have not taken into account the whole life histories of the fungi. The result has been that some of the commoner moulds are known only by tradition and are difficult to recognize from published descriptions, whilst some genera exist only as names and it is impossible to know to which fungi these refer. However, in the last 10 to 20 years there has been intensive study of these fungi with numerous publications and monographs, and identification of many isolates is now a much simpler undertaking.

Here we are only interested in the common moulds. Although taxonomy and nomenclature have changed and monographs have been written it is notable that the actual moulds occurring in the industrial situation have changed little since George Smith first published on them, and his descriptions and photographs require relatively little alteration. This being so, the literature has been updated and we have included line drawings to elucidate the structures which in most cases are clearly visible in the original photographs, especially when the structure is known. G. Smith always stressed that he used photographs which showed pictures of what a worker was likely to see using the sort of equipment he was likely to have in a routine industrial laboratory. At the present time the literature is full of the most beautiful photographs, especially ones involving scanning electron microscopy and interference microscopy. However, no effort has been made to include this type of work in this chapter as it is most unusual for routine laboratories to have these techniques available and in any case the preparation of material is both skilled and time consuming and not suitable at present for routine work.

9.2   Keys to classes and genera of Fungi found as common moulds

Diagrams have been introduced into the keys to assist in rapid identification. The descriptions of individual genera which follow the key are placed in alphabetical order. Conidial (anamorphic) states of teleomorphic fungi have been included where it is considered these are likely to be seen and the key makes no claim to any relationship to classification. It is introduced by a simple key to the main groups of fungi followed by a series of keys initially separated according to Saccardo's spore forms (see p. 103). The two Coelomycete genera treated are included here for convenience. Some genera are keyed in two or more places where it is thought structures might be misleading. As well as the more conventional keys a synoptic key is included as this type of key is becoming increasingly popular. It is less restrictive in the characters which are to be examined and can be entered at any point using the characters which appear most obvious, and in some cases an organism can be identified on one character. (See key, p. 117, for further details.)

It must be pointed out that only 34 genera have been included in the keys to the Hyphomycetes. These are the ones that, in the opinion of the authors, are most likely to be met in the industrial field, but there are many hundreds of other genera. If a match is not obtained it is quite likely that the specimen belongs to one of the other genera and such works as Ellis (1971, 1976), Barron (1968) and von Arx (1970, 1974) should be consulted or the specimen sent to a specialist for examination.

It should be noted that the sketches in the keys are not drawn to the same scale so the relative sizes may be misleading.

Key to classes of Fungi found as common moulds

1.	Mycelium only produced	*Agonomycetes (Mycelia sterilia)*
1'.	Spores produced	2
2.	Spores in closed receptacles	3
2'.	Spores not in receptacles	*Hyphomycetes* and some *Coelomycetes (Melanconiales – Pestalotia* spp. etc.)
3.	Spores in sporangia, mycelium usually coarse and non-septate	*Zygomycetes* (Chap. 4)
3'.	Spores not in sporangia, mycelium septate	4
4.	Spores in asci (groups of eight), freely produced on the mycelium or enclosed in larger bodies	*Ascomycetes* (Chap. 5)
4'.	Spores produced in large numbers from conidiophores, in either globose or flask-shaped receptacles	*Coelomycetes (Sphaeropsidales* (Chap. 8) *Phoma* spp. etc.)

Keys to the more common genera of moulds (Hyphomycetes)

In order to reduce the length and complexity of any one key this separation is dealt with as a series of short keys. The first separation is based on Saccardo's spore forms as being very easy to distinguish. Only four of these spore forms are common. The bold number after a generic name on the right of the key refers to a sketch of the conidia or conidial head.

Dictyospored Genera

(With both longitudinal and transverse septa, only dark spored species are commonly encountered.)

1.	Conidia borne in chains, with an apical rostrum (beak)	*Alternaria* (1)
1'.	Conidia borne singly 2	
2.	True dictyospores	*Stemphylium* and *Ulocladium* (2)
2'.	Multiseptate bodies, lightly coloured, not true conidia	*Papulospora* (3)

(1)

(2)

(3)

Phragmospored Genera

(With transverse septa.)

1.	Conidia hyaline, with basal foot cell, aggregated in slimy masses	*Fusarium* (4)
1'.	Conidia dematiaceous (dark)	2
2.	Centre cells of conidia more swollen and darker than the rest	*Curvularia* (5)
2'.	Conidia straight, with disto (not true) septa	*Helminthosporium* (6)
2".	Conidia fusiform with apical appendages, and darker median cells	*Pestalotia* etc. (7)

(4)

(5)

(6)

(7)

Didymospored Genera

(With one septum.)
Only one species is commonly encountered.

Conidia hyaline, colonies pink .... *Trichothecium roseum* (8)

(8)

Amerospored Genera

(Without septa.)

The amerospored genera form the largest group of fungi which are common moulds, and of these the species producing phialides form a high percentage of those encountered and of particular interest to industry. Two keys are given, one to the phialidic species and one to the other amerospored species.

Phialidic Amerospored Genera

1.	Conidiogenous cells lengthening at apex (annellides) forming a penicillus	2	
1'.	Conidiogenous cells true phialides	3	

(9)

2.	Hyaline	*Scopulariopsis* (9)
2'.	Dematiaceous (dark)	*Doratomyces* (10)

(10)

3.	Hyaline	4
3'.	Dematiaceous	11

4.	Conidia wet, aggregated in slimy masses	5
4'.	Conidia dry, in chains	9

(11)

5.	Conidiogenous structures simple	6
5'.	Conidiogenous structures complex, with a distinct conidiophore	8

(12)

6.	Conidia mostly produced directly on the mycelium	*Aureobasidium* (11)
6'.	Conidia produced on distinct single phialides	7

(13)

7.	Conidia colourless (pale)	*Acremonium* (12)
7'.	Conidia dark, mycelium pale	*Gliomastix* (13)

(14)

8.	Conidial apparatus of mixed complexity from verticillate to penicillate, young conidia sometimes in chains	*Gliocladium* (14)
8'.	Conidial apparatus irregular, conidia usually green	*Trichoderma* (15)
8".	Conidial apparatus regularly verticillate, conidia never green	*Verticillium* (16)

(15)

(16)

9.	Conidiophores arising from specialized foot-cells and terminating in a vesicle	*Aspergillus* (17)
9'.	Conidiophores without foot-cells or vesicles, bearing penicillate structures	10

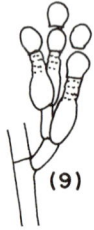

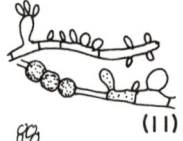

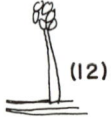

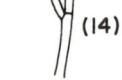

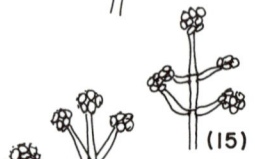

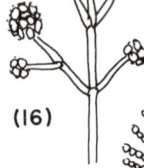

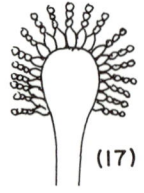

(17)

10.	Conidiogenous cells lengthening at the tip, not true phialides, conidia distinctly truncate at the base.	*Scopulariopsis* (**9**)
10'.	Phialides terminating in long slender tips, never green	*Paecilomyces* (**18**)
10".	Phialides thicker with rounded tip, conidia usually dark	*Memnoniella* (**19**)
10'''.	Phialides with short neck, conidia usually green	*Penicillium* (**20**)
		(young stages of *Gliocladium*)

11.	Conidia in dry chains	12
11'.	Conidia wet, aggregating in slimy masses	14

12.	Conidiogenous cells lengthening	*Doratomyces* (**10**)
12'.	Conidiogenous cells true phialides	13

13.	Phialides in an apical whorl	*Memnoniella* (**19**)
	(if wet in a slimy ball	*Stachybotrys*)
13'.	Single elongate phialides producing endogenously hyaline conidia becoming septate basipetally, and fragmenting	*Wallemia* (**21**)

14.	Conidia borne directly on the hyphae or swellings of the hyphae	*Aureobasidium* (**11**)
14'.	Conidia produced from distinct phialides	15

15.	Phialides hyaline, borne singly on trailing ropes of hyphae, conidia black	*Gliomastix* (**13**)
15'.	Hyphae dark, conidiophores dark and often branched, conidia hyaline to pale brown	*Phialophora* (**22**)
15".	Phialides in apical whorl, with rounded tips, conidia black, in slimy masses	*Stachybotrys* (**23**)

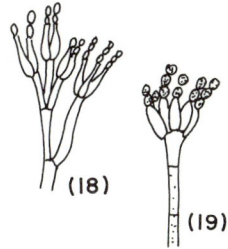

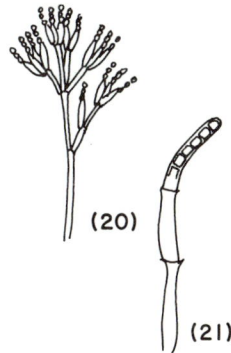

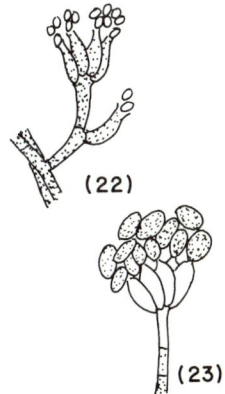

(18)

(19)

(20)

(21)

(22)

(23)

Non-phialidic Amerospored Genera

1.	Conidia and mycelium hyaline (light coloured)	2
1'.	Conidia or mycelium dematiaceous (dark)	11

2.	Conidia wet, forming slimy masses	3
2'.	Conidia dry	4

3.	Conidia produced directly from loci on the conidiogenous hyphae and as chlamydospores	*Aureobasidium* (**11**)

3'.   Conidia produced by fragmentation
     of hyphae, colonies yeast-like ....   *Geotrichum* **(24)**

4.   Conidia produced singly   5
4'.   Conidia produced in chains   7

5.   Conidia produced on pegs or
     narrow denticles   6
5'.   Conidia produced on broad, short,
     hyaline conidiogenous cells, hyaline
     when young, but becoming dark ....   *Humicola* **(25)**

6.   Conidia produced in clusters on
     small points on the narrow
     conidiogenous cell ....   *Sporothrix* **(26)**
6'.   Conidia produced on pegs on the
     swollen apices of specialized branched
     conidiophores, in grape-like clusters,
     conidiophore with distinct metallic
     sheen ....   *Botrytis* **(27)**
6".   Conidia produced from narrow
     hyphae from swollen parts of
     the conidiophore ....   *Myceliophthora* **(28)**

7.   Conidia monoblastic, in acropetal
     chains, produced by budding   8
7'.   Conidia thallic formed from the
     break up of conidiogenous hyphae   9

8.   Conidia profuse, chains branched,
     colonies dusty, common ....   *Monilia* **(29)**
8'.   Few conidia, chains short, also
     thallic conidia which darken
     in age, grows in pickles and
     other acid substrates ....   *Moniliella* **(30)**

9.   Distinct coloured conidiophore,
     conidia fragmenting from tip
     backwards ....   *Oidiodendron* **(31)**
9'.   Conidiophores less distinct, hyaline   10

10.   Conidiophores branched acutely at
     the apex with terminal or lateral
     conidia, and some from the break
     up of the hyphae ....   *Geomyces* **(32)**
10'.   No distinct conidiophores, thallic
     conidia in adjacent or alternate
     segments of the parent mycelium becoming
     thickened and liberated by lysis of
     intervening cells, coloured orange ......   *Sporendonema* **(33)**

11.   Conidia wet ....   *Aureobasidium* **(11)**
11'.   Conidia dry   12

12.   Conidia single ....   *Humicola* **(25)**
12'.   Conidia in chains   13

13. Conidia monoblastic, formed by
     budding in branched acropetal chains,
     dark green with dense tree-like
     heads .... *Cladosporium* (**34**)
13'. Conidiophores penicillate, but
     conidiogenous cells annellidic,
     forming grey synnema .... *Doratomyces* (**10**)
13". Conidia thallic, fragmenting from
     the tip backwards, with
     distinct dark conidiophore .... *Oidiodendron* (**31**)

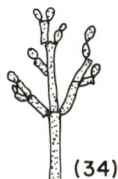

(**34**)

Synoptic key to common mould genera (Hyphomycetes)

The genera covered by the key are numbered. The key consists of a list of characters, many contrasting, in this case based on morphology. At the side of each character the numbers of those genera which show the character are listed. Where there are contrasting characters the absence of a number indicates the contrasting character. As sub-heading to some of the main characters others useful for further separation are listed. Some distinct useful diagnostic characters shown by only one or two genera are also included.

The identity of the organism is found by a process of elimination. The genera are numbered as they appear in the dichotomous key. Where an entry is underlined the character occurs occasionally, is not distinct, or might be misinterpreted.

---

1. Spore form
      Dictyospore   1, 2, 3
      Phragmospore   4, 5, 6, 7
      Didymospore   8
      Amerospore, all the remaining genera   9–34

2. Mycelium hyaline (pale coloured, not so in the contrast)
      4, 8, 9, 11, 12, 13, 14, 15, 16 ,17, 18, 19, 20, 23, 24, 25, 26, 27, 28, 29, 30, 31, 32, 33

3. Conidia hyaline (not so)
      4, 8, 9, 11, 12, 14, 15, 16, 17, 18, 20, 21, 22, 24, 26, 27, 28, 29, 30, 31, 32, 33

4. Conidia in chains (not so)
      1, 2, 8, 9, 10, 14, 17, 18, 19, 20, 21, 24, 29, 30, 31, 32, 33, 34
   Conidia not in chains and as wet spore masses
      4, 7, 11, 12, 13, 14, 15, 16, 22, 23, 24
   Conidia borne singly
      2, 3, 5, 6, 25, 26, 27, 28

5. Conidia produced from phialides (not so)
      4, 9, 10, 11, 12, 13, 14, 15, 16, 17, 18, 19, 20, 21, 22, 23
   Phialides lengthening, that is not true phialides (annellides)
      9, 10
   Phialides
      gradually tapered to long points   12, 13, 14, 16, 18
      with rounded tip   19, 23
      with short neck   15, 17, 20, 22
      endogenous conidia becoming septate basipetally and fragmenting   21
      conidia borne directly on the mycelium through an open cup   11
      single, borne directly on the mycelium   11, 12, 13, 21
      borne as branched heads   4, 9, 10, 14, 15, 16, 17, 18, 19, 20, 22, 23

irregularly branched   15, 18, <u>22</u>
in regular verticils   16
as a terminal verticil   17, 19, <u>20</u>, 23
as terminal verticils on a branched conidiophore   14, 18, 20, 22

6. Non-phialidic genera
Thallic, formed from break up of conidiogenous hyphae   24, 31, 32, 33
   with distinct conidiophore (or not) 31, (coloured) 33
Holoblastic   1, 2, 5, 6, 25, 26, 27, 28, 29, 30, 34
   on narrow pegs   26, 27, 28
   by budding   29, 30, 34

7. Miscellaneous distinctive characters
Conidiophores with swollen apex   17, <u>20</u>, 27
Growing on very acid media   30
Not true conidia, bulbils   3
Conidia curved   4, 5
Conidia rostrate at apex (beaked)   1
Conidia with darker median cells   5, 7
Conidia with a foot cell   4
Conidia with appendages   7

Numerical list of genera for synoptic key

1. *Alternaria*
2. *Stemphylium* and *Ulocladium*
3. *Papulospora*
4. *Fusarium*
5. *Curvularia*
6. *Helminthosporium*
7. *Pestalotia*
8. *Trichothecium*
9. *Scopulariopsis*
10. *Doratomyces*
11. *Aureobasidium*
12. *Acremonium*
13. *Gliomastix*
14. *Gliocladium*
15. *Trichoderma*
16. *Verticillium*
17. *Aspergillus*
18. *Paecilomyces*
19. *Memoniella*
20. *Penicillium*
21. *Wallemia*
22. *Phialophora*
23. *Stachybotrys*
24. *Geotrichum*
25. *Humicola*
26. *Sporothrix*
27. *Botrytis*
28. *Myceliophthora*
29. *Monilia*
30. *Moniliella*
31. *Oidiodendron*
32. *Geomyces*
33. *Sporendonema*
34. *Cladosporium*

9.3   Generic descriptions

Acremonium Link ex Fries
(= *Cephalosporium* Corda) (Figs 9.1, 9.2)
Conidiophores phialidic, hyaline; conidia 0-septate, hyaline, in balls (or chains) in basipetal succession. In a recent monograph of *Cephalosporium*-like genera Gams (1971) showed *Acremonium* to be a more suitable name for the species often described as *Cephalosporium*. He included in the same monograph many species which had previously been classified elsewhere, for example in *Gliomastix* and *Paecilomyces*.

In traditional *Cephalosporium* elongated phialides arise as branches from all parts of the mycelium. The conidia collect in wet balls at the apex of the phialides. Under low magnifications these balls of conidia appear like large spores, but on mounting and examination at high magnifications they are seen to be composed of numerous small conidia. Gams divides the genus into three sections and one series and includes in these divisions species with spores in chains, coloured or encrusted conidia and branched conidiophores. A large number of species have been described and, until Gams' monograph, separation was difficult.

Acremonium species occur quite commonly, even in the spoilage situation and are often seen after a mite infestation. *Acremonium* states are known for several ascomycete genera including *Emericellopsis*, *Chaetomium* and *Nectriopsis*.

Acremonium strictum W. Gams, perhaps the commonest species is very variable, producing white to pale pink, ropy colonies, often slightly wet looking, with single elongate phialides 20–40 μm or more long by 1.5–2(2.5) μm diam., bearing conidia in wet spore balls. Conidia cylindrical, more than twice as long as broad, 3–6(7) × 1–2 μm. There are no chlamydospores or synnemata.

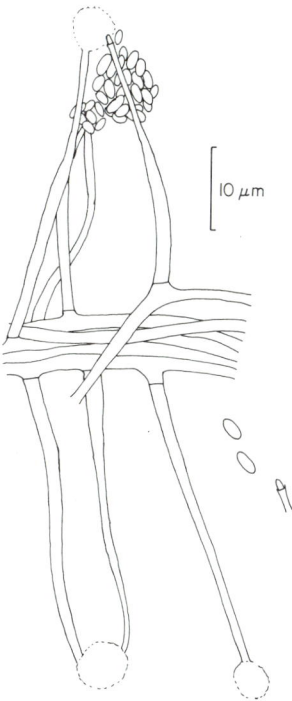

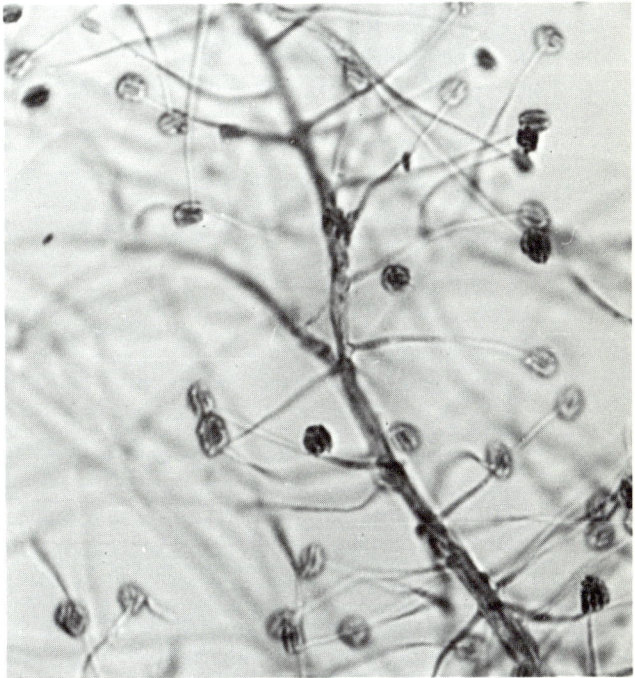

Fig. 9.1 (left)   *Acremonium strictum*   Phialides and spore balls.

Fig. 9.2 (right)   *Acremonium strictum*   Phialides and balls of conidia, ×550.

Species producing white synnemata are placed in the genus *Tilachlidium*.

Acremonium terricola (Miller *et al.*) W. Gams, is common in soil, forms pale pink colonies and produces spindle shaped conidia in chains.

Alternaria Nees ex Wallroth
(Gr. *alteres*, a kind of dumbell) (Figs 9.3, 9.4)
Conidiophores polytretic (or monotretic), sympodial, dark. Conidia dictyo-spores, catenate, dry, dark, ovoid with an apical rostrum (beak). The taxonomy of the genus is difficult, chiefly because of the lack of stable characteristics. The conidia are liable to vary greatly in size, shape and degree of septation even in a single culture, and great variations are seen on different culture media. The majority of the conidia which are pale to dark brown have a definite rostrum, often paler than the body of the spore.

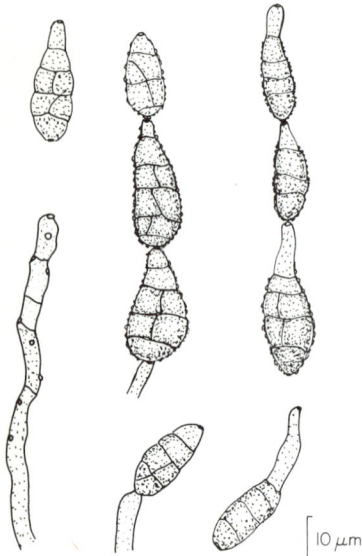

Fig. 9.3 *Alternaria alternata* Conidiophore and conidia in chains.

Comprehensive treatments of the genus are to be found in Ellis (1971, 1976), Simmons (1967) and Neergaard (1945).
    Most of the species grow well in culture but in nature they are found chiefly as parasites of cultivated plants. One species occurs regularly on industrial materials.

Alternaria alternata (Fr.) Keissler. This species has very commonly been known as *Alternaria tenuis* Nees ex Pers. but Simmons (1967) showed that *A. alternata* is the more suitable name. It is sometimes referred to as the *Alternaria alternata* group as the species concept is so broad that it might be a collection of closely related species rather than one single species. However, under normal circumstances when growing amongst other moulds in Petri

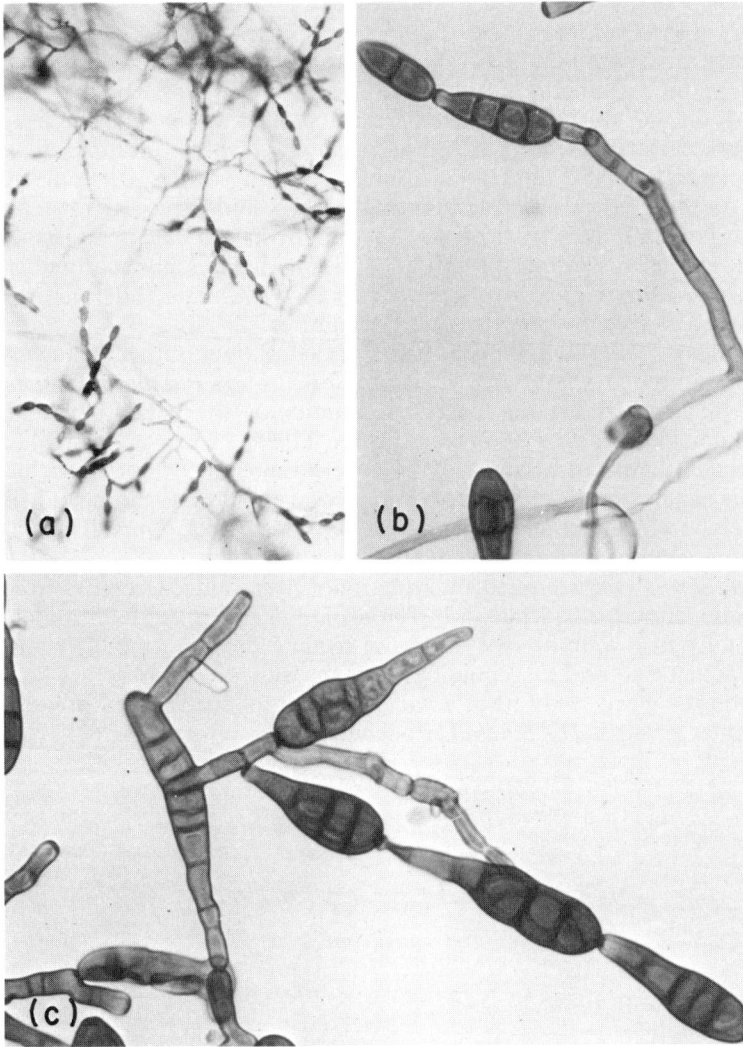

Fig. 9.4 *Alternaria alternata* **(a)** Chains of conidia in living culture, ×100; **(b)** conidiophore with young conidia, ×500; **(c)** chain of mature conidia and proliferating conidia, ×500.

dishes colonies are thin, velvety, almost black, with aerial growth consisting almost entirely of spore chains. In pure cultures aerial mycelium is always more freely produced, but, whilst some isolates continue to spore abundantly, others become sterile after only a few transfers. Spores are produced in chains, straight or branched, of up to a dozen or more spores, are yellow-brown to dark brown, very irregular in shape and size, mostly with short but definite rostrum, 20–63 × 9–18 μm. Found on numerous kinds of organic materials in damp situations.

Aureobasidium Viala & Boyer
(Lat. *aureus*, golden yellow) (Figs 9.5, 9.6)
Conidiophores undifferentiated or short lateral slightly swollen branches; conidia 0-septate, hyaline, enteroblastic, synchronous, endoconidia also produced. Moulds producing yeast-like or slimy colonies at first pale and later becoming dark or dark in sectors are quite common. Of these *Aureobasidium* and allied genera are the ones most usually seen in industry as spoilage organisms. As there is quite a wide range of these moulds which are at least superficially similar it is not surprising that there have been many name changes and confusions in interpretation. The most commonly accepted of these species is *Aureobasidium pullulans* (de Bary) Arnaud, but this was originally called *Dematium pullulans* de Bary and is perhaps best known in industrial circles as *Pullularia pullulans* (de Bary) Berkhout. Other synonyms of interest are *Aureobasidium vitis*, *Exobasidium vitis* and *Pullularia fermentans*. There have been several taxonomic studies of the genus including Cifferi, Ribaldi and Corte (1957) and Cooke (1959). Hermanides-Nijhof (1977) has given a useful account of 14 species and one variety and has explained the separation from *Hormonema*. However, she says more study is needed to elaborate eventual synonymy of the listed taxa. Durrell (1968) undertook an electron microscopic study of conidiogenesis.

Aureobasidium is characterized by synchronously produced conidia from hyaline conidiogenous cells, while in *Hormonema* Lagerberg & Melin conidia arise basipetally from one or two loci from hyaline or dark conidiogenous cells. The actual method of conidiogenesis is difficult to observe, and probably requires more study from a wide selection of species and isolates. Another similar genus is *Exophiala* Carmichael.

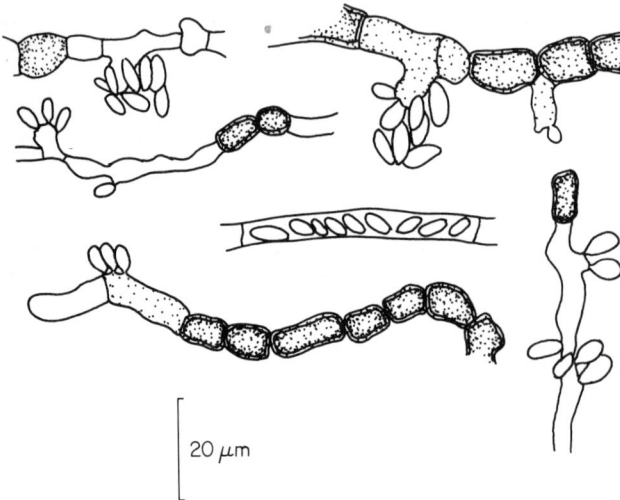

20 μm

Fig. 9.5 *Aureobasidium pullulans* var. *melanigenum*   Dark hyphae and pale conidia.

Aureobasidium pullulans (de Bary) Arnaud is a fairly common mould occurring on damp cellulosic materials. The conidiogenous cells are undifferenti-

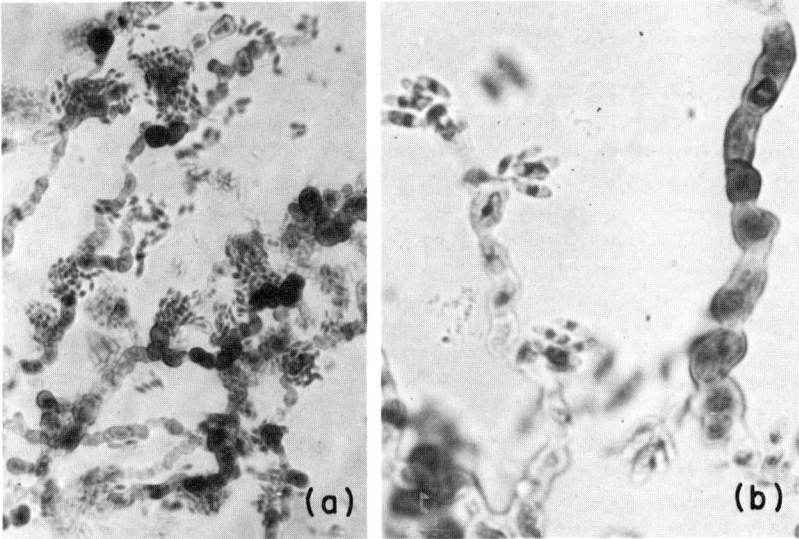

Fig. 9.6 *Aureobasidium pullulans* var. *melanigenum* Slide culture; **(a)** ×500; **(b)** ×1000.

ated or short lateral slightly swollen branches, from which conidia arise synchronously often in great numbers. The conidia are hyaline, smooth ellipsoidal and variable in size, 7–16 × 3.5–7 μm. Secondary smaller conidia are frequently formed from the initial conidia. Endoconidia are sometimes produced in intercalary cells. Hermanides-Nijhof (1977) describes *A. pullulans* as light in colour, yellow, cream, pink or light brown only becoming darker in age due to the production of dark hyphae. She describes a new variety *A. pullulans* var. *melanigenum* Hermanides-Nijhof for those isolates which darken rapidly and in which the dark hyphae thicken and break up into separate dark thick walled cells. This type of isolate is often seen in spoilage situations.

Reynolds (1950) gives a good account of *Aureobasidium* as a spoilage organism especially as a spoiler of painted surfaces. Various ascomycetous and pycnidial fungi also produce *Aureobasidium*-like states, for example *Sclerophoma pythiophila*, making this group of fungi even more difficult to study.

Botrytis Persoon ex Fries
(Gr. *botrys*, a bunch of grapes) (Fig. 9.7)
Conidiophores macronematous, straight, brown, with metallic sheen; conidiogenous cells swollen, polyblastic, synchronous; conidia solitary, dry, usually hyaline, 0-septate; sclerotia often produced.

The name is very descriptive for the conidiophores are irregularly branched at the apices, and the spores are each borne on a little peg on the ends of the swollen final branches. Many species are important plant pathogens, and are the conidial states of *Sclerotinia* species. Only one species is of common occurrence in industry.

The genus is discussed in Ellis (1971, 1976) and Hennebert (1973) who also deals with other related species.

Botrytis cinerea Pers. ex Fries (Lat. *cinereus*, ashy grey), conidial state of *Sclerotinia fuckeliana* (de Bary) Fuckel. Found on a wide variety of materials, and is also parasitic on many kinds of plants. Colonies floccose, pale brownish grey; conidiophores stiff, erect, much branched at the top, and bearing clusters of conidia like bunches of grapes (Fig. 9.7); conidia ovate, pale greyish, 8–12(15) × 6–10 μm; sclerotia beginning to develop after a few days as small, dirty green, mycelial knots, increasing rapidly in size and turning black, eventually irregular in shape, often confluent, and several millimetres long.

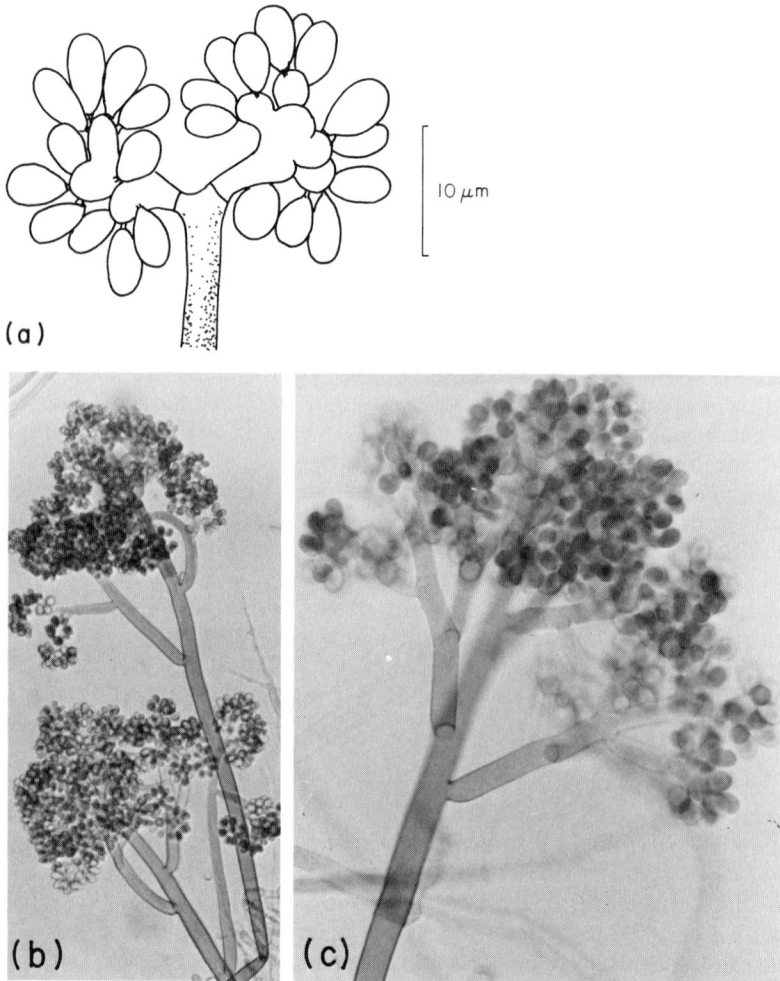

Fig. 9.7 *Botrytis cinerea* **(a)** Conidiophore and conidia; **(b)** typical conidiophores, ×160; **(c)** showing attachment of conidia, ×400.

Cladosporium Link ex Fries
(Gr. *klados*, a branch; hence branched spore chains) (Figs 9.8, 9.9, 9.10)
Conidiophores dematiaceous, macronematous, branched; conidia polyblastic, 0–3 septate, with ramo-conidia (branched or branch) in acropetal chains.

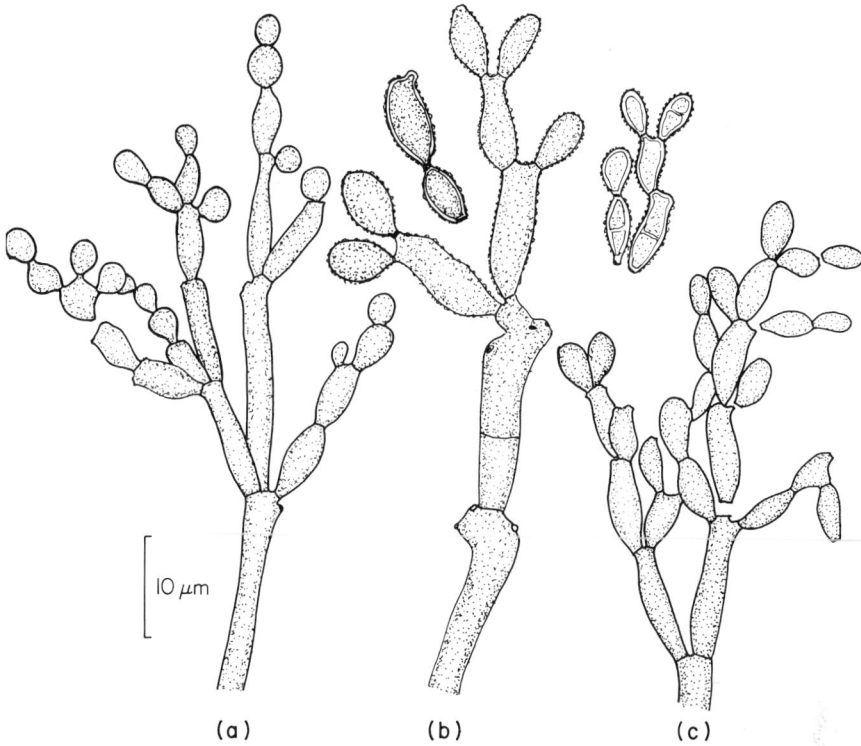

Fig. 9.8   *Cladosporium species*   Conidiophore and conidia. **(a)** *C. sphaerospermum*; **(b)** *C. herbarum*; **(c)** *C. cladosporioides*.

Fig. 9.9   Conidiophores and conidia of **(a)** *Cladosporium sphaerospermum*; **(b)** *C. herbarum*; **(c)** *C. cladosporioides*; all ×440.

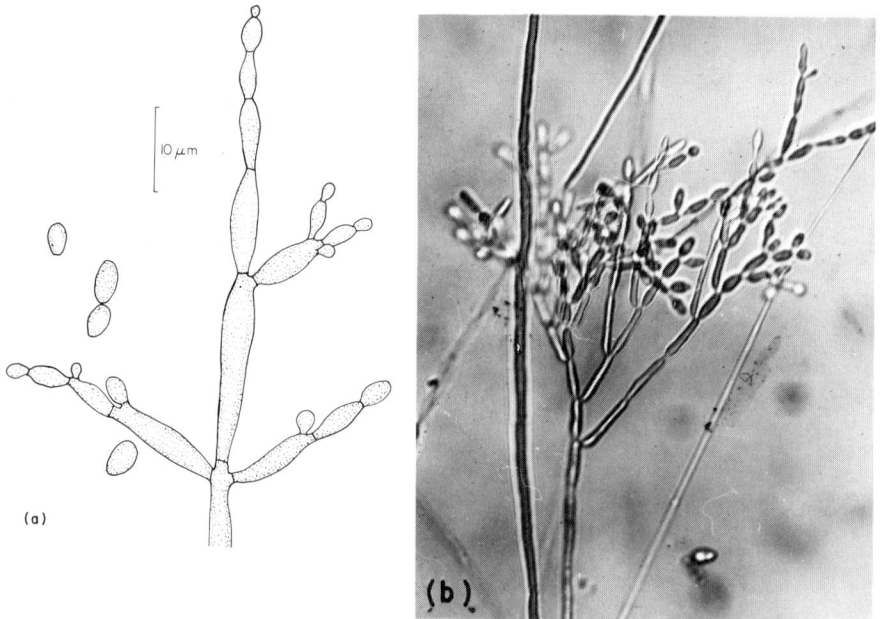

Fig. 9.10 *Cladosporium resinae* **(a)** Drawing; **(b)** growing in paraffin, ×500.

Members of the genus *Cladosporium* are very common and have been given many names. A large number of them are plant pathogens, but a few species are saprophytic and cause spoilage problems. The literature tends to be scattered and they have been difficult to place. De Vries (1952) produced a useful monograph, and Ellis (1971, 1976) described and keyed out most (43) of the common species.

Originally the name *Cladosporium* was restricted to forms with septate spores only and *Hormodendrum* was used for forms with non-septate spores, but it is now recognized that most isolates produce a high percentage of one-celled conidia.

Four species occur frequently in industrial situations (see Key). Other species are mostly plant pathogens.

Key to four common *Cladosporium* species

1.	Conidial scars prominent	2
1′.	Conidial scars not prominent	*C. resinae*
2.	Conidiophore often with terminal and intercalary swelling, nodose, 6–8 μm thick and verruculose	*C. herbarum* and others
2′.	Conidiophores not nodose	3
3.	Conidia mostly spherical or subspherical, 3–4.5 μm diam.	*C. sphaerospermum*
3′.	Conidia not spherical or subspherical, usually saprophytic	*C. cladosporioides*

C. herbarum (Pers.) Link ex S. F. Gray (Lat. *herbarum*, of plants). An exceedingly common organism, found on dead herbaceous and woody plants, textiles, rubber, leather, paper, and foodstuffs of all kinds. Petri dishes exposed in the open air usually trap hundreds of spores, far more than of any other genus of moulds. It grows over a wide range of temperatures, and has been reported fairly frequently as infecting meat in cold storage.

In culture this species is easy to recognize. It grows somewhat restrictedly, thick velvety, varying from deep rich green to dark grey-green in colour, reverse a characteristic opalescent blue-black or greenish black. Examination of living cultures, under a low magnification, shows the spores to occur in large, tree-like clusters. The sporing structures are very brittle, breaking up completely, when mounted, into spores and rod-like fragments of conidiogenous cells and mycelium. Under the microscope all parts of the fungus are dark-coloured, greenish brown to dark brown. Examination of young cultures or slide cultures will show that the first formed conidia (blastospores) increase by budding, producing chains with the youngest spore at the top (acropetal chains), eventually forming tree-like masses of much-branched chains. The conidiophores usually show a swelling at the original point of attachment of the chains. The conidia are distinctly verruculose, with scars at one or both ends and young spores mostly one celled.

C. sphaerospermum Penz. This very common species has been isolated from air, soil, foodstuffs, paint and textiles and similar situations. The conidia are mostly globose to subglobose, 3–4.5 μm diam., dark brown and verrucose.

C. cladosporioides (Fresen.) de Vries A very common cosmopolitan species which has been isolated from air, soil, textiles and paint. It is often used for testing resistance of materials to mould damage.

C. resinae (Lindau) de Vries (Lat. *resinae*, of resin) (now known as *Hormoconis resinae* (Lindau) Arx & de Vries), imperfect state of *Amorphotheca resinae* Parbery.

It is best known as an inhabitant of wood impregnated with creosote or coal-tar, both of which it can utilize as sources of carbon. Although it grows well on laboratory culture media it is remarkable that, on creosoted timber, the growth stops short on reaching untreated wood. During recent years the fungus has become a nuisance in the petroleum industry, and in the fuel tanks of aircraft. Most storage tanks contain a small amount of water under the fuel – the fungus grows at the interface, utilizing water from below and carbon compounds from above. The tangled masses of mycelium which are formed may block supply pipes, control valves, etc. and metabolites liberated may cause corrosion. See Hendy (1964), Parbery (1969, 1971 and others), Scott (1971) and Sheridan (1972 and others) and more recent work by Park (1975) and Neihof and Bailey (1978).

As *C. resinae* is still an acute economic problem work on control is very active. Infection is usually controlled by quality control of fuel, strict servicing and regular inspection followed by rapid treatment with biocides. Tank cleaning programmes are very expensive and to be avoided if possible by diagnosing the problem early. Suitable biocides are difficult to find as many ordinary ones would damage the delicate aeroengines. Biobor and ethylene glycol have been found effective. According to Neihof and Bailey (1978) diethylene glycol monomethyl ether appears to be the most promising replacement for the currently used additive 2-methoxyethanol.

Curvularia Boedijn
(Name refers to shape of spores) (Fig. 9.11)
Conidiophores polytretic, dark coloured, macronematous and producing conidia in a sympodial manner; conidia phragmospores, centre cells dark and swollen.

Both mycelium and conidia are dark coloured, the latter borne on more or less erect conidiophores. The conidia are produced at the apex of the conidiophores sympodially until a cluster is formed. The cluster may be tight, or more open, but the conidiophores from which spores have been shed always appear knobbly or tortuously bent along the spore-bearing part. The conidia have 3 or 4 cross septa, and are mostly curved or bent about the third cell from the base, this cell being broader and darker in colour than the others. The genus has been discussed and species described by Boedijn (1933) and Ellis (1966, 1971, 1976). All species are difficult to maintain in good condition in culture, many isolates becoming sterile after one or two transfers.

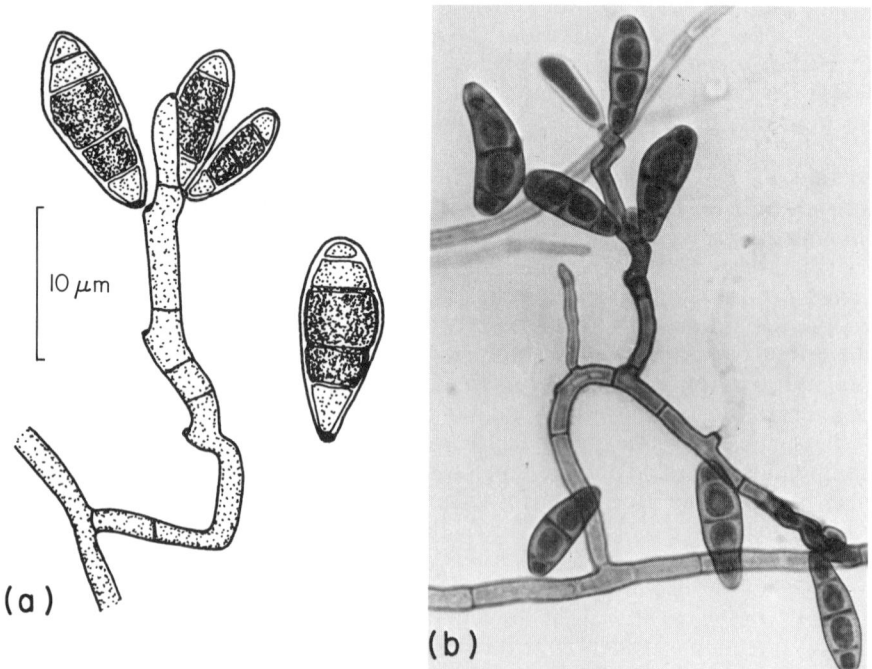

Fig. 9.11 *Curvularia palescens* **(a)** Conidiophore and conidia; **(b)** showing sympodial growth of conidiophore, ×490.

C. lunata (Wakker) Boedijn (Lat. *lunatus*, like the crescent moon), perfect state *Cochliobolus lunatus* Nelson & Haasis is very common and appears on a wide variety of substrates. Conidiophores usually short, up to about 100 μm long and 2–4 μm diam., septate; conidia 20–30 × 8–16 μm, mostly 23 × 11 μm.

C. geniculata (Tracy & Earle) Boedijn (Lat. *geniculatus*, knotty), perfect state *Cochliobolus geniculatus* Nelson is also very common and occurs on many different substrata. Conidiophores fairly long, 300–900 μm, widening upwards and 3.5–5 μm diam., at tip; conidia forming dense clusters 4-septate, with thick cell much swollen and very dark, 20–45 × 7–14 μm, mostly 24 × 9 μm.

Doratomyces Corda
(Gr. *doratos*, gen. of *dory*, a spear) (Figs 9.12, 9.13, 9.14)
Conidiogenous cells annellides in branched or penicillate heads aggregated into synnemata; conidia dark, globose to elliptical, 0-septate, with truncate base, in chains.

The genus *Stysanus* Corda, described in previous editions of this book, has been shown by Morton and Smith (1963) to be a later synonym of *Doratomyces* Corda.

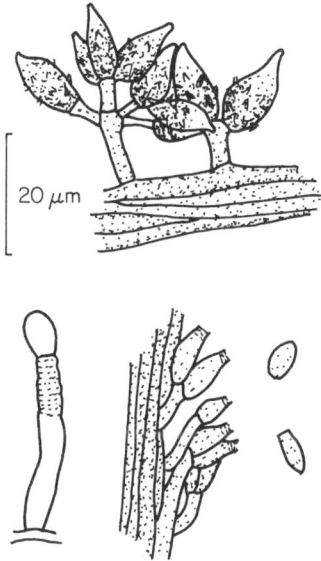

Fig. 9.12 *Doratomyces stemonitis* Both types of conidium.

D. stemonitis (Pers. ex Fr.) Morton & G. Smith (*Stemonitis*, a genus of Myxomycota with similar fructifications) is not uncommon, particularly in soil. Colonies a distinctive dark sooty grey, consisting of a mixture of simple fascicles and definite synnemata (as in Fig. 9.14); synnemata consisting of a

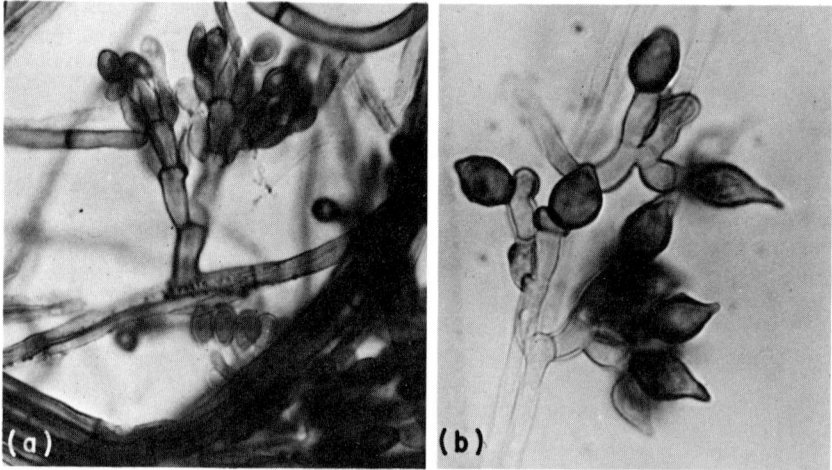

Fig. 9.13 *Doratomyces stemonitis* **(a)** Dissected coremium showing penicillate sporing structures, ×1000; **(b)** second conidial form belonging to *Echinobotryum,* ×1000.

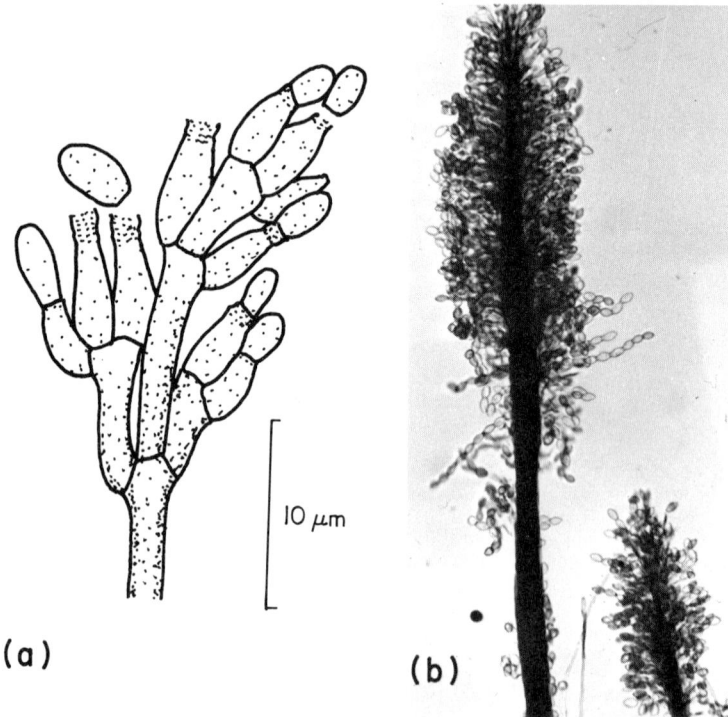

Fig. 9.14 *Doratomyces microsporus* **(a)** Conidial head, showing annellophores; **(b)** synnema, ×225.

stalk, which is a tight bundle of dark, septate hyphae, and a head, where the conidiophores spread out and terminate in penicillate structures showing a marked resemblance to those of *Scopulariopsis* (cf. Figs 9.13, 9.40, 9.41). The conidia, which are borne in chains, are ovate to lemon-shaped, almost colourless to pale brown, about 7 μm in long axis. On continued cultivation in the laboratory the production of large synnemata is often progressively reduced.

As in *Scopulariopsis* the conidium bearing cells are annellides, not phialides. A number of authors have expressed the opinion that the two genera should be regarded as one, since the production of synnemata has, in recent years, been regarded with little favour as a basis for classification. The situation is, however, complicated by the fact that some isolates of *Dorato-myces*, with the typical spore form indistinguishable from *D. stemonitis*, produce in addition a second spore form, which belongs to the genus *Echinobotryum* Corda (Fig. 9.13).

Note: The fungus pictured as *Stysanus stemonitis* in Editions 1–5 of this book, Figs 84 and 85, has been found to be a species of the related genus *Trichurus*. This differs from *Doratomyces* in producing sterile hairs amongst the penicillate structures of the spore head.

Fusarium Link ex Fries
(Lat. *fusus*, a spindle; describes shape of spores) (Figs 9.15, 9.16)
Conidiogenous cells hyaline, enteroblastic, mono- or polyphialidic; conidia, *Fusarium* species produce several types of conidia, microconidia hyaline, 0-, 1- or 2-septate, small; macroconidia hyaline, curved (sickle-shaped) phragmospores, with a foot cell bearing some kind of heel (this is the most distinctive character of the genus); chlamydospores may also be present, borne terminally, intercalary or on the macroconidia.

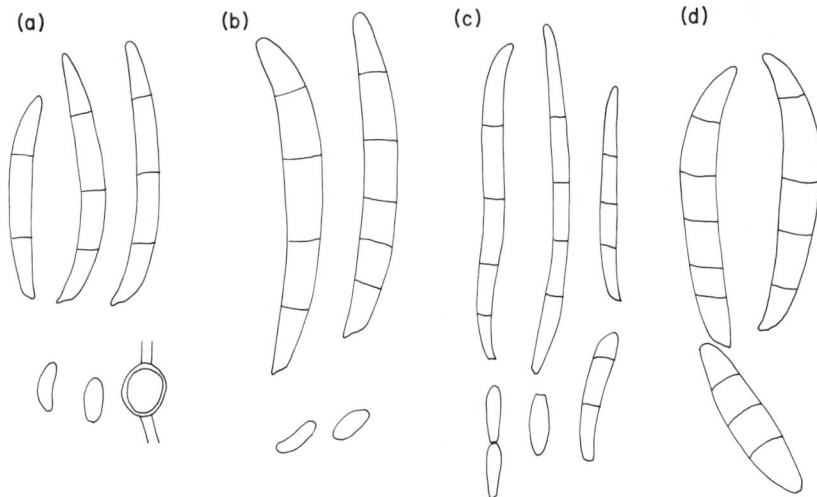

Fig. 9.15 *Fusarium* species   Conidia. **(a)** *F. oxysporum;* **(b)** *F. solani;* **(c)** *F. moniliforme;* **(d)** *F. culmorum.*

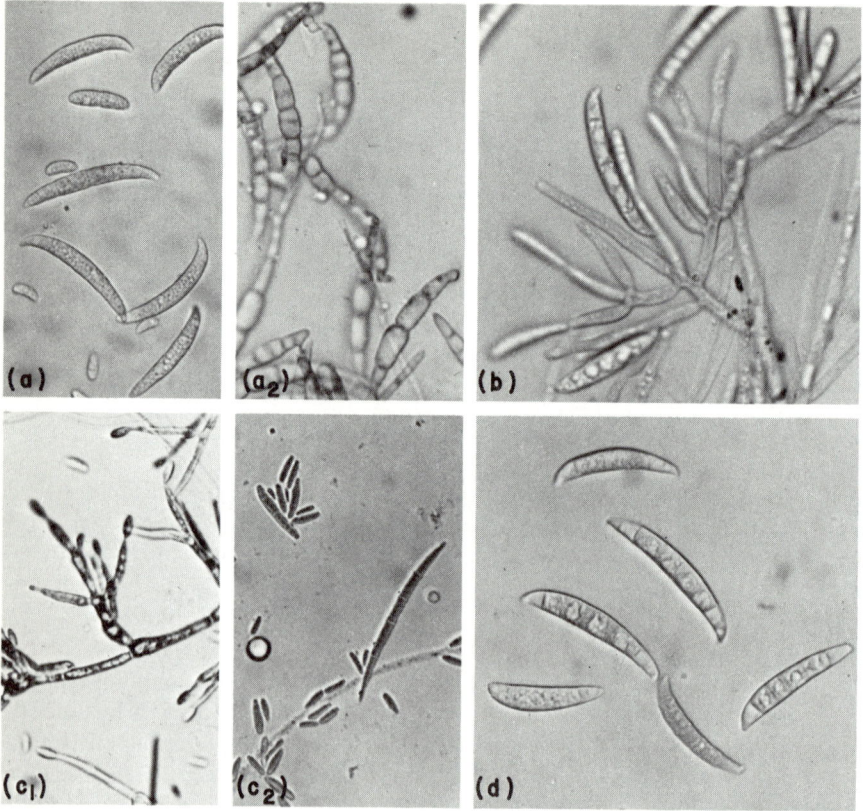

Fig. 9.16 Conidia of **(a)** *Fusarium oxysporum;* **(b)** *F. solani;* **(c)** *F. moniliforme;*
(d) *F. culmorum;* all ×470.

The perfect states have been found for many of the species and fall in
the pale coloured genera of the Pyrenomycetes, for example *Nectria, Hypo-
myces, Gibberella* and *Calonectria.*

The genus has a very wide host range, occurring in soil, as a plant
pathogen, on man and as a spoilage organism in industrial situations, even
occurring in chemicals, water and aircraft fuel. Recently it has become of
increasing importance as producing mycotoxins in foodstuffs. It has been
particularly studied as a plant pathogen, while in the industrial situation
identification has usually only been taken to genus level. This is no doubt due
to the difficulty encountered in identification of cultures. One of the main
causes of difficulty in identification is the ability of *Fusarium* species for rapid
change, variability with substrate, and rapid loss of ability to spore. It is
therefore necessary to adopt standard conditions of culture to obtain consistent results. The conditions standardized include media, temperature, light
and length of growth, and vary from author to author. The use of single spore
cultures has also helped in stabilization and either microconidia or macroconidia are suitable.

Besides being variable, *Fusarium* is a large genus with a considerable range of characters. It was Appel and Wollenweber (1910) and Wollenweber (1913) who made the first breakthrough in classification and divided the genus into groups or sections. These studies culminated in Wollenweber and Reinking's monograph, *Die Fusarien* (1935) which remained for years the standard work, though Wollenweber published further papers. However, there were still considerable problems in identification and many papers and monographs have followed, with great simplification, some authors think too much simplification, by Snyder and Hansen (1941, 1945), Toussoun and Nelson (1968), and Snyder *et al.* (1957). Other monographs of interest were produced by Gordon (1952 and others), Messiaen and Cassini (1968) and Booth (1971). In the latter, Booth stresses the importance of the conidiogenous cell in identification, a concept not usually studied in *Fusarium*. In 1977, Booth produced a useful laboratory guide to the identification of the major species.

When identifying *Fusarium* species Booth (1971, 1977) stressed the importance of single spore cultures, standard methods of cultivation (he used potato sucrose agar, pH 6.5 at 25°C for 4 days) for the study of conidiogenesis, micro- and macroconidia, the presence of chlamydospores and growth rate.

Identification of *Fusarium* species is still a subject for the specialist, and of more importance to the plant pathologist. However, four of the commonest species are mentioned below.

F. oxysporum Schlecht, is about the most variable species of *Fusarium*, and of world wide distribution. It occurs in soil and occasionally as a spoilage organism, but most frequently as a plant pathogen. Growth is moderate, white, peach, to salmon pink or violet. Microconidia are oval to cylindrical or even curved and produced on simple short phialides. Macroconidia, 3–5 septate, $27–60 \times 3–5$ μm. Chlamydospores are produced.

F. solani (Mart.) Sacc., is slower growing than *F. oxysporum*, producing greyish white to blue or bluish brown colonies. Microconidia are cylindrical to oval and produced from long phialides which may arise from complex branched conidiophores. Macroconidia are 1–5 septate. Globose chlamydospores are produced.

It is of world wide occurrence, found in soil and occasionally as a spoilage fungus, but is most important as a plant pathogen. The perfect state is *Nectria haematococca* Berk. & Br.

F. moniliforme Sheldon forms colonies white to peach or salmon pink later developing vinaceous to violet colours. The fusoid to clavate microconidia, $5–12 \times 1.5–2.5$ μm are sometimes 1-septate, and produced in chains from long lateral phialides. Macroconidia are rare in some isolates, 3–7 septate, $25–60 \times 2.5–4$ μm. No chlamydospores are produced.

The perfect state is *Gibberella fujikuroi* (Sawada) Wollenw. It is biochemically interesting especially for the production of gibberellins. It is an important plant pathogen.

F. culmorum (W. G. Smith) Sacc. grows very rapidly, producing abundant red pigment and becoming reddish brown. No microconidia are produced.

Macroconidia are uniform in shape and size, 3–5 septate, 27–50 × 5–7.5 μm, produced from simple phialides arising from complexly branched conidiophores. Oval to globose chlamydospores are produced.

It occurs frequently in soil and is an important plant pathogen of cereals.

Geomyces Traaen
(Gr. *ge*, the earth; *mykes*, a fungus) (Figs 9.17, 9.18)
Conidiophores present, hyaline, branched acutely at the apex; conidiogenous cells thallic; conidia formed terminally or laterally on short pedicels, intergrading with arthrospores produced in alternate segments of the mycelium, becoming enlarged and liberated by lysis of intervening cells.

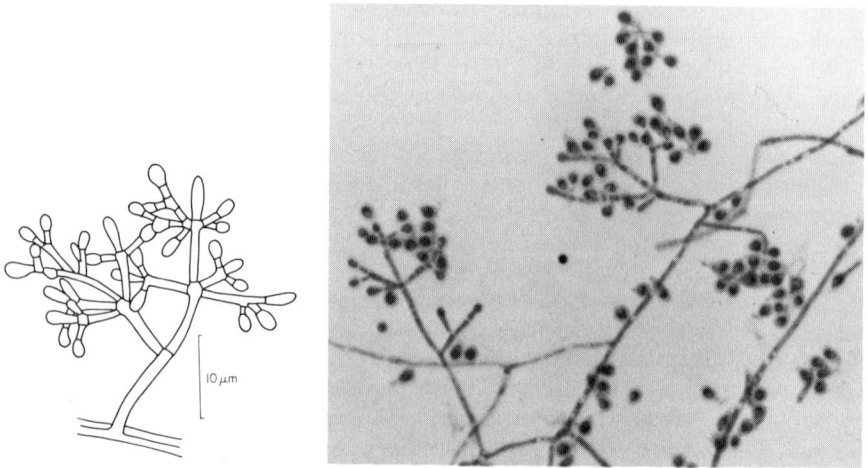

Fig. 9.17 (left) *Geomyces pannorus* Conidiophore and conidia.

Fig. 9.18 (right) *Geomyces cretacea* Conidiophore and conidia, ×550.

Isolates producing light coloured or hyaline arthrospores are frequently isolated from soil and decaying materials. Until recently opinions as to classification have differed. Carmichael (1962) included many of these in his monograph of *Chrysosporium*. In previous editions of this book they were referred to *Aleurisma* and *Sporotrichum*. However, in a recent monograph Sigler and Carmichael (1976) use the name *Geomyces* Traaen and make the new combination *Geomyces pannorus* (Link) Sigler & Carmichael, for the most common species encountered. This is very variable in colour and some workers would divide it up into several species, but for most purposes it is convenient to regard it as one large and variable species. It produces slow growing dusty colonies, spreading in age, compacted almost velvety, ranging in colour from pure white to greyish through buff to orange brown, reverse colourless, to clear yellow to orange-yellow to brownish. Spores hyaline, barrel shaped, ovate to pear shaped, smooth to rough, mostly 2 × 3 μm but ranging from 2–4 × 2–5 μm. The very smallness of the conidia helps to distinguish it from the species of *Chrysosporium*.

It is reported from a variety of sources and under the name of *Sporotrichum carnis* Brooks & Hansford from meat in cold storage at temperatures as low as −6°C.

Geotrichum Link ex Pers
(Gr. *ge*, the earth; *thrix*, hair; hence forming mycelium close to the substrate) (Fig. 9.19)
Conidiogenous cells thallic, micronematous, hyaline; conidia 0-septate, hyaline, arthrospores, produced by random septation and disjunction at a double septum, slimy, thin walled, smooth, cylindrical but sometimes oblong or subglobose.

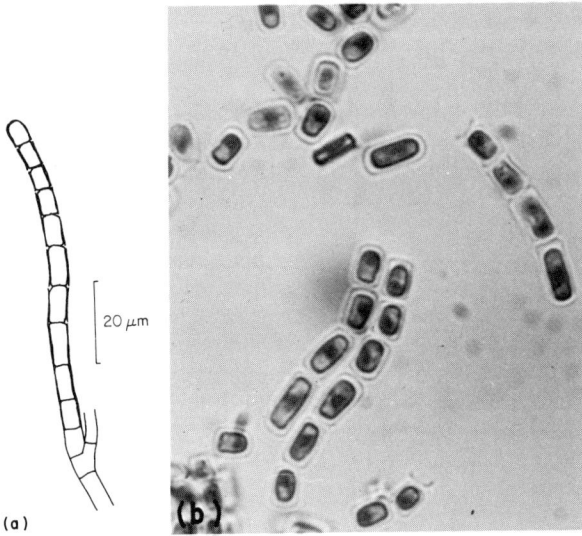

Fig. 9.19 *Geotrichum candidum* **(a)** Fragmenting mycelium; **(b)** fragmenting mycelium and arthrospores, ×550.

G. candidum Link (Lat. *candidus*, white). This species grows well on wort agar, less well on synthetic media; colonies thin, spreading, creamy white, soft and somewhat yeast-like in texture; in young colonies fairly long mycelial strands present, showing characteristic dichotomous branching at the colony margin, mycelium breaking up into arthrospores as colonies age; arthrospores cylindrical with rounded ends. This species is common on all types of milk products. It has been known for a long time as *Oospora lactis* (Fresenius) Sacc., but it is unlike other species of the genus *Oospora* and has been identified as Link's species. It is also common, and sometimes creates quite a problem in polluted water (Cooke, 1954; Tubaki, 1962).

Carmichael (1957) has made a detailed study of the morphology and occurrence of this fungus, and provided a long synonymy. Sigler and Carmichael (1976) included the genus in their study on Hyphomycetes with arthroconidia and accepted *Endomyces geotrichum* Butler & Petersen as the name for its perfect state. Von Arx (1972) also used this name. However, Redhead and Malloch (1977) renamed the telemorph as *Galactomyces geotrichum* (Butler & Petersen) Redhead & Malloch (= *Dipodascus geotrichum*). Von Arx (1977) was not happy about this and included the common rapidly growing non-ascosporic fungus found on milk products as *Geotrichum candidum* in his key to the accepted species of *Dipodascus* and *Geotrichum*.

Gliomastix Gueguen
(Gr. *gloios*, mucilage; *mastix*, a whip) (Fig. 9.20)
Conidiogenous cells monophialidic; conidia 0-septate, catenate or aggregated in slimy heads, usually pigmented.

This genus was monographed by Dickinson (1968) and described by Ellis (1971, 1976). It was treated as a section of *Acremonium* by Gams (1971). The commonest species in the spoilage situation is *G. murorum*.

G. murorum (Corda) Hughes (Lat. *murorum*, of walls), is frequently found in soil and isolated from textiles and wood. Colonies at first white or pale pink, turning black in patches as conidia develop, and eventually black all over; conidiophores merely elongated phialides, arising as lateral branches from trailing hyphae or ropes of hyphae; conidia developing successively from the tips of the phialides, forming, at least during the rapidly growing period, slimy balls, in old cultures forming whip-like twisted chains; ripe conidia dark coloured, $2.5-5.5 \times 2-4.5$ μm, with the colouring matter aggregated in small granules which tend to become detached in fluid mounts. It is usually assumed that the conidiogenous cells are phialides but the structure at the apex of the cells is so fine and obscured by granules that it is difficult to study the exact structure. The actual structure is still a matter of discussion and opinions differ.

Helminthosporium Link ex Fries
Drechslera Ito, *Bipolaris* Shoemaker and *Exosporium* Link ex Fries
(Gr. *helmins, helminthos*, a worm; hence worm-like spores) (Figs 9.21, 9.22)
Conidiophores dark coloured, bearing conidia in bunches at the apex; conidia dark, apparently phragmospores but usually distoseptate.

The genus *Helminthosporium* has until recent times included a large number of species, the best known of which are parasites of cereals and grasses. Many of them, however, grow well as saprophytes, and are not uncommon on plant debris. They are sometimes isolated from soil, but are rarely seen in the industrial situation. However, the genus is well known and of considerable economic importance.

In culture some species grow normally, producing typical dark coloured hyphae and abundant conidia. Others appear to be quite sterile, and tend to produce mostly hyaline or pale coloured hyphae. In classical keys it is described as producing spores singly on the ends of the conidiophores whereas examination of sporing cultures will usually show clusters of conidia, at least at low magnifications, resembling bunches of bananas. The mode of conidial production varies. The conidia are dark, obclavate to cylindrical with rounded apex and slightly truncate base, with apparent transverse septa, but

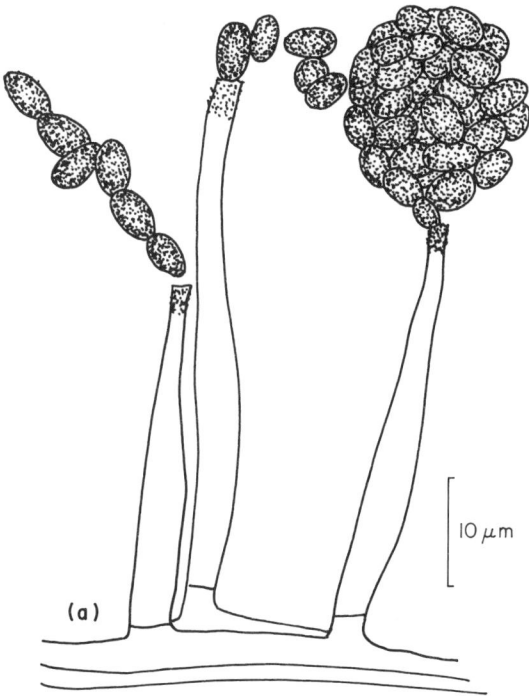

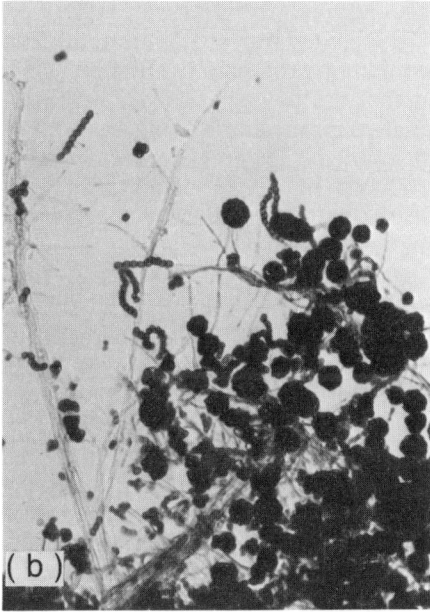

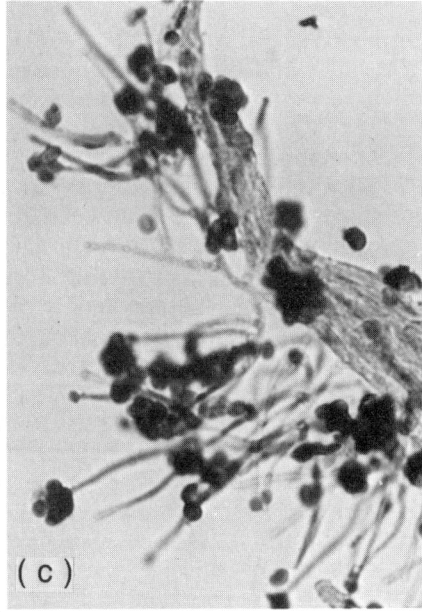

Fig. 9.20 *Gliomastix murorum* **(a)** Phialides and conidia; **(b)** conidia in balls and chains, ×500; **(c)** phialides and mature conidia, ×500.

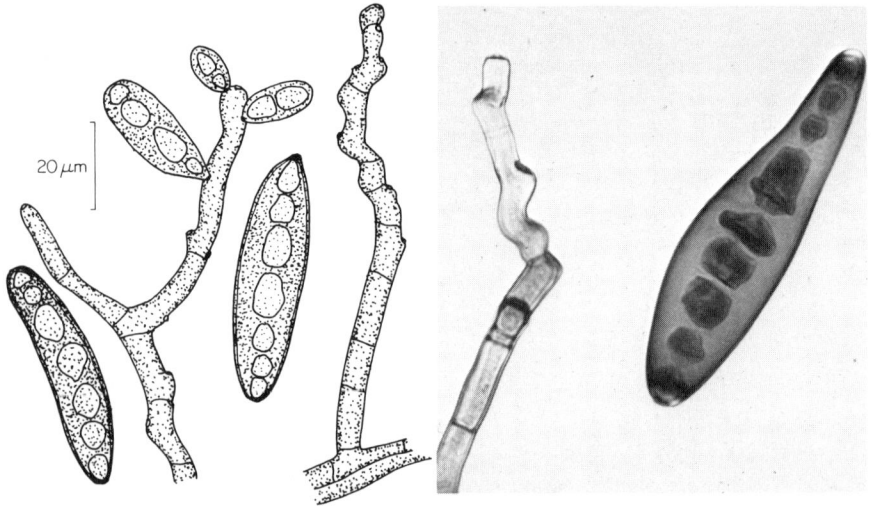

Fig. 9.21 (left)   *Helminthosporium cynodontis*   Sympodial conidiophores and conidia.

Fig. 9.22 (right)   *Helminthosporium (Cochliobolus sativus)*   Conidiophore and conidium, ×500.

in the majority of species these are distosepta, which do not reach across the conidium from wall to wall.

The genus as interpreted by the older monographers, Drechsler (1923) and Nisikado (1929), was not homogeneous. It included fungi with different kinds of conidia and different methods of conidium production. Hughes (1953) when he examined type material found that the type species, *H. velutinum* Link ex Fries produces its conidia terminally and in whorls below the apex of stiff conidiophores whereas in the majority of species previously included in *Helminthosporium* the conidia are produced in a sympodial fashion. In addition perfect states have been found for quite a number of species, but these belong to several different genera.

The situation is not yet fully clarified and numerous papers have been published in which the taxonomy of this group of fungi has been discussed, including Ellis (1961), Shoemaker (1959), Luttrell (1963, 1964), Leonard and Suggs (1974), Chidambaram *et al.* (1973) and Barron (1968) who gives a short survey of the situation. Ellis (1971, 1976) describes and keys many species and allocates them to genera. He keeps the name *Helminthosporium* for the type species and for species with terminal conidia, while placing other species according to their method of spore production, putting the common sympodial forms in *Drechslera* (which is the most likely genus to occur in industry) and *Exosporium*. However, Ellis (1976) under *Drechslera* states 'There remain so many critical species which are not known to have perfect states or which even the experts have difficulty in deciding whether they should be in *Drechslera*, *Bipolaris* or some other genus that the continued use of *Drechslera* for all graminicolous species previously in *Helminthosporium* is recommended.'

Humicola Traaen
(Lat. *humus*, soil; *colo*, to inhabit (Fig. 9.23)
Conidiophores micronematous, hyaline; conidiogenous cells, short, hyaline, monoblastic, determinate, terminal; conidia 0-septate, large, dark, spherical, single, dry and usually smooth. A sparse hyaline phialidic state with very small 0-septate conidia is also produced.

Colonies are hyaline becoming greyish to dark brown or black in age. The conidia are dark, more or less globose, smooth and produced singly at the ends of short hyaline conidiogenous cells. The phialidic state is often very sparse and may not be seen.

Two species are common, both occurring in soil, on wood and wood products, and on plant debris. They are strongly cellulolytic.

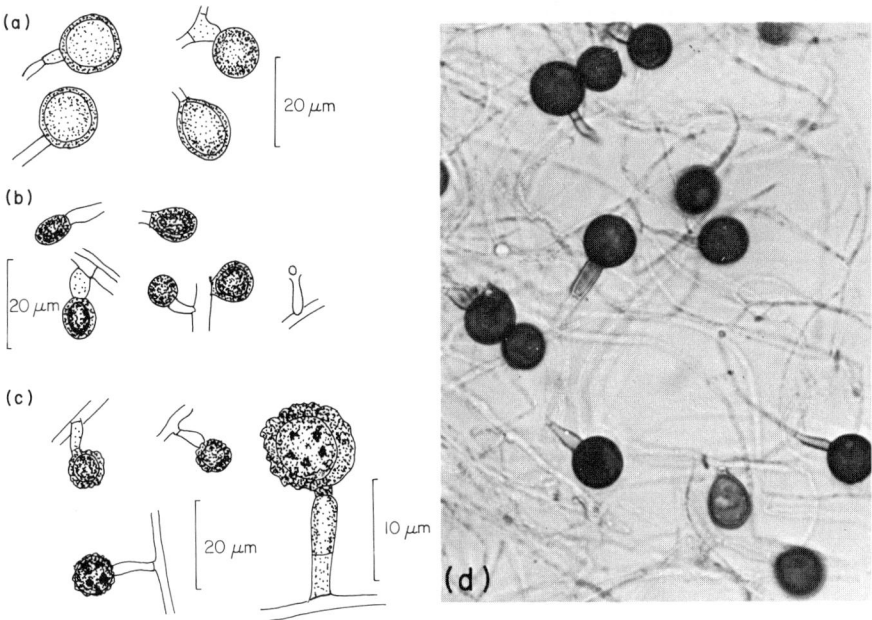

Fig. 9.23 *Humicola* species **(a)** *H. grisea;* **(b)** *H. fuscoatra;* **(c)** *Thermomyces lanuginosus;* **(d)** *H. grisea;* ×490.

H. fuscoatra Traaen has smaller conidia, 6–11 μm diam.

H. grisea Traaen has conidia 12–17 μm diam.
Similar species which are thermophilic and do not produce a phialidic state are usually referred to the genus *Thermomyces. T. lanuginosus* Tsiklinsky has dark globose conidia, which are very warted and is quite common. *T. stellatus* (Bunce) Apinis has lobed conidia, occurs on mouldy hay and is not so common.

Ellis (1971) and Barron (1968) discuss and describe these and similar genera.

Isaria Persoon ex Fries
(Gr. *is*, a fibre) (Fig. 9.24)
Species which have been referred to this genus are best known as parasites of insects, but occasionally typical isarioid colonies are isolated in the laboratory, forming long white synnemata, cylindrical, occasionally branched along the whole length.

Fig. 9.24 *Isaria* sp.   Culture on agar slant, showing coremia, natural size.

The fungi which have been assigned to the genus *Isaria* include species with very different methods of spore production. Hence, in the opinion of most present-day mycologists, the genus has no real standing, but merely represents growth forms belonging to various better defined genera.

Monilia Persoon ex Fries
(Lat. *monile*, a necklace of beads) (Figs 9.25, 9.26)
Conidiophores not differentiated; conidiogenous cells holoblastic; conidia hyaline, 0-septate, produced by budding in acropetal chains.
   This rather heterogeneous genus is characterized by the production of long branched chains of bead-like conidia, produced in acropetal succession, i.e. with the youngest conidium at the apical end of the chain. *M. fructigena* and *M. cinerea*, imperfect states of *Sclerotinia*, are well known parasites of tree fruits, forming greyish pustules consisting chiefly of masses of spores, and producing soft rot.

M. sitophila (Montagne) Saccardo (Gr. *sitos*, corn; *phileo*, to love). Best known as the red bread mould. From time to time outbreaks of the infection

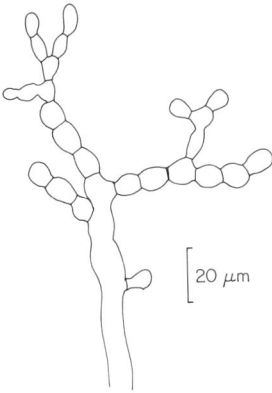

Fig. 9.25 *Monilia sitophila* Mycelium and conidia.

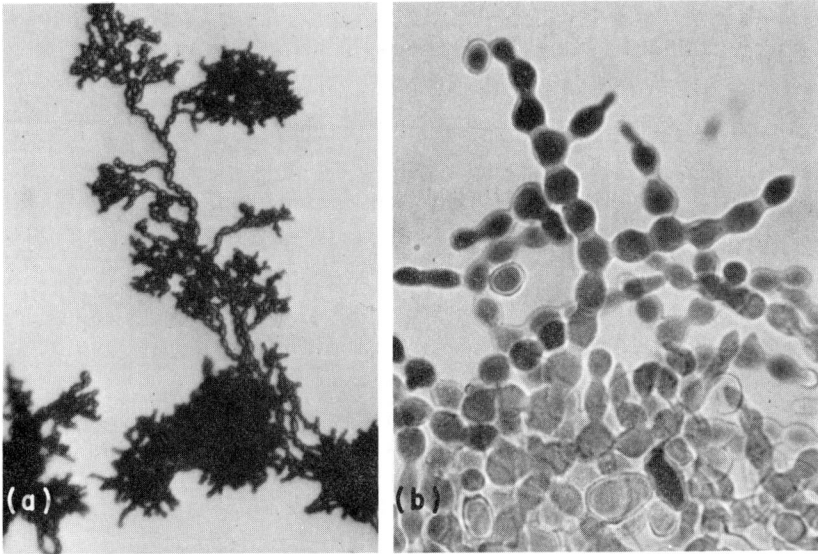

Fig. 9.26 *Monilia sitophila* **(a)** Spore masses as seen in culture tube, ×100; **(b)** ×550.

occur in bakeries but, more often, the fungus attacks wrapped and, particularly, sliced bread, and is noticed only after the bread has left the bakery. In an amazingly short time the bread becomes covered with a characteristic pink, loose-textured growth. According to Shear and Dodge (1927) it also attacks sugarcane bagasse in store, forming long pink festoons from the bales. It is sometimes found as an aerial contaminant and can then be a great nuisance in a culture room, since it spreads with extreme rapidity on suitable media and often sheds spores outside Petri dishes in which it is growing.

It grows best on organic culture media, forming loose floccose masses of a pale pink to salmon-pink colour. Ovate conidia are formed at the ends of aerial hyphae; these increase by budding, and eventually form enormous irregular masses. Later the mycelium breaks up to some extent at the septa. Shear and Dodge (1927) have shown that *M. sitophila* is the imperfect state of *Neurospora sitophila*, but, since the latter is heterothallic, it is seldom that perithecia are found. They have also shown that there are several species of *Neurospora* with very similar conidial states. Of these, *N. tetrasperma* has asci containing only four ascospores and is homothallic, but single strains of the others cannot be recognized with certainty unless they are grown in a series of cultures along with authentic + and − strains of the various species.

Moniliella Stolk & Dakin
(Figs 9.27, 9.28)
Conidia hyaline, 0-septate, monoblastic in acropetal chains or thallic arthro-spores sometimes thickening to form chlamydospores.

In 1966 Stolk and Dakin described a new genus of Hyphomycetes occurring in very acid syrups such as pickles. It produces hyaline to pale coloured conidia in monoblastic acropetal chains as in *Monilia* or *Cladosporium*, but the hyphae tend also to fragment forming thallic arthrospores which some-times thicken up to form chlamydospores.

M. acetoabutans Stolk & Dakin
is the commonest species. The colonies are restricted hyaline to pale grey or even darker, forming blastospores in chains, $5–11 \times 3–8~\mu m$; cylindrical thallospores, often quite long, $6–15 \times 3–5.5~\mu m$ and dark brown chlamydospores, smooth walled, globose to ellipsoidal, $8–12 \times 7–11~\mu m$.

The fungus is unique in its tolerance to acetic acid especially associated with sugar as in sweet pickle, mint sauce and similar pickle products containing vinegar, to which it often gives a characteristic fruity or ester-like odour reminiscent of ethyl acetate (Dakin and Stolk, 1968).

Myceliophthora Cost.
(Figs 9.29, 9.30)
Conidiogenous cells hyaline, poorly differentiated, with conidia borne directly on hyphae or on ampuliform swellings of the hyphae; conidia 0-septate, hyaline, blastospores, borne on short or long pedicels, usually determinate, occasional secondary blastospores produced.

A recent study has been made by van Oorschot (1977). One species occurs in the spoilage situation and then only under extra warm conditions. *Myceliophthora thermophila* (Apinis) Oorschot was originally described by Apinis as *Sporotrichum* and later transferred to *Chrysosporium* by Klopotek (1974).

Myceliophthora thermophila (Apinis) Oorschot
produces white cottony to pale to pinkish brown colonies, growth is rapid at 30°C, with a maximum growth temperature of 54°C, and a minimum growth temperature of 24°C. Blasto-conidia are borne terminally or laterally on short or long pedicels on trailing hyphae or ampuliform swellings of the hyphae. Occasional secondary conidia

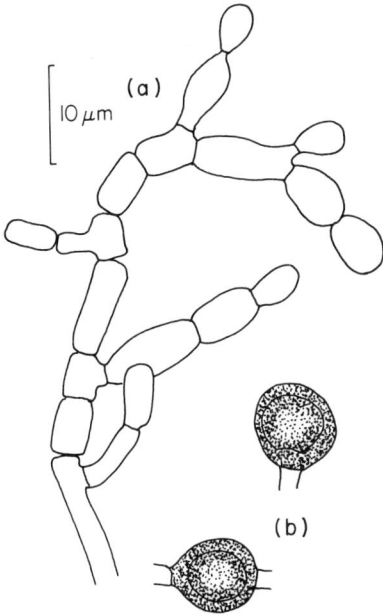

Fig. 9.27 *Moniliella acetoabutans* **(a)** arthrospores; **(b)** chlamydospores.

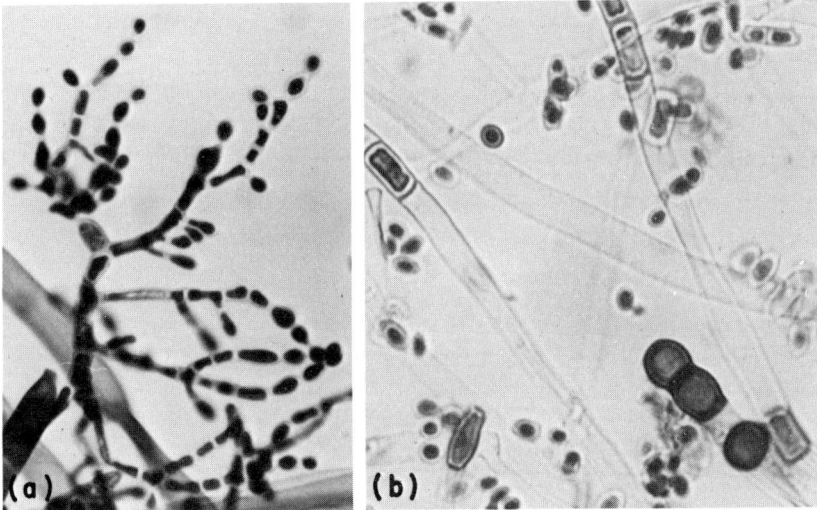

Fig. 9.28 *Moniliella acetoabutans* **(a)** Arthrospores, ×700; **(b)** chlamydospores, ×550.

are produced. Conidia are hyaline, 4–11 × 3–4.5 μm, smooth or when first isolated finely roughened.

This moderately thermo-tolerant species occurs from time to time in decomposing vegetable materials especially straw and was used by Barnes *et al.* (1972) in the microbial upgrading of waste paper.

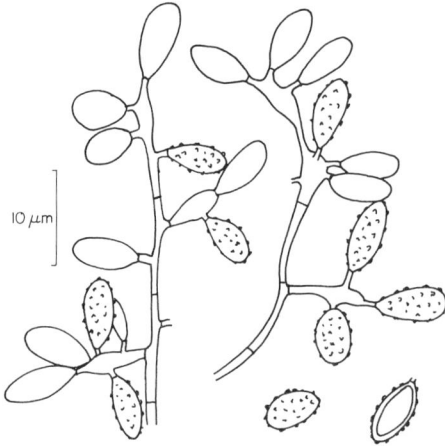

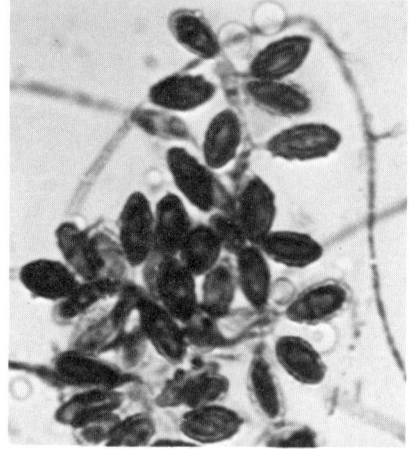

Fig. 9.29 (left)   *Myceliophthora thermophila*   Rough and smooth conidia and swollen conidiogenous cells.

Fig. 9.30 (right)   *Myceliophthora thermophila*   Rough conidia, ×1300.

M. lutea Cost. produces yellow to brown or occasionally dark green colonies and is the cause of the 'vert de gris' disease of cultivated mushrooms.

Oidiodendron Robak
(oidio-, like *Oidium*; Gr. *dendron*, a tree) (Figs 9.31, 9.32)
Conidiophores distinct, dark stipe, branching at the apex; conidiogenous cells thallic, integrated, terminal on branches and fragmenting from the tip backwards; conidia hyaline, 0-septate, arthroconidia in basipetal chains.

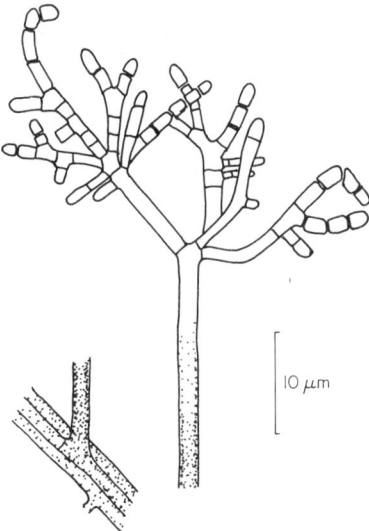

Fig. 9.31 *Oidiodendron griseum*
Conidiophore and conidia.

Fig. 9.32 *Oidiodendron griseum*   Conidiophore and conidia. **(a)** ×250; **(b)** ×550.

Superficially this genus has the appearance of a miniature *Cladosporium*, but differs in the method of spore production. The erect conidiophores branch freely in a tree-like manner and then the branches gradually split up, from the tip backwards, into conidia. All the species are very slow growing, and hence are easily missed when mouldy materials are plated out. Robak (1932) described four species all from Norwegian wood pulp, Barron (1962) monographed the genus and Ellis (1971, 1976) keys out ten species. Sigler and Carmichael (1976) include it in their monograph of some Hyphomycetes with arthroconidia and describe *Myxotrichum* species as the perfect states of some species.

The genus seems common on wood pulp and paper products. The commonest species are *O. tenuissimum* (= *O. fuscum* Robak) and *O. griseum* Robak.

O. tennuissimum (Peck) Hughes. Colonies very slow growing on all media, brownish grey, powdery with reverse gradually turning dark brown; all parts of the fungus very small; conidiophores brown, septate, branched at the apex, fragmenting to form chains of globose to elliptical conidia, dark, distinctly verruculose, 1.5–4 × 1.5–2.5 µm, linked by narrow connectives to look like a chain of beads.

O. rhodogenum Robak produces a red pigment and can be the cause of red staining of wood pulp.

Papulospora Preuss
(Lat. *papula*, pimple, pustule) (Figs 9.33, 9.34)
No true conidia are formed, but bulbils are produced.
Reproduction is by structures known as 'bulbils' or papulospores which are

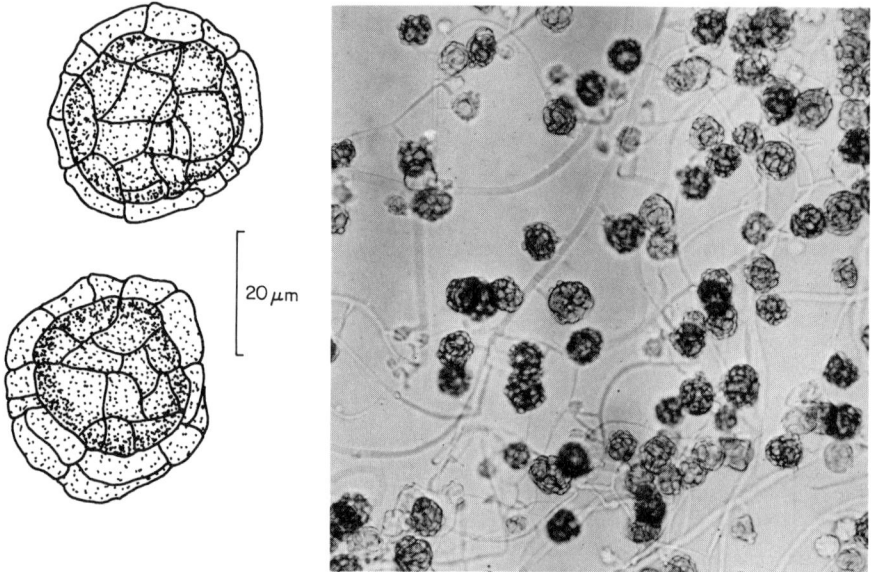

Fig. 9.33 (left)   *Papulospora* sp.   Bulbils.

Fig. 9.34 (right)   *Papulospora coprophila*   Bulbils, ×150.

multicellular structures like miniature sclerotia, usually differentiated into central and sheathing cells. Some species have been shown to be the imperfect states of Ascomycetes or Basidiomycetes, whilst in other cases no connections are known. Care should be taken when examining them to be sure that no other conidial form is produced as several genera have species which produce miniature sclerotia as well as a conidial form, for example *Diheterospora* (see Barron and Onions, 1966) which has an *Acremonium* conidial state.

Species belonging to *Papulospora* are found on dung, in soil, on rotting plant debris, on paper and on textiles. One species, *P. byssina* Hotson & Fergus (Fergus, 1971), is of importance as the brown plaster mould of commercial mushroom beds. A large number of species have been described and the reader is referred to a number of papers by Hotson (1912, 1917, 1929, 1942) and more recently to the paper by Weresub and LeClair (1971) on *Papulospora* and bulbiferous basidiomycetes.

Pestalotia Notaris, **Monochaetia** (Sacc.) Allescher., **Pestalotiopsis** Steyaert and **Truncatella** Steyaert
(F. Pestalozzi, Italian mycologist and often spelt *Pestalozzia*). (Figs 9.35, 9.36)
Conidia dark phragmospores with hyaline end cells bearing two or more appendages.

These genera of the Coelomycetes occur in acervulae and even in culture produce conidia in black pustules. The conidia are very distinctive having 3–5,

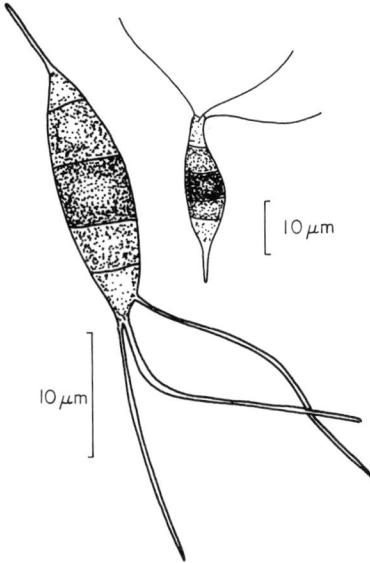

Fig. 9.35 (left)   *Pestalotiopsis*   Conidia.

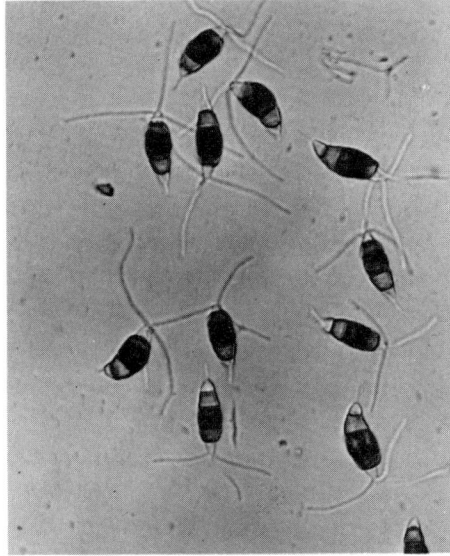

Fig. 9.36 (right)   *Pestalotiopsis*   Conidia, ×500.

mostly 4, cross septa, with terminal cells, colourless or nearly so and median cells dark coloured and the lowest of the median cells paler than the others. The lower end of the conidium bears an appendage and the upper end is furnished with a crest of 2–5 long colourless appendages. These appendages are not easily seen in lactophenol mounts, but show up well in water.

Determination of species is not easy and there has been controversy regarding the systematics. Most workers appear at present to follow the monograph of Guba (1961), who divided the genera according to the number of septa.

These genera usually occur on plants and are of little interest to the industrial mycologist, but they have been occasionally isolated from soil and as causing destruction of cotton and jute fabrics.

Phialophora Medler
(Gr. bearing phialides) (Figs 9.37 and 9.38)
Conidiogenous cells dark coloured phialides, which arise directly from aerial hyphae or as part of simple branched or penicillate structures; conidia produced singly, 0-septate, hyaline and collecting in wet balls.

This difficult genus has aroused considerable interest in recent years. It occurs on wood and woody products, some species from man and one, *P. radicicola*, is an important plant pathogen, often known as its perfect state *Gaeumannomyces graminis*. Some species have alternative means of producing conidia and have been referred to *Rhinocladiella* by Schol-Schwarz (1968). Taxonomy of the group has been confusing with monographs by van

Beyma (1943), Moreau (1963), Wang (1965), Nicot and Caillat (1967), Schol-Schwarz (1968, 1970) and Cole and Kendrick (1973).

The position has now been largely clarified by de Hoog and Hermanides-Nijhof (1977) in a series of papers on the black yeasts, in which *Phialophora mansonii* becomes *Exophiala mansonii* and in which *Aureobasidium* is also included. It is only the true *Phialophora* species which are of industrial interest and for general identification of the common species Ellis (1971, 1976) proves most useful. Identification is based on the shape of the conidia and on the size and shape of the collarettes at the apex of the phialides.

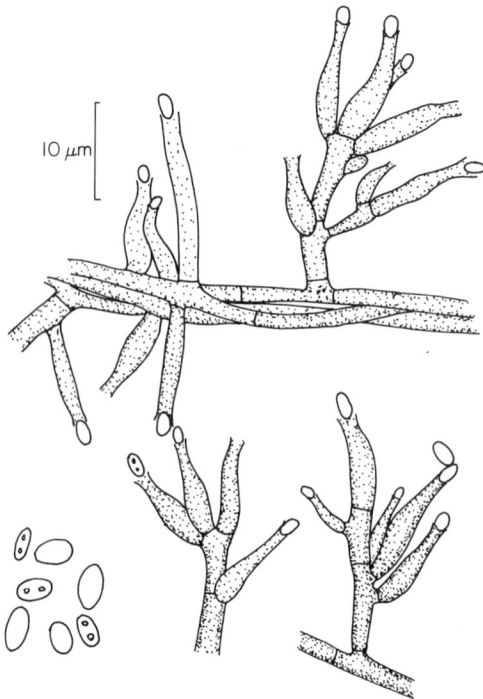

Fig. 9.37 (left)   *Phialophora fastigiata*   Phialides and conidia.

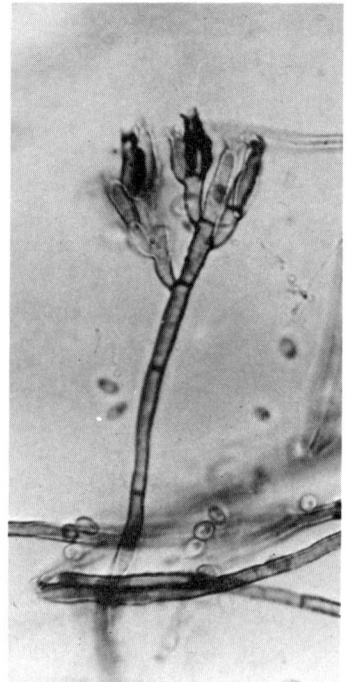

Fig. 9.38 (right)   *Phialophora melinii*   Showing typical phialides, ×1000.

P. fastigiata (Lagerb., Lundberg & Melin) Conant is the most common species. Colonies are fairly restricted, brown to almost black, matted floccose or almost velvety, conidiophores variable, sometimes branched but often bearing single phialides. Phialides pale brown with slightly swollen base tapering to a broad neck and producing a not very distinct collarette. Conidia colourless to very pale brown, variable in shape, often ellipsoidal, 3–7 × 1.5–2.5 µm.

Phoma Sacc. (nom. conser.)
(Probably a corruption of Gr. *Phyma*, a wart or pustule) (Fig. 9.39)
Coelomycete producing pycnidia. Conidia hyaline, 0-septate, elliptical to oval, produced from phialides.

There is an enormous number of genera and species of the Coelomycetes producing pycnidia. The majority of these occur on leaves and stems. The classification of the group is undergoing rapid development with the study of methods of conidium production. As some existing genera are being shown to exhibit several methods of conidium production it is clear that the situation needs clarification and additional papers are being added almost daily. Sutton (1973) gives a good survey and key to the species. A relatively simple key to common species from soil is given by Dorenbosch (1970).

In culture, most species produce a fair amount of floccose aerial mycelium, white at first then darkening, with pycnidia formed slowly, and best in the light, mostly scattered and close to the surface of the medium. Pycnidia are more or less flask-shaped, sometimes irregular, with short necks, dark brown, fairly thin-walled and easily crushed, emitting a mass of hyaline, ovate to elongate, 0-septate conidia. Only one species, *P. herbarum* is a common saprophyte and occurs frequently in the industrial situation causing spoilage of butter, paint, cement and rubber (Boerema, 1964).

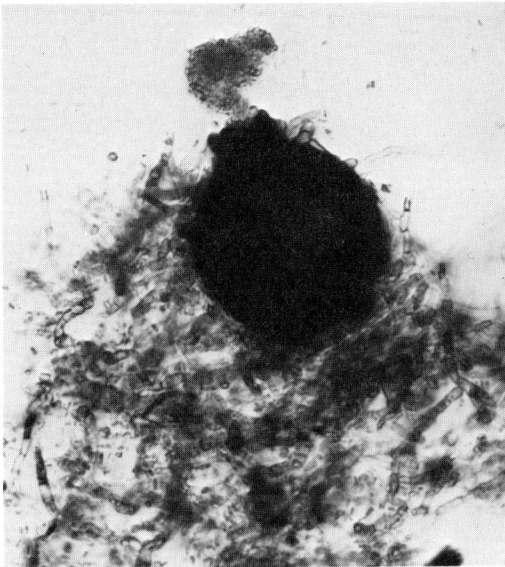

Fig. 9.39   *Phoma violacea*   Mounted in water, extruding conidia, ×250.

P. herbarum Westend. This is probably the most common *Phoma* species and as such has had numerous synonyms (Boerema, 1964, 1970), of which *P. oleracea* Sacc., *P. hibernica* Grimes, O'Connor and Cummins, *P. pigmentivora* Massec and *P. violacea* (Bertel) Eveleigh are best known, the latter epithets referring to its habit of growing on painted surfaces, on which it

produces unsightly pink to purple spots, often several centimetres in diameter. Pycnidia detached or clustered, almost globose with scarcely any neck,
125–150 μm diam., phialo-conidia ellipsoid, 4–6 × 2–2.5 μm. Never produces
chlamydospores.

Scopulariopsis Bainier

(Gr. *opsis*, like; hence like *Scopularia*, another and older genus of fungi. The
latter name is derived from Lat. *scopula*, a little broom) (Figs 9.40 and 9.41)
Conidiogenous cells annellides, borne singly or in branched or penicillate
heads; conidia globose to elliptical, 0-septate, with truncate base, in chains.
    Colony colour varies from white through various shades of brownish yellow
to deep brown, greyish brown, or practically black; it is never green, this
point being one of the reasons given by Thom (1930) and Raper and Thom
(1949) for separating the genus from *Penicillium*. Actually the genus is not
related to *Penicillium* at all, the occasional production of more or less
penicillately branched conidiophores being the only point of resemblance.
The conidium producing cell is not a phialide but, as shown by Hughes
(1953), an annellide (Fig. 9.41c).

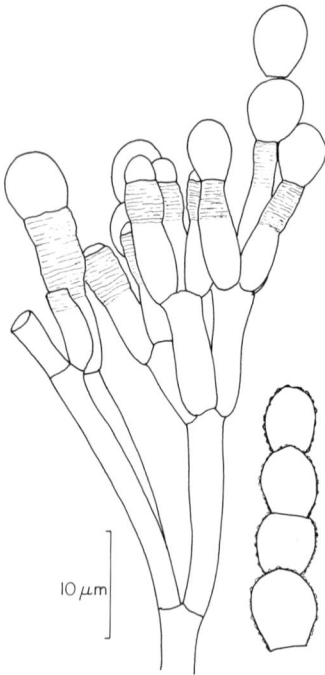

Fig. 9.40 *Scopulariopsis brevicaulis* Annellophores and truncate conidia.

    Fruiting structures in this genus are very irregular, varying from definitely
penicillate structures to single annellides sessile on aerial hyphae. Conidiophores when present are always short. The most distinctive feature of the
genus is the conidium, which varies in form from almost spherical to

lemon-shaped, but always has, at the point of attachment, a thickened ring with central pore (Fig. 9.41b). The conidia may be rough or smooth.

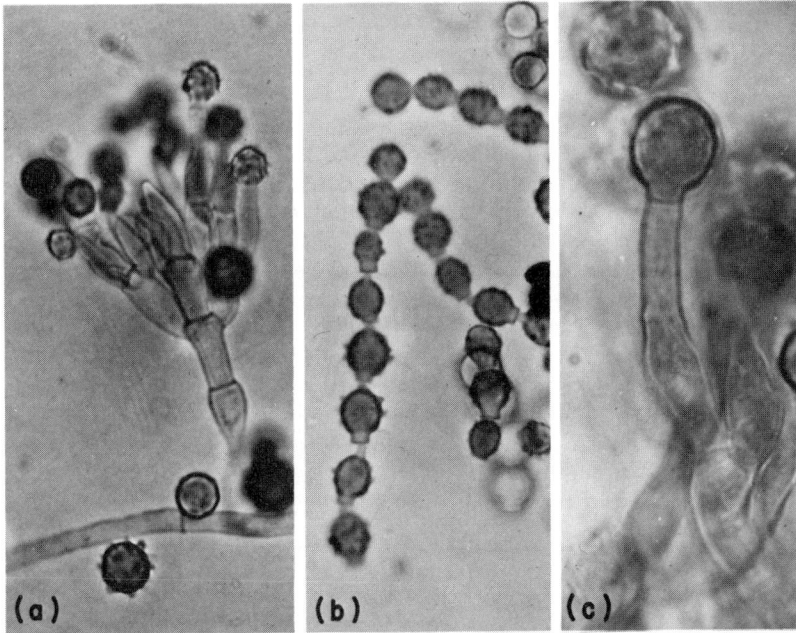

Fig. 9.41 *Scopulariopsis brevicaulis* **(a)** Penicillus, ×1000; **(b)** conidia showing typical thickend ring, ×1000; **(c)** annellide, ×2000.

The perfect state, known for several species, is *Microascus* Zukal. The perithecia are black and carbonaceous, usually almost globose, but always with a definite opening (the ostiole) through which the ascospores are forcibly expelled when ripe, often forming golden-brown, long, twisted, more or less columnar masses, usually termed cirrhi.

Morton and Smith (1963) made a detailed study of the genus, its perfect state *Microascus*, and the related genus *Doratomyces* (= *Stysanus*).

S. brevicaulis (Sacc.) Bainier (Lat. brevis, short; caulis, a stalk). Synonym *Penicillium brevicaule* Saccardo. Colonies usually thin and smooth velvety at first, furrowed, greyish white then yellowish brown, becoming overgrown with loosely floccose to funiculose hyphae, conidia lemon-shaped, coarsely roughened, 6–7 μm in long axis.

S. brevicaulis is by far the commonest species encountered. It is found growing on all kinds of decomposing organic matter and, unlike many moulds, flourishes on substrates containing a high percentage of protein, such as meat and ripening cheese. It is also found as a human parasite, causing a serious infection of the nails.

A special point of interest is that most species of *Scopulariopsis* (and of *Paecilomyces* too) can liberate arsenic in the form of very poisonous gaseous compounds from any substrate containing even only a trace of this element. When grown on ordinary gelatine, for example, the garlic-like odour common to arsine and its alkyl derivatives is distinctly noticeable. In the past there have been one or two serious cases of arsenic poisoning due to the growth of *S. brevicaulis* on wallpapers coloured with paris green.

Sporendonema Desmazières ex Fries
(Gr. *endon*, within; *nema*, a filament) (Figs 9.42, 9.43)
Conidiogenous cells micronematous, thallic; conidia arthrospores, produced in adjacent or alternate segments of the parent mycelium, becoming thick walled and liberated by lysis of intervening cells. *Sporondonema* has been confused with *Wallemia*, but Sigler and Carmichael (1976) have made a clear separation of these and other hyaline arthrospore producing genera.

S. casei Desmazières ex Fries (Lat. *casei*, of cheese). Found principally on cheese rind where it forms orange coloured colonies, in culture it grows best at temperatures not exceeding 18°C, spreading slowly at first then more rapidly, usually orange to reddish, with aerial chains of rounded conidia which appear to be formed by fragmentation of the hyphae. This species was formally known as *Oospora crustacea* (Sull.) Sacc.

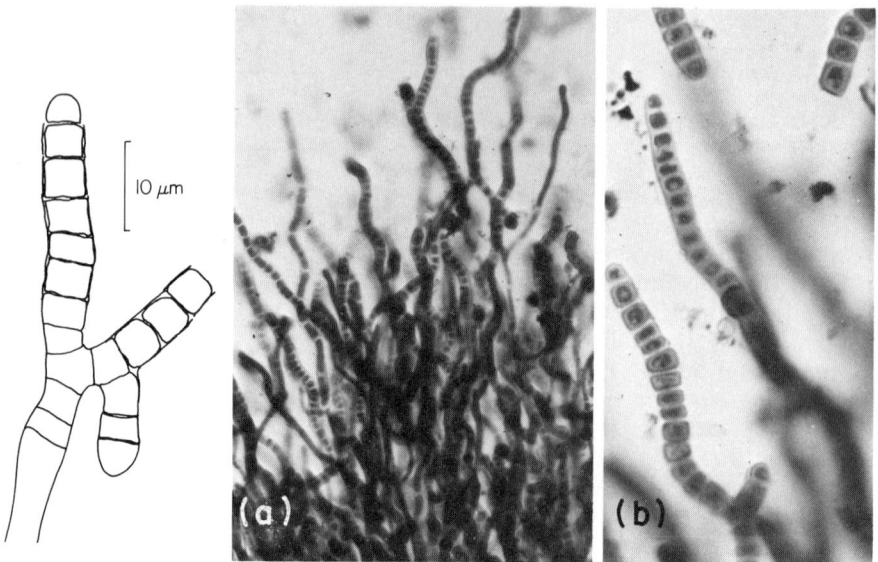

Fig. 9.42 (left)   *Sporendonema casei*   Arthroconidia.

Fig. 9.43 (right)   *Sporendonema casei*   Chains of conidia from slide culture; **(a)** ×250; **(b)** ×550.

S. purpurascens (Bonorden) Mason & Hughes produces pink colonies and is common on mushroom beds.

Sporothrix Hektoen & Perkins
(Gr. *thrix*, hair; i.e. hyphae covered with spores) (Figs 9.44, 9.45)
Conidiogenous cells hyaline, simple with denticulate tips; conidia 0-septate, hyaline.

Conidia are borne, often in small clusters, on short branches from trailing hyphae, each spore being produced on a small peg. The best known species is *S. schenckii* Hektoen & Perkins, which causes the human disease known as sporotrichosis. The fungus is also known as a saprophyte, occurring chiefly on wood, and it is probable that most human infections originate from handling infected timber. Figure 8.44 shows spore production in *S. schenckii*.

The name *Sporotrichum*, commonly used for this species by medical mycologists, is incorrect in this context, since the fungus is quite different from Link's genus (see Carmichael, 1962).

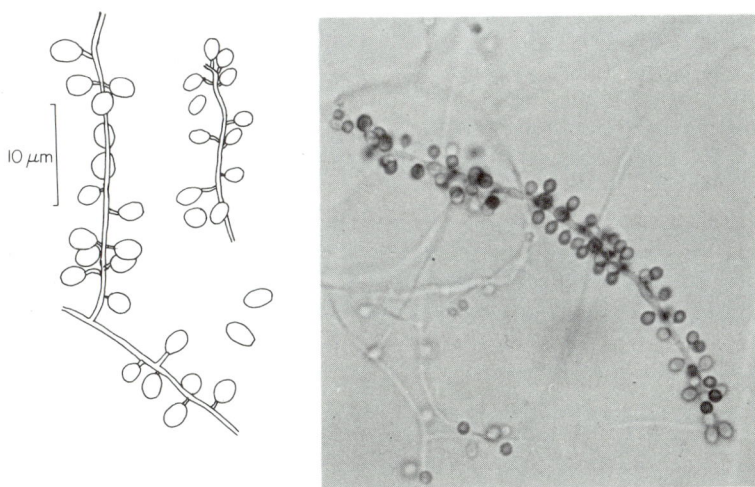

Fig. 9.44 (left)   *Sporothrix schenkii*   Conidiophores and conidia.

Fig. 9.45 (right)   *Sporothrix schenkii*   Slide culture, ×250.

Stachybotrys Corda and *Memnoniella* Hohnel
(Gr. *stachy*, progeny; *botrys*, a bunch of grapes) (Figs 9.46 and 9.47)
Conidiophores dark to hyaline usually present; conidiogenous cells phialides in a whorl at the apex of the stipe, with a rounded tip with a small opening; conidia 0-septate, dark (in some species hyaline conidia are produced and these were previously referred to *Hyalostachybotrys*) aggregated in wet masses (*Stachybotrys*), in chains (*Memnoniella*).

These genera have been discussed by Bisby (1943, 1945), monographed by Verona and Mazzuchetti (1968) and most species keyed out and described by Ellis (1971, 1976). Two species are particularly common and found on paper, textiles, seeds, soil and dead plants. They attack and decompose cellulose and during the second world war caused spoilage of sandbags, fibre boards, tentage, etc. in tropical and sub-tropical countries. Growth in culture is most satisfactory if cellulose, for example filter paper, is included.

S. atra Corda (Lat. *ater*, black). Vegetative hyphae hyaline; conidiophores at first hyaline, then dark, bearing a terminal whorl of phialides; phialides thick, rounded at the top, hyaline at first then brownish, producing slimy spores which collect in a dense, irregular mass; spores ellipsoidal to subglobose, often slightly curved, dark brown to black, smooth when young, but mostly becoming rough in old cultures, 8–12 × 5–7 μm.

M. echinata (Riv.) Galloway (Lat. *echinatum*, spiny). This is very similar to *S. atra* but the conidia are produced in chains, phialides 8–10 × 3–4 μm, conidia dark brown to black, rough, globose to ovate, 3.5–6 μm.

   In the previous edition Smith included this as *Stachybotrys echinata* (Rivolta) G. Smith as he considered the difference between the two genera was very slight. However, most workers at present still separate the two genera and retain the name *Memnoniella echinata*.

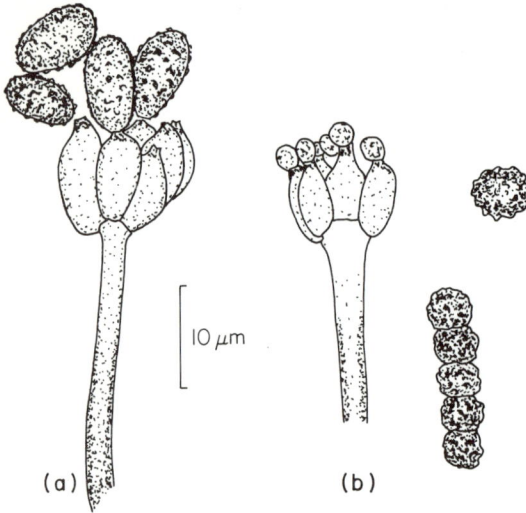

Fig. 9.46   (a) *Stachybotrys atra*   Conidial head. **(b)** *Memoniella echinata* Conidial head and conidia in chain.

***Stemphylium* Wallroth and *Ulocladium* Preuss**
(Gr. *stemphylon*, a mass of pressed grapes; from the many celled spores) (Figs 9.48, 9.49, 9.50)
Separation of these two genera with dark dictyospores has proved difficult, and studies were made by Wiltshire (1938) and Neergaard (1945). Simmons (1967) separated them and *Alternaria* by studying the method of conidia production. He figured the differences and described seven species of *Ulocladium*. Ellis (1971, 1976) describes species of both genera.

Fig. 9.47 **(a)** *Stachybotrys atra* Conidiophores borne on rope of hyphae, ×300; **(b)** conidial head ×375. **(c)** *Memnoniella echinata* Conidiophore and conidial head. ×600.

Stemphylium. Conidiogenous cells monotretic, dark, with slightly swollen tips, secondary conidia or conidiogenous cells may proliferate through the initial pore; conidia dark dictyospores, spherical, ovoid or obovoid.

The type species of the genus *S. botryosum* Wallr., the perfect state of *Pleospora herbarum*, is the only common saprophyte likely to be encountered and though typically occurring on dead herbaceous plants and wood is isolated from air, paper and cellulosic materials.

Ulocladium. Conidiogenous cells dark, polytretic, may become sympodially geniculate, secondary conidia may be produced by germination of individual cells of the conidium to produce an apical or lateral secondary conidiogenous cell or 'false beak' and may thus produce a chain of conidia by repeated production of secondary conidia; conidia dark, dictyospores, smooth or rough, obovoid, i.e. the portion of the conidium proximal to the conidiogenous cell is tapered and narrower than the distal portion. The obovoid, non-rostrate conidia distinguish *Ulocladium* from *Alternaria* with its ovoid distally tapered or rostrate conidia.

Several species of *Ulocladium* are isolated from time to time from wood, paper, paint, textiles and soil as well as from dead herbaceous plants.

U. chartarum (Preuss) Simmons (Lat. *charta*, paper) is common on paper and in spoilage situations. Conidia with 1–5 often 3 transverse septa and 1–5 longitudinal septa, commonly in chains of 2–10, ellipsoidal to obovoid, broadly rounded, often with short false beaks, smooth to rough, 18–38 × 11–21 μm.

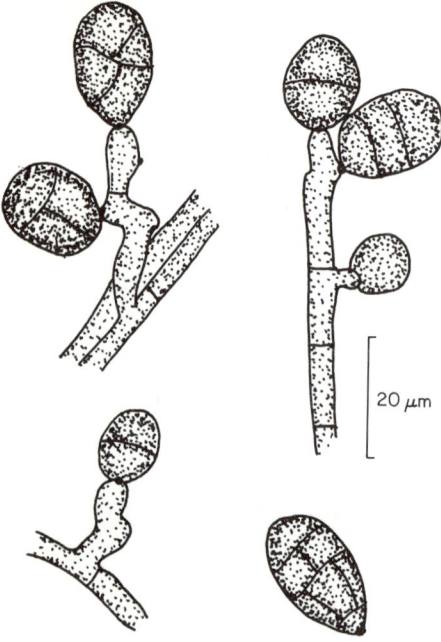

Fig. 9.48   *Ulocladium chartarum*   Conidia.

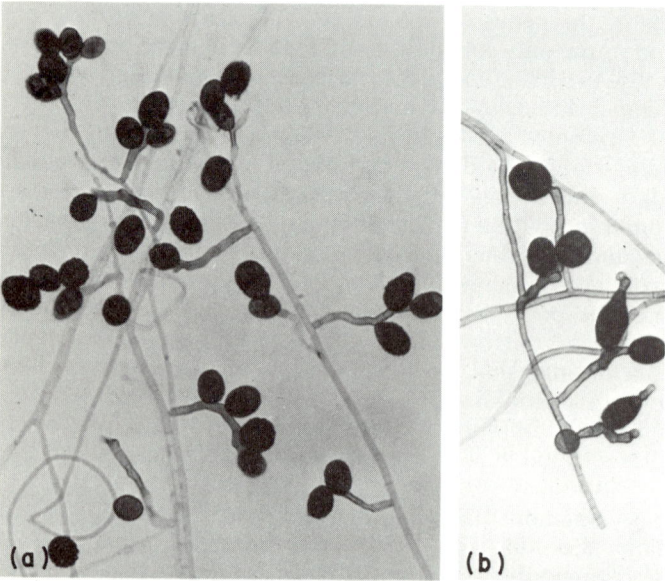

Fig. 9.49   **(a)** *Ulocladium atrum*   Conidiophores and conidia, ×200; **(b)** *U. chartarum*   ×250.

U. consortiale (Thum) Simmons, with conidia mostly solitary but some in chains, smooth, to inconspicuously roughened, 18–34 × 7–15 μm is also common.

U. lanuginosum (Harz) Simmons, usually has solitary conidia, obovoid, and smooth, 25–32 × 18–21 μm. It attacks cellulose and is found on coarse fabrics, straw board and the like.

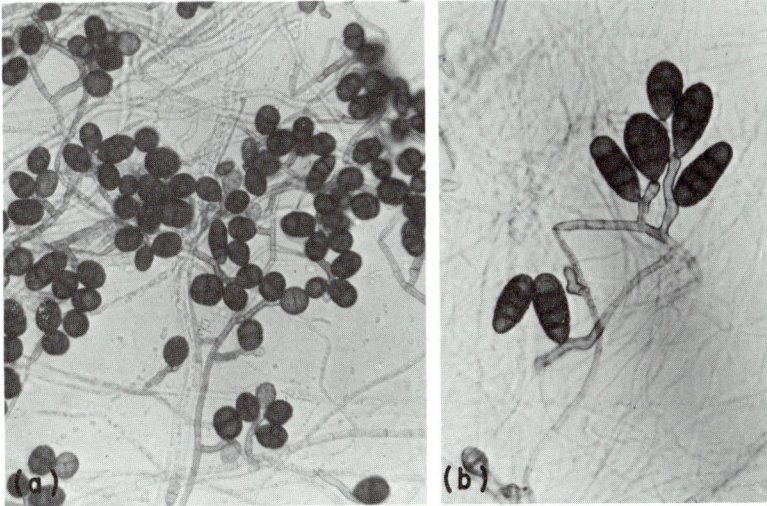

Fig. 9.50   *Ulocladium lanuginosum*   Conidiophores and conidia. **(a)** ×250; **(b)** ×500.

Trichoderma Persoon ex Fries
(Gr. *thrix*, hair; *derma*, skin) (Figs 9.51 and 9.52)
Conidiophores bearing branches or phialides irregularly or in verticils at a wide angle or at right angles to the main stipe; conidiogenous cells hyaline, phialides; conidia 0-septate, hyaline, globose to elongate with truncate base, collecting in wet masses, usually green.

This genus is common on woody products and in soil. It has become of considerable economic interest in recent years on account of its strong cellulose decomposing action, both as a spoilage organism and for use in various schemes for upgrading of cellulose waste.

Although first described by Persoon in 1794 and being common in the field (it can often be seen forming sheets of green growth on damp logs or fungi) there has been and still is confusion concerning its classification. *Trichoderma* species have been shown to be the imperfect states of various *Hypocrea* species, but relationships have not been worked out for all species so it is convenient to treat them as a form genus of the Hyphomycetes.

Bisby (1939) considered the genus to be monotypic, as when a large number of isolates were examined there were no sharp differences to be observed, and he referred all isolates to *T. viride* Pers. ex Fr. This has resulted in almost indiscriminate use of this name for all isolates of

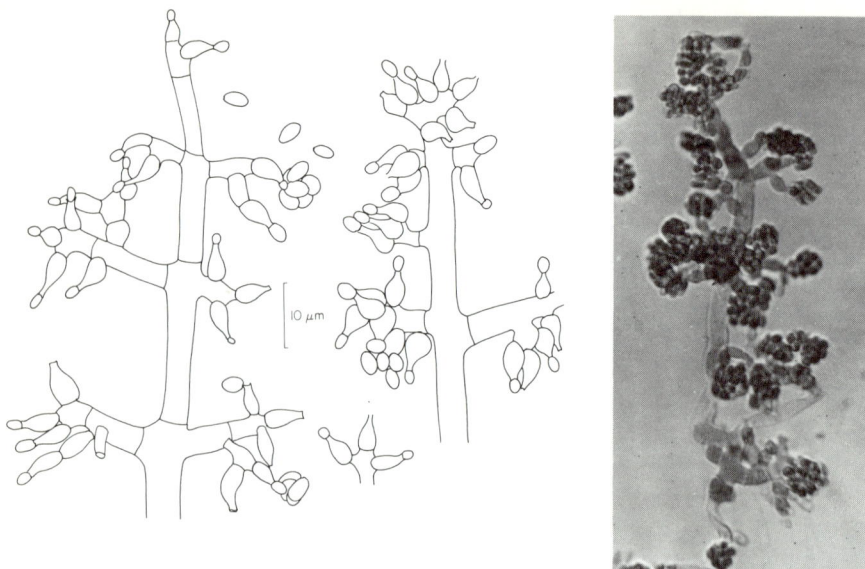

Fig. 9.51 (left)   *Trichoderma harzianum*   Showing conidiophores, phialides and conidia.

Fig. 9.52 (right)   *Trichoderma* sp.   Conidiophore with balls of conidia, ×500.

Trichoderma. Rifai (1969) has shown in a recent monograph that the situation is not nearly so simple. He still found a continuum of characters from one species to another, but described nine species aggregates. His system is usable though he himself considers that it requires modification.

Rifai includes two white species, *T. piluliferum* and *T. polysporum* and excludes all forms with branches forming a penicillus or an acute angled arrangement of branches and phialides as belonging to *Gliocladium*, even when they may also show a verticillate structure.

Most isolates of *Trichoderma* have colonies that spread rapidly forming a somewhat thin mycelial layer with irregular shaped patches of verdigris green, this being the colour of ripe conidia in mass.

Three species aggregates with smooth conidia, no sterile tips to the conidiophores, and the phialides borne in definite verticils are particularly common.

T. koningii Oud., produces ellipsoidal conidia; *T. aureoviride* Rifai produces obovoid conidia with a truncate base and often produces a strong yellow pigment in the colony reverse; *T. harzianum* Rifai, produces almost globose conidia.

T. viride Pers. ex S. F. Gray is not very common and can be distinguished from all other species by having rough conidia. *T. reesii* Simmons is a recently described species which shows particularly active cellulase decomposition and is being used in upgrading of cellulosic wastes as it has a high level of cellulose productivity coupled with non-utilization of $-NO_3$.

Trichothecium Link ex Fries (Figs 9.53, 9.54)
Conidia 1-septate, hyaline, in basipetal short chains at the apex of simple
conidiophores.

T. roseum (Pers.) Link ex Fries (Lat. *roseus*, pink); synonym *Cepalothecium
roseum* Corda.

This is the only important species of the genus, causes a rot of apples in
storage and is found, somewhat less frequently, on other substrates. Colonies
somewhat thin, floccose, wide-spreading, white at first, then slowly clear
pink; conidiogenous cells erect, bearing terminal clusters of conidia directly
attached to the tip; spores 2-celled, roughly ovate with a nipple-like projection at the point of attachment, 18–20 × 8–10 μm.

Fig. 9.53   *Trichothecium roseum*   Conidiophore and conidia, showing arrangement of conidia.

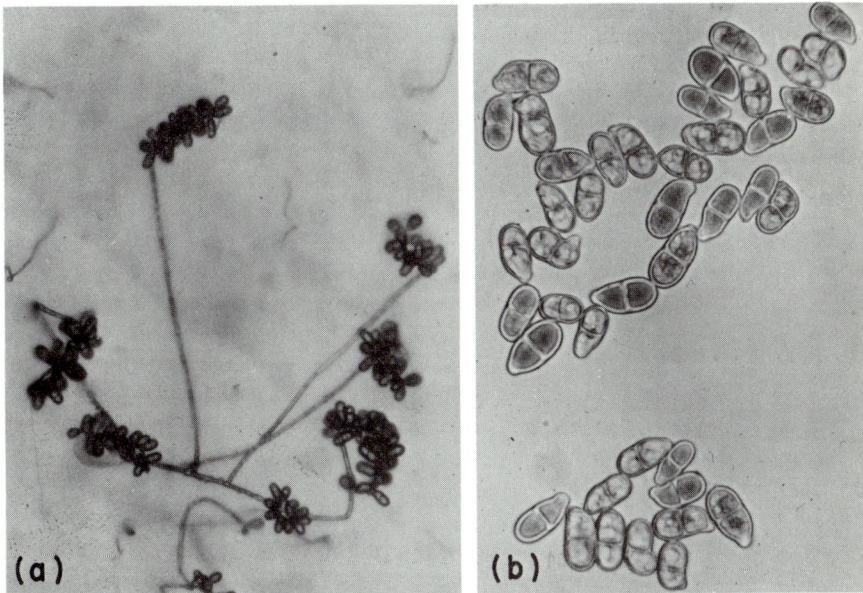

Fig. 9.54   *Trichothecium roseum*   **(a)** Conidiophore as seen in Petri dish culture, ×200;
(b) conidia, ×500.

The actual method of conidium formation is distinctive and placement of the genus in the present classification is a matter for discussion but it has been well described by Ingold (1956), and the genus has been described by Rifai and Cook (1966). The conidia form a chain, with each conidium disposed so that the long axis is almost horizontal, alternate conidia being turned in the opposite direction. The conidia are attached to one another through slight thickenings in each side of the base.

Verticillium Nees ex Wallroth
(Lat. *verticillus*, a whorl of branches) (Figs 9.55 and 9.56)
Conidiophores, conidia produced in slimy masses from phialides which are borne in whorls (verticils) directly or on branches at intervals up the length of the conidiophore; conidiogenous cells, phialides, hyaline; conidia hyaline, 0-septate.

The systematic position of this genus is sometimes difficult as the conidiophores of some species are hyaline where others are dark. Thus Ellis (1971) includes the genus in his work on dematiaceous Hyphomycetes and Gams (1971) describes 18 species in his monograph on *Cephalosporium*-like fungi, but he also includes species with spores in chains. Neither of these authors mention the *Verticillium alboatrum* group of species which are well known plant pathogens with usually hyaline conidiophores but also some pigmented mycelium as hyphae, sclerotia or chlamydospores.

Verticillium species are wide spread in soil, as pathogens of other fungi, insects and plants.

V. lateritium (Ehrenberg) Rabenhorst (Lat. *lateritius*, brick coloured) Synonym *Acrostalagmus cinnabarinus* Corda; *V. cinnabarinum* (Corda) Reinke & Berthold.

This is not uncommon, particularly in soil. Colonies thin spreading, dull brick red; conidiophores long stiff, mostly several times branched; conidia ellipsoidal, about 3×1.5 μm.

It is the conidial state of an Ascomycete, *Nectria inventa* Pethybridge.

V. alboatrum Reinke & Berth. (Lat. *albus*, white; *ater*, black) is a serious parasite of many plants, causing wilt. Colonies at first white, then turning black in patches (hence the name); conidiophores usually much simpler than in the last species, often being reminiscent of *Cephaolosporium (Acremonium)*. Cultures are difficult to maintain in good condition, usually becoming white. There are several other very similar species often classed with this in a group known as the *V. alboatrum* group. Isaac in a series of papers (1949, 1953, 1955) sorted them into distinct species.

Wallemia Johan-Olsen
Conidiogenous cells usually unbranched phialides with a dark collarette, with a protuberant upper part which becomes septate basipetally and finally fragments to form conidia, frequently proliferating through the collarette; conidia fragmenting, modified arthroconidia, hyaline, 0-septate, cubical, becoming subspherical and slightly rough.

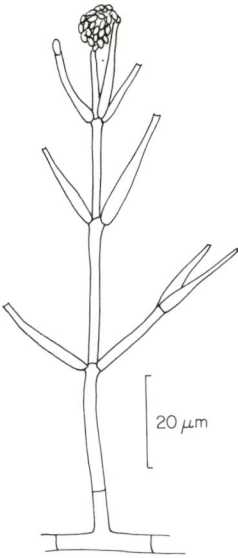

Fig. 9.55 *Verticillium lateritium* Conidiophore with verticils of phialides.

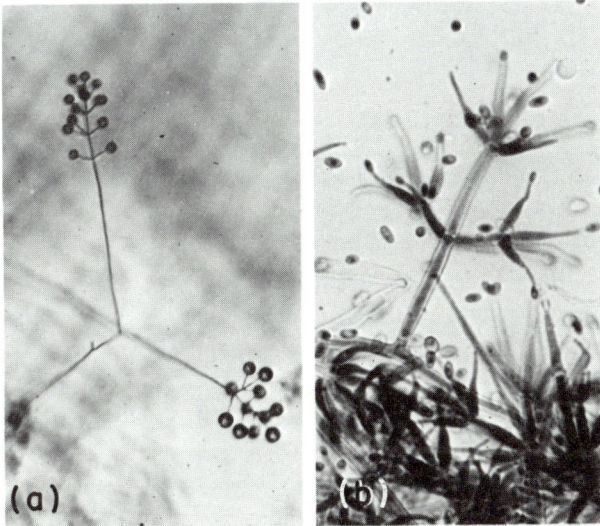

Fig. 9.56 **(a)** *Verticillium theobromae*, ×100. **(b)** *Verticillium lateritium*, ×500.

An electron microscope study of this unusual method of spore production was made by Hill (1974).

W. sebi (Fries) von Arx (Lat. *sebi*, of tallow). A species with many synonyms, most often known as *Sporendonema sebi*. It is very xerophilic and chiefly found on highly sugared or salted materials such as foodstuffs, jam, bread, cakes, salted fish, milk products, fats and tobacco and has been isolated from air, hay, textiles and man. It forms characteristic 'buttons' in sweetened condensed milk. In culture it grows very slowly, forming velvety chocolate-coloured colonies, with long rows of spores, which round up and become free at maturity, and are mostly 2–2.5 μm, but occasionally up to 4 μm or even 5 μm in diameter.

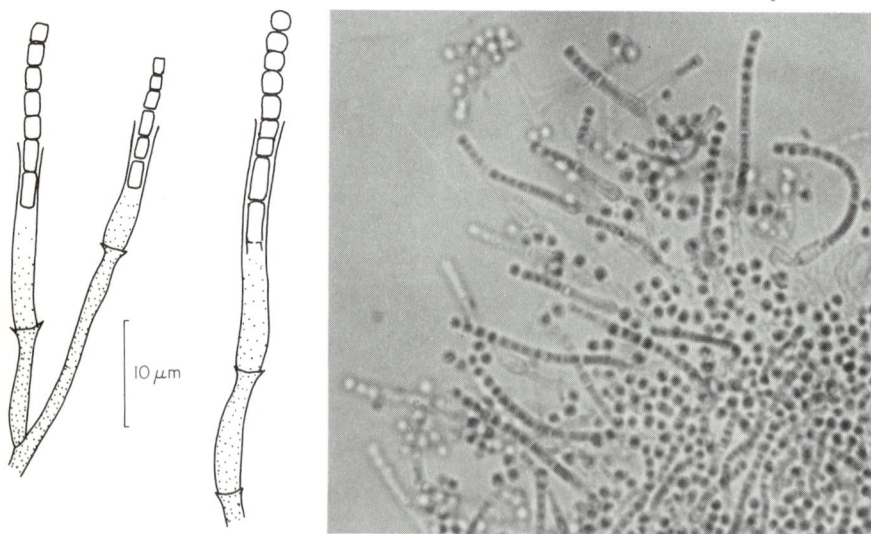

Fig. 9.57 (left) *Wallemia sebi* Phialides with endogenous fragmenting conidia.

Fig. 9.58 (right) *Wallemia sebi* Phialides with conidia, ×1000.

9.4 References

APPEL, O. and WOLLENWEBER, H. W. (1910). Grundlageneiner Monographie der Gattung *Fusarium* Link. *Arb. biol. Bund. Anst. Land-u. Forstw.*, **8(1)**, 1–209.

ARX, J. A. VON (1970). *The genera of fungi sporulating in pure culture.* J. Cramer, Lehre.

ARX, J. A. VON (1972). On *Endomyces, Endomycopsis* and related yeast-like fungi. *Antonie van Leeuwenhoek*, **38**, 289–309.

ARX, J. A. VON (1973). Some further observations on *Sporotrichum* and some similar fungi. *Persoonia*, **7**, 127–30.

ARX, J. A. VON (1974). *The genera of fungi sporulating in pure culture.* 2nd edition. J. Cramer, Vaduz.

ARX, J. A. VON (1977). Notes on *Dipodascus, Endomyces* and *Geotrichum* with the description of two new species. *Antonie van Leeuwenhoek*, **43**, 333–40.

BARNES, T. G., EGGINS, H. O. W. and SMITH, E. L. (1972). Preliminary stages in the development of a process for the microbial upgrading of wastepaper. *Int. Biodet. Bull.*, **8**, 112–6.

BARRON, G. L. (1962). New species and new records of *Oidiodendron. Canad. J. Bot.*, **40**, 589–607.

BARRON, G. L. (1968). *The Genera of Hyphomycetes from Soil.* Williams & Wilkins, Baltimore.

BARRON, G. L. and ONIONS, A. H. S. (1966). *Verticillium chlamydosporum* and its relationship to *Diheterospora, Stemphiliopsis* and *Paecilomyces. Canad. J. Bot.*, **44**, 861–9.

BEYMA, F. H. VAN (1943). Beschreiburg der in Centraalbureau voor Schimmel-cultures vor handenem Arten der Gattung *Phialophora* Thaxter und Margarinomyces Laxa, nebst Schlüssel zu ihrer Bestimmung. *Leeuwenhoek med. Tijdschr.*, **9**, 51–76.

BISBY, G. R. (1939). *Trichoderma viride* Pers. ex Fries, and notes on *Hypocrea. Trans. Br. mycol. Soc.*, **23**, 149–68.

BISBY, G. R. (1943). *Stachybotrys. Trans. Br. mycol. Soc.*, **26**, 133–43.

BISBY, G. R. (1945). *Stachybotrys* and *Memnoniella. Trans. Br. mycol. Soc.*, **28**, 11–12.

BOEDIJN, K. (1933). Uber einige phragmosporen Dematiazeen. *Bull. Jard. bot. Buitenzorg.* Ser. 3, **13**, 120–34.

BOEREMA, G. H. (1964). *Phoma herbarum* Westend., the type species of the form-genus *Phoma* Sacc. *Persoonia*, **3**, 9–16.

BOEREMA, G. H. (1970). Additional notes on *Phoma herbarum. Persoonia*, **6**, 15–48.

BOOTH, C. (1971). *The genus Fusarium.* Commonwealth Mycological Institute, Kew.

BOOTH, C. (1977). *Fusarium: laboratory guide to the identification of the major species.* Commonwealth Mycological Institute, Kew.

CARMICHAEL, J. W. (1957). *Geotrichum candidum. Mycologia*, **49**, 820–9.

CARMICHAEL, J. W. (1962). *Chrysosporium* and some other aleuriosporic Hyphomycetes. *Canad. J. Bot.*, **40**, 1137–73.

CHIDAMBARAM, P., MATHUR, S. B. and NEERGAARD, P. (1973). Identification of seed-borne *Drechslera* species. *Friesia*, **10**, 165–207.

CIFERRI, R., RIBALDI, M. and CORTE, A. (1957). Revision of 23 strains of *Aureobasidium pullulans* (de Bary) Arn. (= *Pullularia pullulans*). *Atti Ibt. bot. Univ. Pavia*, Ser. 5, **14**, 78–90.

COLE, G. T. and KENDRICK, B. (1973). Taxonomic studies of *Phialophora. Mycologia*, **65**, 661–88.

COOKE, W. B. (1954). Fungi in sewage and polluted water. II. Isolation technique. *Sewage and Industrial Wastes*, **26**, 661–74.

COOKE, W. B. (1959). An ecological life history of *Aureobasidium pullulans* (de Bary) Arnaud. *Mycopathologia*, **12**, 1–45.

DAKIN, J. C. and STOLK, A. C. (1968). *Moniliella acetoabutans*: some further characteristics and industrial significance. *Journal of Food Technology*, **3**, 49–53.

DE VRIES, G. A. (1952). *Contribution to the Knowledge of the Genus Cladosporium Link ex Fr.* Baarn: Uitgeveij and Drukkerij, Hollandia.

DICKINSON, C. H. (1968). *Gliomastix* Guéguen, *Mycological Papers*, **115**, 1–24.

DORENBOSCH, M. M. J. (1970). Key to nine ubiquitous soil-borne *Phoma*-like fungi. *Persoonia*, **6**, 1–14.

DRECHSLER, C. (1923). The graminicolous species of *Helminthosporium* I. *Journal of Agricultural Research*, **24**, 641–740.

DURRELL, L. W. (1968). Studies of *Aureobasidium pullulans* (de Bary) Arnaud. *Mycopathologia et Mycologia Applicata*, **35**, 113–20.

ELLIS, M. B. (1961). Dematiaceous hyphomycetes III. *Mycological Papers*, **82**, 1–55.

ELLIS, M. B. (1966). Dematiaceous hyphomycetes VII. *Curvularia, Brachysporium* etc. *Mycological Papers*, **106**, 1–57.

ELLIS, M. B. (1971). *Dematiaceous Hyphomycetes*. Commonwealth Mycological Institute, Kew.

ELLIS, M. B. (1976). *More Dematiaceous Hyphomycetes*. Commonwealth Mycological Institute, Kew.

FERGUS, C. L. (1971). The temperature relationships and thermal resistance of a new thermophilic *Papulospora* from mushroom compost. *Mycologia*, **63**, 426–31.

GAMS, W. (1971). *Cephalosporium artige Schimmelpilze (Hyphomycetes)*. G. Fischer, Stuttgart.

GORDON, W. L. (1952). The occurrence of *Fusarium* species in Canada. II. Prevalence and taxonomy of *Fusarium* species in cereal seed. *Canad. J. Bot.*, **30**, 209–51.

GUBA, E. F. (1961). *Monograph of Monochaetia and Pestalotia*. Harvard Univ. Press, Cambridge, Massachusetts.

HENDY, N. S. (1964). Some observations on *Cladosporium resinae* as a fuel contaminant and its possible role in the corrosion of aluminium alloy fuel tanks. *Trans. Br. mycol. Soc.*, **47**, 467–75.

HENNEBERT, G. L. (1973). *Botrytis* and *Botrytis*-like genera. *Persoonia*, **7**, 183–204.

HERMANIDES-NIJHOF, E. J. (1977). The black yeasts and allied hyphomycetes: *Aureobasidium* and allied genera. *Studies in Mycology* No. 15, CBS, 141–77.

HILL, S. T. (1974). Conidium ontogeny in the xerophilic fungus *Wallemia sebi*. *J. stored Prod. Res.*, **10**, 209–15.

DE HOOG, G. S. (1977). The black yeasts and allied hyphomycetes. *Rhinocladiella* and allied genera. *Studies in Mycology* No. 16, CBS, 1–140.

DE HOOG, G. S. and HERMANIDES-NIJHOF, E. J. (1977). The black yeasts and allied Hyphomycetes. Survey of black yeasts and allied hyphomycetes. *Studies in Mycology* No. 15, CBS, 178–222.

HOTSON, J. W. (1912). Cultural studies of fungi producing bulbils and similar propagative bodies. *Proc. Amer. Acad. Arts and Sci.*, **48**, 227–306.

HOTSON, J. W. (1917). Notes on bulbiferous fungi with a key to described species. *Botany Gazette*, **44**, 265–84.

HOTSON, J. W. (1929). *Papulospora* atra n. sp. *American Journal of Botany*, **16**, 219–20.

HOTSON, J. W. (1942). Some species of *Papulospora* associated with roots of *Gladiolus* bulbs. *Mycologia*, **34**, 391–9.

HUGHES, S. J. (1953). Conidiophores, conidia and classification. *Canad. J. Bot.*, **31**, 577–659.

INGOLD, C. T. (1956). The conidial apparatus of *Trichothecium roseum*. *Trans. Br. mycol. Soc.*, **39**, 460–4.

ISAAC, I. (1949). A comparative study of pathogenic isolates of *Verticillium*. *Trans. Br. mycol. Soc.*, **32**, 137–57.

ISAAC, I. (1953). A further comparative study of pathogenic isolates of *Verticillium*. *V. nubilum* Pethybridge and *V. tricorpus* sp. nov. *Trans. Br. Mycol. Soc.*, **36**, 180–95.

ISAAC, I. and DAVIES, R. R. (1955). A new hyaline species of *Verticillium*. *V. intertextum* sp. nov. *Trans. Br. mycol. Soc.*, **38**, 143–56.

KLOPOTEK, A. VON (1974). Revision der thermophilen *Sporotrichum*-Arten *Chrysosporium thermophilum* (Apinis) comb. nov. und *Chrysosporium fergusii* spec. nov. = Status conidiales von Corynasus thermophilus (Fergus & Sinden) comb. nov. *Arch. Microbiol.*, **98**, 365–9.

LEONARD, K. J. and SUGGS, E. G. (1974). *Setosphaeria prolata*, the ascigerous state of *Exserohilum prolatum*. *Mycologia*, **66**, 281–97.

LUTTRELL, E. S. (1963). Taxonomic criteria in *Helminthosporium*. *Mycologia*, **55**, 643–74.

LUTTRELL, E. S. (1964). Systematics of *Helminthosporium* and related genera. *Mycologia*, **56**, 119–32.

MESSIAEN, C.-M. and CASSINI, R. (1968). Recherches sur les Fusarioses IV. La systématique des *Fusarium*. *Annles des Epiphyties*, **19**, 387–454.

MOREAU, C. (1963). Morphologie comparée de quelques *Phialophora* et variations du *P. cinerescens* (Wr.) van Beyma. *Rev. Mycol.*, **28**, 260–76.

MORTON, F. J. and SMITH, G. (1963). The genera *Scopulariopsis* Bainier, *Microascus* Zukal, and *Doratomyces* Corda. *Mycological Papers*, **86**, 1–96.

NEERGAARD, P. (1945). *Danish species of Alternaria and Stemphylium*. Oxford University Press, London.

NEIHOF, R. A. and BAILEY, C. A. (1978). Biocidal properties of anti-icing additives for aircraft fuels. *Appl. Environ. Microbiol.*, **35**, 698–703.

NICOT, J. and CAILLAT, M. (1967). Étude morphologique d'une souche africaine de *Phialophora richardsiae* (Nann.) Conant. *Rev. Mycol.*, **32**, 28–40.

NISIKADO, Y. (1929). Studies of the *Helminthosporium* diseases of Gramineae in Japan. *Ber. Ohara Inst.*, **4**, 111–26.

OORSCHOT, C. A. N. VAN (1977). The genus *Myceliophthora*. *Persoonia*, **9**, 401–8.

PARBERY, D. G. (1969). *Amorphotheca resinae* gen. nov. sp. nov: the perfect state of *Cladosporium resinae*. *Australian Journal of Botany*, **17**, 331–57.

PARBERY, D. G. (1971). Biological problems in jet aviation fuel and the biology of *Amorphotheca resinae*. *Material und Organismen*, **6**, 161–207.

PARK, P. B. (1975). Biodeterioration in aircraft fuel systems. In *Microbial Aspects of the Deterioration of Material*. LOVELOCK, D. W. and GILBERT, R. J. (Eds). Society of Applied Bacteriology, Technical Series 9, 106–26.

RAPER, K. B. and THOM, C. (1949). *A manual of the Penicillia*. Williams & Wilkins, Baltimore.

REDHEAD, S. A. and MALLOCH, D. W. (1977). The endomycetaceae: new concepts, new taxa. *Canad. J. Bot.*, **55**, 1701–11.

REYNOLDS, E. S. (1950). *Pullularia* as a cause of deterioration of paint and plastic surfaces in South Florida. *Mycologia*, **42**, 432–48.

RIFAI, M. A. (1969). A revision of the genus *Trichoderma*. *Mycological Papers*, **116**, 1–56.

RIFAI, M. A. and COOKE, R. C. (1966). Studies on some didymosporus genera of nematode-trapping Hyphomycetes. *Trans. Br. mycol. Soc.*, **49**, 147–67.

ROBAK, H. (1932). Investigations regarding fungi on Norwegian ground wood pulp and fungal infections at wood pulp mills. *Nyt. Mag. Naturn.*, **71**, 185–330.

SCHOL-SCHWARZ, M. B. (1968). *Rhinocladiella*, its synonym *Fonsecaea* and its relation to *Phialophora*. *Antonie van Leeuwenhoek*, **34**, 119–52.

SCHOL-SCHWARZ, M. B. (1970). Revision of the genus *Phialophora* (Moniliales). *Persoonia*, **6**, 59–94.

SCOTT, J. A. (1971). Microbiological contaminants of aircraft fuel tanks – airframe considerations. *Society of Automative Engineers Publication 710438.* 7pp.

SHEAR, C. L. and DODGE, B. O. (1927). Life histories and heterothallism of the red bread mould fungi of the *Monilia sitophila* group. *Journal of Agricultural Research*, **34**, 1019–42.

SHERIDAN, J. E. (1972). Studies on the 'kerosene fungus' *Cladosporium resinae* (Lindau) de Vries. Part III. Morphology, taxonomy and physiology. *Tuatara*, **19**, 130–65.

SHOEMAKER, R. A. (1959). Nomenclature of *Drechslera* and *Bipolaris*, grass parasites segregated from *Helminthosporium*. *Canad. J. Bot.*, **37**, 879–87.

SIGLER, L. and CARMICHAEL, J. W. (1976). Taxonomy of *Malbranchea* and some other Hyphomycetes with arthroconidia. *Mycotaxon*, **4**, 349–488.

SIMMONS, E. G. (1967). Typification of *Alternaria*, *Stemphylium* and *Ulocladium*. *Mycologia*, **59**, 67–92.

SNYDER, W. C. and HANSEN, H. N. (1941). The species concept in *Fusarium* with special reference to section *Martiella*. *Amer. J. Bot.*, **28**, 738–42.

SNYDER, W. C. and HANSEN, H. N. (1945). The species concept in *Fusarium* with special reference to *Discolor* and other sections. *Amer. J. Bot.*, **32**, 657–66.

SNYDER, W. C., HANSEN, H. N. and OSWALD, J. W. (1957). Cultivars of the fungus *Fusarium*. *Journal of Madras University*, Section B, **27**, 185–92.

STOLK, A. C. and DAKIN, J. C. (1966). *Moniliella*, a new genus of Moniliales. *Antonie von Leeuwenhoek*, **32**, 399–409.

SUTTON, B. C. (1973). Coelomycetes. In *The Fungi*, Vol. 4A. AINSWORTH, G. C., SPARROW, E. K. and SUSSMAN, A. S. (Eds). pp. 513–82. Academic Press, London.

THOM, C. (1930). *The Penicillia*. Williams & Wilkins, Baltimore.

TOUSSOUN, T. A. and NELSON, P. E. (1968). *A pictorial guide to the identification of Fusarium species*. Pennsylvania State University Press.

TUBAKI, K. (1962). Studies on the slime-forming fungus in polluted water. *Trans. mycol. Soc. Japan*, **3**, 29–35.

VERONA, O. and MAZZUCHETTI, G. (1968). I generi *Stachybotrys* et *Memnoniella*. *Publ. Ente naz. Cellul. Carta*. Rome.

WANG, C. J. K. (1965). *Fungi on Pulp and Paper (Phialophora 69–79)*. State University College of Forestry, Syracuse, New York.

WERESUB, L. K. and LECLAIR, P. M. (1971). On *Papulospora* and bulbiferous basidiomycetes *Burgoa* and *Minimedusa*. *Canad. J. Bot.*, **49**, 2203–13.

WILTSHIRE, S. P. (1938). The original and modern conceptions of *Stemphylium. Trans. Br. mycol. Soc.*, **21**, 211–38.

WOLLENWEBER, H. W. (1913). Studies on the *Fusarium* problem. *Phytopathology*, **3**, 24–50.

WOLLENWEBER, H. W. and REINKING, O. A. (1935). *Die Fusarien*. Paul Parey, Berlin.

10
Aspergillus

Species of the great group *Aspergillus* . . . together with the Penicillia and Mucors, furnish the 'weeds' of the culture room.

Thom and Church, *The Aspergilli*, 1926.

10.1 Introduction

Species of the genus *Aspergillus* Micheli ex Fries form a very large proportion of all the moulds encountered in industrial work. They are to be found almost everywhere on any conceivable type of substratum, and a good working knowledge of the genus is essential to anyone undertaking serious work on problems of mouldy deterioration. They are of particular importance as spoilage organisms of food. Many species will grow at very low water activity and are found attacking jam, cakes, tobacco, and are often the primary invaders in moulding of cereals. Not only do they attack the food, but also produce toxins, which are harmful to man and animals (see Tables 10.1 and 10.2). Thus *Aspergillus* species are one of the major producers of mycotoxins, in particular aflatoxin (see p. 196). The activities of the Aspergilli, however, are not entirely destructive in nature, for several species have had their fermentative powers harnessed for commercial purposes. In the East, strains of *A. oryzae* have long been used for the saccharification of rice starch in the

Table 10.1 Species of *Aspergillus* most commonly recorded as producing mycotoxins

Species	Toxin produced
A. flavus and A. parasiticus	Aflatoxins
A. clavatus	Patulin and sterigmatocystin
A. versicolor	Ochratoxins and sterigmatocystin
A. nidulans	Sterigmatocystin
A. ochraceus	Ochratoxin A
A. terreus	Patulin

Table 10.2 Species of *Aspergillus* producing diseases of man and animals

A. flavus
A. fumigatus
A. nidulans
A. niger
A. terreus

production of sake and similar potable liquors. Strains belonging to the same group are used in the manufacture of soy sauce and of the enzymic mixtures sold under the names 'Takadiastase' and 'Polyzime'. The black Aspergilli, long known to be capable of producing considerable amounts of oxalic acid from sugar, are now used for the successful commercial production of citric acid, following Currie's discovery that, by suitable adjustment of conditions, the formation of oxalic acid can be inhibited and the production of the more valuable citric acid encouraged. The same group of Aspergilli are also used for the production of gluconic acid, and of gallic acid from tannin. A comprehensive work on the activities of the Aspergilli is given in Smith and Pateman (1977).

10.2 Generic diagnosis

The following is a generic diagnosis sufficiently broad to cover all the species of any importance.

Aspergillus Micheli ex Fries
(Lat. *aspergillum,* a mop for distributing holy water) (Fig. 10.1)
Mycelium colourless or bright or pale coloured, or bearing surface concretions of colouring matter, never dematiaceous (black or smoky brown), septate, partly submerged or partly aerial; fertile branches (conidiophores) arising from, and more or less perpendicular to, specialized thick-walled enlarged mycelial cells (foot-cells), mostly non-septate, smooth, roughened or pitted, frequently enlarging towards the apex and terminating in a swelling

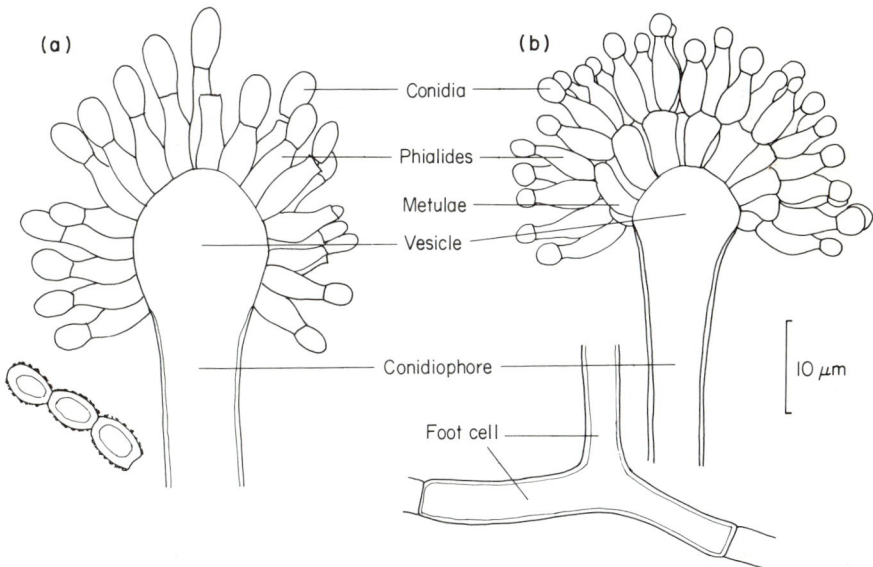

Fig. 10.1 (a) *Aspergillus sejunctus*   Conidial head with phialides only.
(b) *A. nidulans* Conidial head with both metulae and phialides.

(the vesicle) which may be variously globose, sub-globose, club-shaped, hemispherical, or a mere thickening of the stalk; vesicle bearing, from whole surface or from upper part only either phialides or more or less cylindrical intermediate cells (metulae) which themselves bear clusters of phialides; phialides or metulae produced simultaneously from surface of vesicles; conidia produced successively from the tips of the phialides, thus forming unbranched chains, variously shaped or coloured; ripe conidial heads of various shapes – globose, radiate, club-shaped or columnar – varying much in size in different species; sclerotia formed by several species but of minor diagnostic importance; perithecia of the perfect state produced by some species, thin-walled, breaking up to liberate the ascospores.

10.3   Separation of *Aspergillus* from *Penicillium*

Some of the smaller and more delicate species bear a strong resemblance to certain of the monoverticillate Penicillia, but there are two criteria by which such may be separated. Thom and Church (1926) and Thom and Raper (1945) regard the presence of foot-cells as the important distinguishing feature and, in spite of the fact that Smith (1933) has described two species of *Penicillium* with foot-cells to the conidiophores, the criterion is quite valid for the borderline species of the two genera. Figure 10.2(a) shows a typical foot-cell with the stalk and head arising from it, and this may be compared with several photographs of Penicillia in the next chapter. Some mycologists consider the method of production of phialides, simultaneous in *Aspergillus* (see Fig. 10.2b), successive in *Penicillium*, to be of primary generic

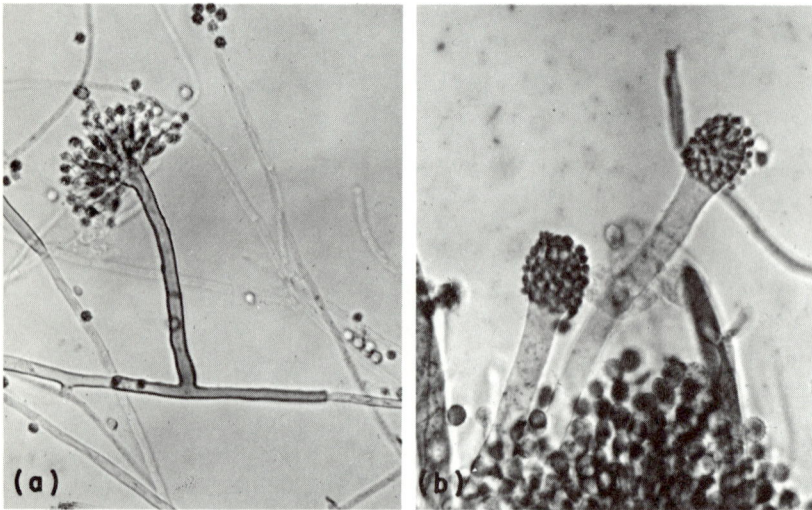

Fig. 10.2   (a) *Aspergillus ustus*   Conidiophore with foot-cell, ×500. **(b)** *A. repens*   Showing 'simultaneous' production of phialides, ×500.

significance. Both distinctions are equally to be trusted, but it is sometimes very difficult to demonstrate the presence of foot-cells in the borderline species of *Aspergillus*, owing to their compact habit of growth and the persistence of tangled masses of mycelium in microscopic mounts. On the other hand it is usually easy to find, especially near the edge of a growing colony, partially developed heads showing various stages in the growth of the phialides.

It is interesting to note that although separation of taxa in *Aspergillus* is usually based on other characters, one author (Singh, 1973) studied the foot-cells produced by different species and found many of them quite characteristic of the species.

Sterigmatocystis

Various authorities in the past have considered *Aspergillus* to include only the species with phialides borne directly on the vesicle, and have included all the species with metulae as a separate genus, *Sterigmatocystis*. However, the production of metulae is not constant and some species such as *A. flavus* and *A. tamarii* produce both kinds of head, with and without metulae respectively and it is not unusual to find heads with both phialides and metulae borne directly on the vesicle.

10.4   The perfect states of *Aspergillus*

As noted in Chapter 5, de Bary showed the connection between the ascomycetous genus *Eurotium* and the *Aspergillus glaucus* group, and many species of the group are to be found mentioned in the literature under the name *Eurotium*. Other species or species aggregates have been shown to have perfect states, for example *A. nidulans* and *Emericella* Berk. & Br. and *A. fumigatus* and *Sartorya* Vuill. When the perfect state is present and known it is correct to use the name of the perfect state (see Chapter 3). However, species of the form genus *Aspergillus* do not always produce the perfect state regularly or easily so it is convenient to study them together as the form genus. This practice has been followed by Raper and Fennell (1965) and it is proposed to do so here. However, other authors have felt that the perfect state names should be used and have looked for, and found, ascospores for many of the species. As these perfect states vary considerably they belong in different genera. The use of the names *Eurotium*, *Emericella* and *Sartorya* was revived by Benjamin in 1955. Subramanian (1972) reviewed the position and revived old names or introduced new ones for the perfect states of many of the species. His work just predated the paper by Malloch and Cain (1972) on Ascomycetes with *Aspergillus*, *Paecilomyces* and *Penicillium* imperfect states. Since then further ascosporic states have been found by Wiley and Simmons (1973). Table 10.3 lists the common groups of *Aspergillus* species with the names of the perfect states associated with them.

Table 10.3   Perfect (teleomorphic) names of *Aspergillus*

Teleomorph	Anamorph
Chaetosartorya Subramanian	*A. cremeus* group
Dichlaena Montagne & Durien	? *A. ornatus* group
? near *Patromyces*	
Emericella Berk.	*A. nidulans* group
Eurotium Link ex Fr. = *Edyuillia*	
and *Gymnoeurotium*	*A. glaucus* group
Fennellia Wiley & Simmons	*A. flavipes* group
Hemicarpenteles Sarbhoy & Elphick	
= *Sclerocleista*	*A. paradoxus* of *A. ornatus* group
Neosartorya Malloch & Cain	
= *Sartorya* (invalid) =	
? *Hemisartorya* (? invalid)	*A. fumigatus* group
Petromyces Malloch & Cain =	
Synclerstostroma (invalid)	*A. alliaceus* of *A. ochraceus* group or
	A. cremeus group
Warcupiella Subramanian	*A. spinulosus* of *A. ornatus* group
Saitoa Rajendram & Muthappa	*A. japonicus* of *A. niger* group

10.5   Determination of species

The taxonomy of the genus is, up to a point, not difficult owing to the excellent scheme of classification given in the monograph by Thom and Church (1926). The later manual, by Thom and Raper (1945) followed the same classificatory scheme with little modification. The most recent revision of the genus, by Raper and Fennell (1965), makes a number of improvements in the classification, and takes into account the large number of new names introduced since 1945. Since then no major reclassification has been undertaken but studies of individual groups within the genus have been made. These will be discussed when the groups are described. Samson (1979) has made a survey of species described since 1965.

One factor which facilitates identifications is the wide range of colony colour exhibited, the genus being probably unique amongst the common Hyphomycetes in this respect. The colours, moreover, are comparatively stable, and are correlated with morphology and biochemical characteristics.

One of the most important factors in Thom and Church's classification of the taxonomy of this important genus, is their conception of what may be termed the 'group species'. Some species are fairly definite and of stable morphology. A single strain may be described in definite terms and other strains encountered will fit the description with considerable accuracy, justifying the conclusion that here we have a true species. In other cases, as in the series represented by *A. versicolor, A. candidus, A. flavus* and others, different strains, whilst clearly related, differ somewhat widely in minor details. Anyone who collects only a few strains belonging in such a series may observe differences which seem to justify the bestowal of several specific epithets. However, when scores or hundreds of strains are examined, the sharp lines of demarcation disappear and instead of a few definite species, there is a perfectly graduated series such that no one could describe any particular strain in terms which would permit of its certain identification by another who had not the same series of strains before him. A key to the common 'group series' is given below. Delimination of species in such a group is also complicated by the fact that a single strain may cover a considerable range in the series in response to variations in cultural conditions, or even in successive cultures grown under the same conditions. Modern technology, for example scanning electron microscopy, allows examination of fine structures not previously possible and detailed studies are being undertaken of these 'group species' to discover if they are true continua or can now be separated into distinct species, and to relate this morphology with biochemical characters such as the ability to produce mycotoxins.

The specialist who is interested in one particular group will collect a number of strains and will know them sufficiently well to be able to separate them on grounds dictated by his own requirements. For most purposes, however, it is sufficient to be able to identify any particular culture as a member of one of the group species. After all, this so-called 'lumping' of strains is only the equivalent of the recognition of variation amongst individuals in any one species of flowering plant or animal.

10.6   Key to the common 'group series' of *Aspergillus*

See pages 174–5.

10.7   Descriptions of 'group series' of *Aspergillus*

A. *clavatus* group

(Lat. *clavatus,* club-shaped) (Figs 10.3, 10.4)
No perfect state known.

Growth rapid, velvety, forming fairly dense felt, bluish grey-green; conidial heads clavate when young, splitting into divergent columns; conidiophores coarse, smooth, hyaline and can be very long; vesicles elongate club-shaped; phialides borne directly on the vesicles over the entire surface, short, densely packed; conidia elliptical, smooth.

A. clavatus Desmazières is the commonest species with conidiophores up to 4 mm long and conidia 3.5–4.5 × 2.5–3 μm. It is quite often seen as a spoilage organism, on dung and in soil and can grow in strongly alkaline conditions.

Other species in the group tend to be larger or smaller than *A. clavatus*. *A. giganteus* Wehner produces some conidiophores up to 5 cm long and giant heads besides a basal growth of smaller heads.

Key to the common 'group series' of *Aspergillus*

Uniseriate

Vesicles club shaped, heads clavate .... *A. clavatus*

Vesicles otherwise
- Heads globose, radiate, bright yellow cleistothecia .... *A. glaucus*
- Heads loosely to definitely columnar, conidia cylindrical at first .... *A. restrictus*
- Conidial heads compactly columnar
  - Conidia only ... *A. fumigatus*
  - Cleistothecia ... *A. fischeri*

Mixed uniseriate and biseriate

- Conidial heads globose radiate splitting in age
  - Brown to black .... *A. niger*
  - White to cream .... *A. candidus*
  - Yellow to ochraceous ...... *A. ochraceus*
- Conidial heads radiate or forming poorly defined columns
  - Yellow green, rough conidiophores .... *A. flavus*
  - Yellow brown, smooth conidiophores .... *A. wentii*
- Conidial heads large, conidiophores constricted below vesicles — Rare

Biseriate

Conidial heads green

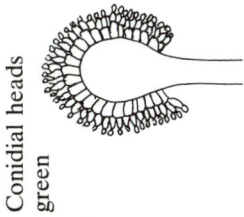

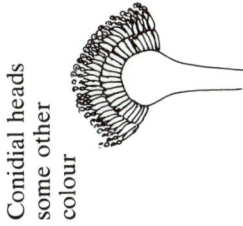

Heads radiate, stipes uncoloured {
One type of head .... A. *versicolor*
Two types of head .... A. *janus*
}

Heads loosely columnar, cleistothecia usually present, stipes brownish .... A. *nidulans*

Conidial heads some other colour

Heads radiate, conidiophores brown, conidia drab olive to dull brown .... A. *ustus*

Heads loosely columnar, conidia white to pale avellaneous, stalks yellow brown to yellow to colourless .... A. *flavipes*

Heads compactly columnar, conidia cinnamon to orange brown to pale buff, conidiophores colourless .... A. *terreus*

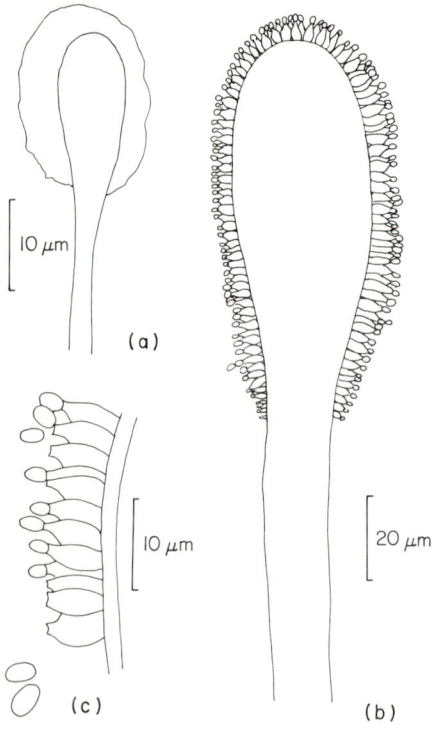

Fig. 10.3 *Aspergillus clavatus* **(a)** Outline of conidial head; **(b)** conidial head showing phialides; **(c)** phialides and conidia.

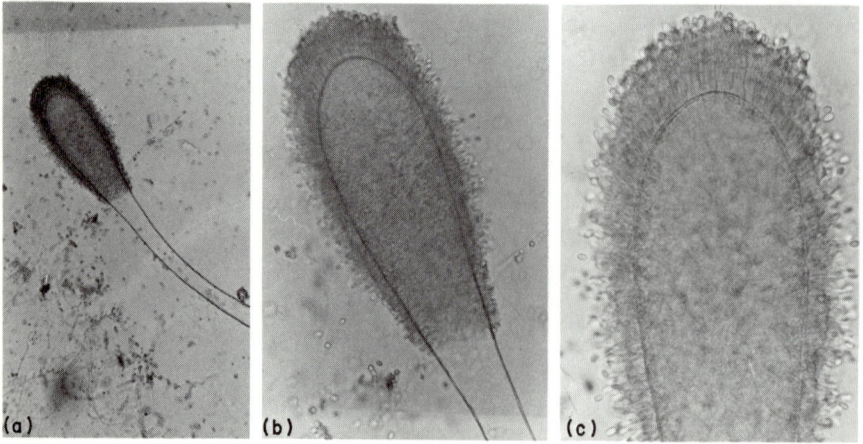

Fig. 10.4 *Aspergillus clavatus* **(a)** Conidial head, ×80; **(b)** conidial head, ×200; **(c)** phialides and conidia, ×320.

A. *glaucus* group

(Lat. *glaucus*, grey-green) (Figs 10.5–10.9)
Perfect state *Eurotium* Link ex Fries.

Growth conidia bright bluish green becoming dull green to brownish green; cleistothecia yellow to orange. Colour sometimes masked by red and brown pigments of encrusted mycelium. Conidial heads loosely radiate; conidiophores smooth, septate, thin walled, often collapsing in mounts; vesicles varying from mere thickenings of the conidiophores to more or less globose; phialides borne directly on the vesicle, covering the whole of it or only the upper part, usually broad; conidia usually spiny, rarely smooth, mostly greater than 5 μm long, and up to 15 or 24 μm at low temperatures, oval to pear shaped; cleistothecia yellow to orange, globose, 80–250 μm diam. according to species, containing 8 spored asci, ascospores irregularly arranged, asci breaking up as spores mature; ascospores colourless, biconvex, usually with a furrow marking the line along which the wall slits on germination, with sides of the furrow rounded or raised into crests or frills according to species.

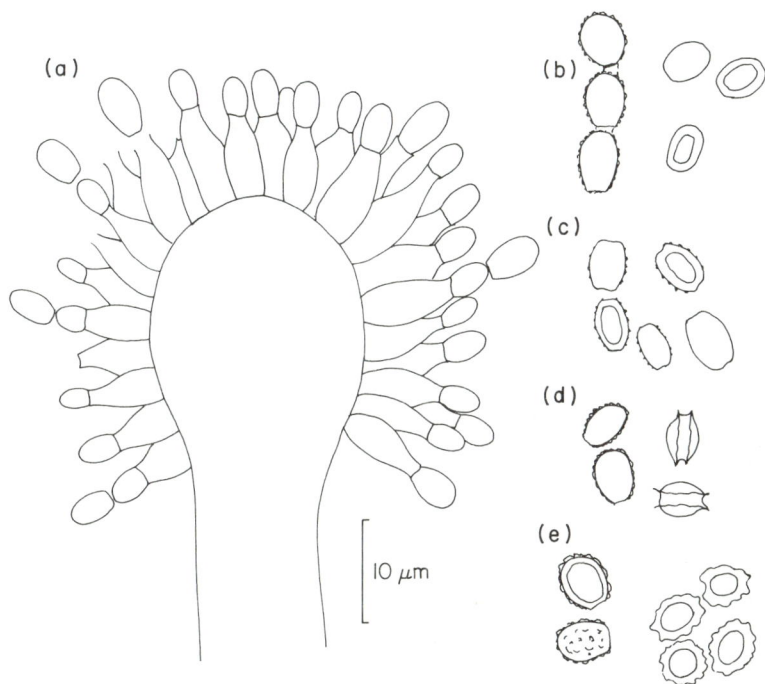

Fig. 10.5 *Aspergillus glaucus* group *A. repens* **(a)** *A. repens* Conidial head. **(b)** *A. repens* Conidia and ascospores. **(c)** *A. sejunctus* Ascospores. **(d)** *A. chevalieri* conidia and ascospores. **(e)** *A. amstelodami* conidia and ascospores.

Several species belonging to this group are amongst the most commonly occurring and most destructive of all moulds. They require less moisture for germination of spores and subsequent growth than practically all other moulds, or, what amounts to the same thing, they will grow on substrata of high osmotic concentration or to use the modern term of 'low water activity'. Thus they are commonly found in pure culture on jams which contain only a little less than the safe percentage of sugar. They grow on textiles which contain very little more moisture than the amount usually considered to be normal for any particular material. They attack tobacco, and, in general, seem to be capable of infecting any organic material which is in equilibrium with an atmosphere of relative humidity greater than 70%, and are frequently the first invaders in a moulding situation, for example in grain stores.

The various species can usually be recognized as belonging to this group – complete identification is more difficult – even in their natural habitats, without isolation in pure culture, by the presence of both bluish green or greyish conidial heads and bright yellow perithecia, the latter being easily visible to the naked eye.

The group was, for a long time, a difficult one for taxonomists. Simplification has come only after it has been recognized that satisfactory delimitations of species must be based on adequate study of a very large number of isolates, and that a too narrow conception of what constitutes a species results only in confusion.

Mangin (1909) was the first to make a comparative study of a number of isolates from various sources. He followed earlier workers in regarding size and markings of ascospores as of primary significance in separating species, and this approach is still followed. Thom and Church (1926) adopted Mangin's scheme of classification and fitted into it species described by Bainier and Sartory (1911, 1912). Thom and Raper (1941) reappraised the situation and include this work in their monograph of 1945, and this was again modified by Raper and Fennell in their work *The genus Aspergillus* (1965). This latter classification is adequate for most purposes. However, as the genus is more a continuum of isolates than a series of distinct species there were still problems. Locci (1972) included *Eurotium* species in his scanning electron microscope studies of the ascosporic Aspergilli. Blaser (1976) monographed the genus *Eurotium*. He used standard culture techniques and divided the genus, according to morphology of the ascospores and surface and colour of the conidia, into three groups, *E. herbariorum*, *E. amstelodami* and *E. echinulatum*.

However, whatever they are called, the common species remain the same. The following key is based on Raper and Fennell (1965). It includes four small spored species, i.e. those with ascospores 6 μm or less in long axis, and one large spored species. A great many species are omitted. The large spored species are rarely seen. Thom and Raper (1941) point out that the large spored species have much lower temperature optimums than the small spored forms, and suggest that they might be isolated more frequently if the incubation temperature were 15–20°C instead of the usual 24–25°C. They are also more osmophilic than the other species, so this may be another reason why they are not isolated by the usual methods.

A. repens (Corda) Saccardo (Lat. *repens*, creeping; refers to the stolon-like hyphae at the edges of colonies). Perfect state *Eurotium repens* Corda (Blaser accepts the name and places it in his section *E. herbariorum*).

Probably the commonest and most cosmopolitan member of the group,

Key to the *Aspergillus glaucus* group

1.	Ascospores 8–10 μm in long axis	*A. echinulatus* (Delacroix) Thom & Church (*Eurotium echinulatum* Delacroix)
1'.	Ascospores 6 μm or less in long axis	2
2.	Ascospores with curved surfaces smooth or nearly so	3
2'.	Ascospores with curved surfaces rough	*A. amstelodami* (Mangin) T. & C. (*Eurotium amstelodami* Mangin)
3.	Equatorial ridges (crests) lacking; furrow nil or very shallow and inconstant	*A. repens* (Corda) Sacc. (*Eurotium repens* Corda)
3'.	Ridges low and rounded; furrow broad and shallow	*A. ruber* (Konig, Spieckermann & Bremer) T. & C. (*Eurotium rubrum* Konig, Spieckermann & Bremer)
3''.	Ridges thin, crest-like; spore like a pulley	*A. chevalieri* (Mangin) T. & C. (*Eurotium chevalieri* Mangin)

forming characteristic colonies, the surface being an intimate mixture of dirty green and yellow, with reverse and medium gradually becoming dirty brown on malt agar and yellow to brownish red on Czapek agar containing 20% sucrose (the medium favoured by Thom and Raper); conidial heads large, with conidia mostly 5–6.5 μm in diam., rough; cleistothecia mostly 75–100 μm diam.; ascospores 4.8–5.6 × 3.8–4.4 μm, smooth, with usually a trace of a furrow, but very indistinct, and no crests.

A. ruber (Konig, Spieckermann & Bremer) Thom & Church (Lat. *ruber*, red). Perfect state *Eurotium rubrum* Konig, Spiecermann & Bremer. Blaser considers it a synonym of *E. herbariorum* Link ex Fries, which he thinks has priority. However, there is some doubt as to the actual validty of *E. herbariorum* for which Malloch and Cain created a neotype in 1972. The name *Aspergillus sejunctus* Vuillemin probably has priority over the name *A. ruber*. However, the latter is very descriptive and well used.

It is found very frequently on all kinds of substrata; colonies rusty red, the colour being due to incrustations of pigment on the mycelium, with the reverse deep red to intense red brown; conidial heads pale blue-green turning greyish; conidia elliptical, rough, 5–6.5 μm in long axis; cleistothecia abundant, 80–120 μm diam.; ascospores 5.2–6.0 × 4.4–4.8 μm, with broad shallow furrow.

A. chevalieri Mangin (F. Chevalier, French mycologist). Perfect state *Eurotium chevalieri* Mangin (Blaser accepts the name and places it in his group *E. amstelodami*).

Colonies show patches of pale blue-green conidial heads and areas of orange hyphae in which cleistothecia are embedded, with reverse orange-red; conidial heads radiate, fairly large; conidia almost globose, spiny, 4.5–5.5 μm diam.; cleistothecia 100–140 μm diam.; ascospores smooth walled, with thin prominant crests, 4.6–5.0 × 3.4–3.8 μm.

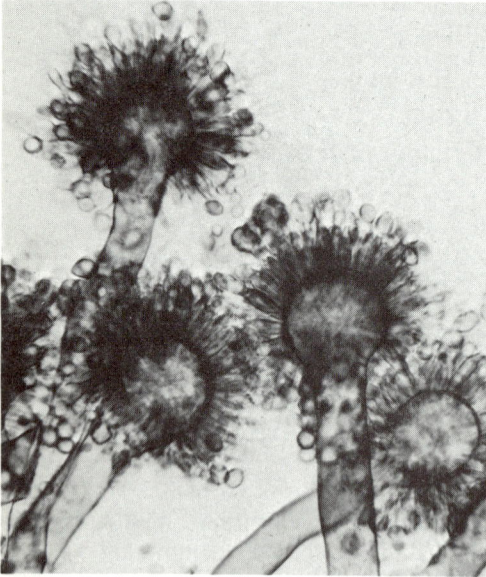

Fig. 10.6 (left)   *Aspergillus glaucus* group *(A. chevalieri* var. *intermedius)*   Conidial heads with globose vesicles, ×500.

Fig. 10.7 (right)   *Aspergillus chevalieri*   Proliferation of phialides to form secondary heads, ×550.

A. amstelodami (Mangin) Thom & Church (Lat. of Amsterdam). Perfect state *Eurotium amstelodami* Mangin. (Blaser accepts the name and places it in his group *E. amstelodami.*)

Colonies usually deep rich green, speckled with sulphur yellow, with reverse almost colourless or at most pale yellow; conidial heads radiate or more or less columnar, large; conidia sub-globose, delicately roughened, mostly about 4 μm diam.; cleistothecia up to 140–160 μm diam.; ascospores rough all over, with prominent furrow and rounded crests. Occasional cultures vary from type in being almost entirely cleistothecial, with conidial heads produced tardily and sparingly.

A. echinulatus (Delacr.) Thom & Church (Lat. *echinulatus,* with little spines). Perfect state *Eurotium echinulatum* Delacroix. (Blaser accepts the name and places it in his group *E. echinulatum.*)

This is the most common of the large-spored species. Colonies slow growing, mottled with a mixture of blue-green conidial heads and orange-red

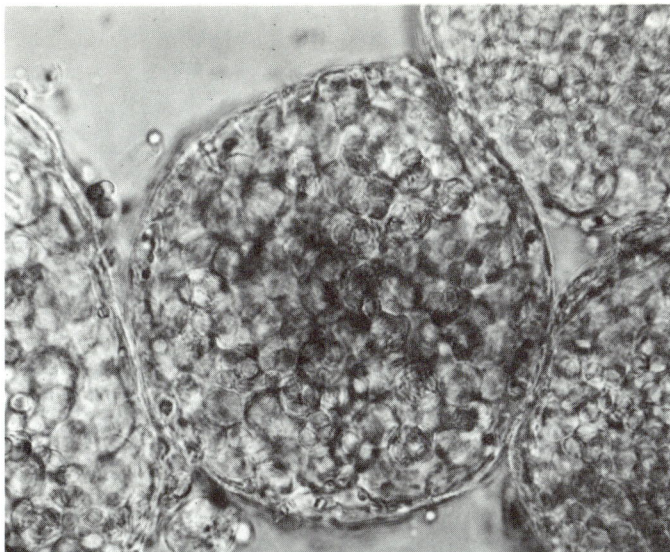

Fig. 10.8 *Aspergillus amstelodami* Perithecia, ×600.

cleistothecial masses, with reverse pale to deep reddish brown; conidiophores long, up to about 1 mm, and mostly broadening upwards; conidia elliptical, pear-shaped or sub-globose, rough, up to 10 μm in long axis; cleistothecia embedded in red hyphae growing close to the surface of the medium, mostly 150–160 μm diam.; ascospores large up to 10 μm long by 6–7.5 μm diam., convex surface rough, with broad shallow furrow and definite crests.

A. *restrictus* group

No perfect state known.

Growth very sparse, conidia very dark green; conidiophores short, smooth, hyaline, occasionally greenish at the apex of the conidiophores; vesicles varying from convex enlargements of the apex of the conidiophores to small globose swellings; phialides borne directly on the vesicle on the upper two thirds or third of the vesicle; conidia hyaline, elliptical to barrel shaped, rough, often adhering in chains or columns in mounts.

Species belonging to this group are of common occurrence but are easily missed in plates from infected material owing to their very slow rate of growth. Some of them have been described as species of *Penicillium* and certainly bear a striking resemblance to certain of the monoverticillate members of that genus. They are very resistant to antiseptics. Like the A. *glaucus* group, they are able to grow on comparatively dry materials and are common on textiles which contain little more than the normal amount of moisture. They are frequently the first invaders in cereal stores and grow on the glass of optical instruments in the hot wet tropics, often etching the glass. They are easy to recognize by their very long slender columnar heads,

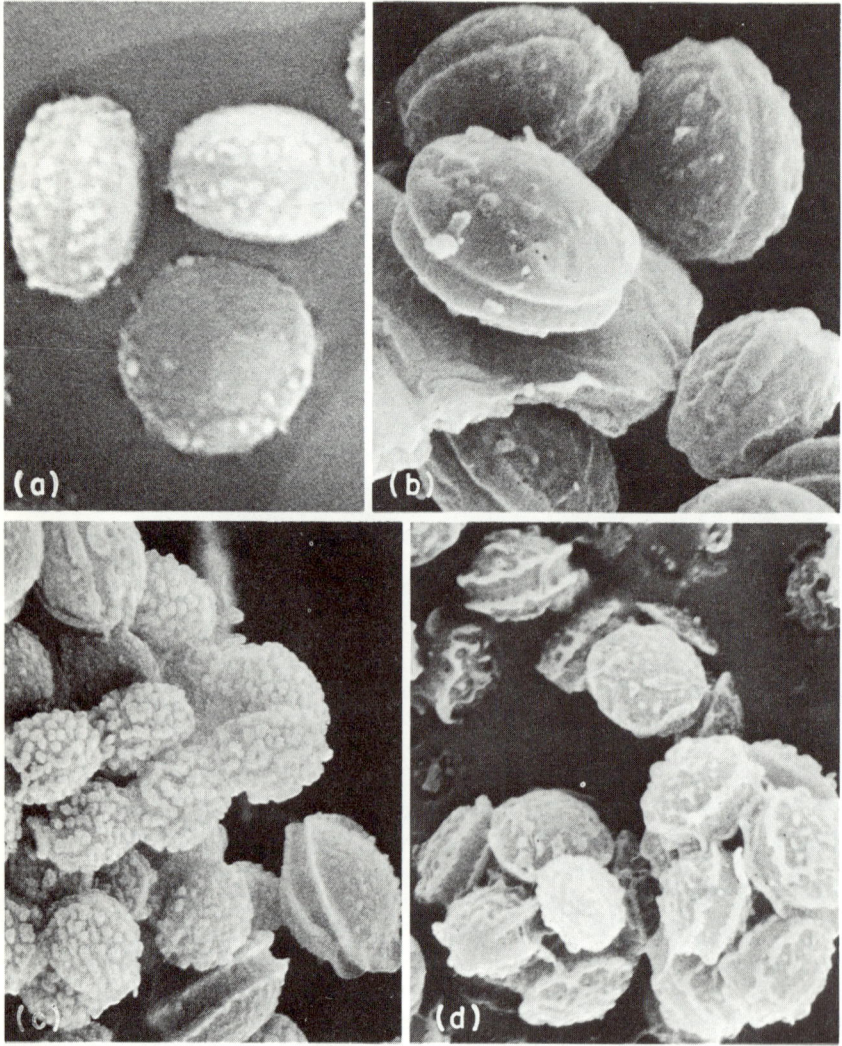

Fig. 10.9 *Aspergillus glaucus* group   Stereoscan photographs of ascospores. **(a)** *A. repens,* ×9000; **(b)** *A. sejunctus,* ×10800; **(c)** *A. chevalieri,* ×7200; **(d)** *A. amstelodami,* ×8000.

composed of chains of fairly large, ovate to barrel shaped, rough conidia. They resemble *A. fumigatus* in disposition of phialides and shape of heads, and the *A. glaucus* group in the size and markings of conidia.

Five species (shown in the key to the *A. restrictus* group) are regarded as distinct by Raper and Fennell, but only two are common.

A. caesiellus Saito has been isolated very infrequently. It is less osmophillic than the other species of the series and has conidia adhering in long columns like *A. restrictus.*

Key to the *Aspergillus restrictus* group

1.	Heads columnar; vesicles small, fertile on upper surface only	2
1'.	Heads radiate when young, tardily loosely columnar; vesicles fertile over upper half to two thirds	*A. penicillioides*
2.	Colonies growing moderately well on standard Czapek agar	*A. caesiellus*
2'.	Colonies growing very slowly on Czapek agar	3
3.	Colonies dark olive green; spore columns long, often twisted	*A. restrictus*
3'.	Colonies light grey green; columns shorter and more delicate	4
4.	Conidia elliptical at first, often remaining so, 4.0–4.5 × 3.0–3.5 μm	*A. conicus*
4'.	Conidia sub-globose to pyriform, mostly 3.0–3.5 μm diam.	*A. gracilis*

A. conicus Blochwitz is found occasionally in soil. Colonies rather pale grey-green at first, often becoming wet and slimy, with reverse in dark green shades; conidia rough.

A. gracilis Bainier grows slowly on all media, pale grey-green, with reverse greenish; spore chains in very slender columns; conidia delicately roughened, small.

A. penicillioides Spegazzini (Gr. *eidos*, like; hence like *Penicillium*) (Fig. 10.10). Differs from the other members of the series in that the spore heads are radiate rather than columnar. Grows scarcely at all on Czapek agar; for normal growth requires a medium containing a high percentage of sugar. Conidia at first elliptical, becoming barrel-shaped to globose, very rough, 3.0–4.0 μm or even 5.0 μm diam. This often occurs on optical instruments and was described from this source as *A. vitricola*. From this habitat the phialides and conidia are often very swollen and enlarged. It is quite common in the spoilage situation.

A. restrictus G. Smith (Lat. *restrictus,* close, tight; from the habit of growth) (Figs 10.11, 10.12). Colonies dull green to brownish green; vesicles more or less conical; heads forming very long compact columns; conidia almost colourless and cylindrical when young, becoming pear-shaped and dark brownish green, 4.5–6.0 × 3–4 μm, in occasional strains larger, up to 10 μm long. The commonest of the four species and appears to be cosmopolitan.

A. fumigatus group

(Lat. *fumigatus*, smoky) (Figs 10.13–10.15)
Perfect state *Sartorya* Vuill.

Growth spreading, dark, smoky green becoming darker, more or less velvety, occasionally floccose; sporing tardily with young heads very bluish

184   *Aspergillus*

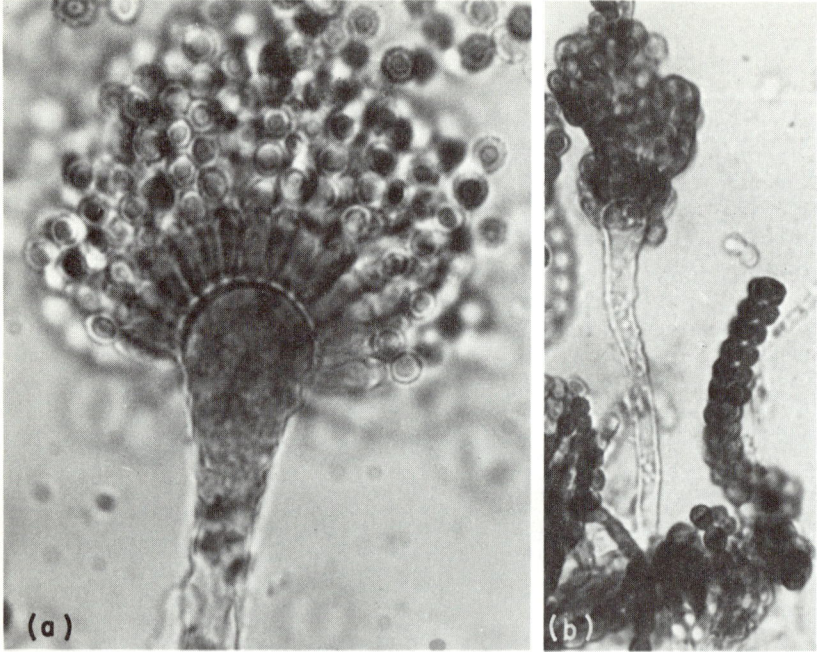

Fig. 10.10   *Aspergillus penicilloides*   **(a)** Conidial head showing rough conidia, ×1300; **(b)** column of conidia from old culture, ×250.

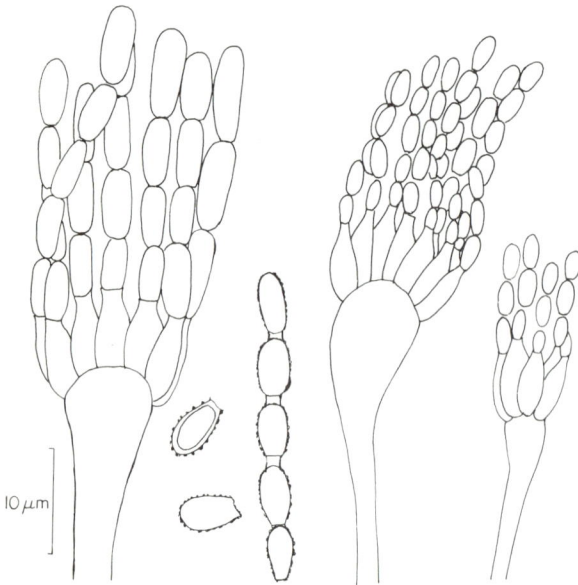

10 μm

Fig. 10.11   *Aspergillus restrictus*   Conidial heads and conidia.

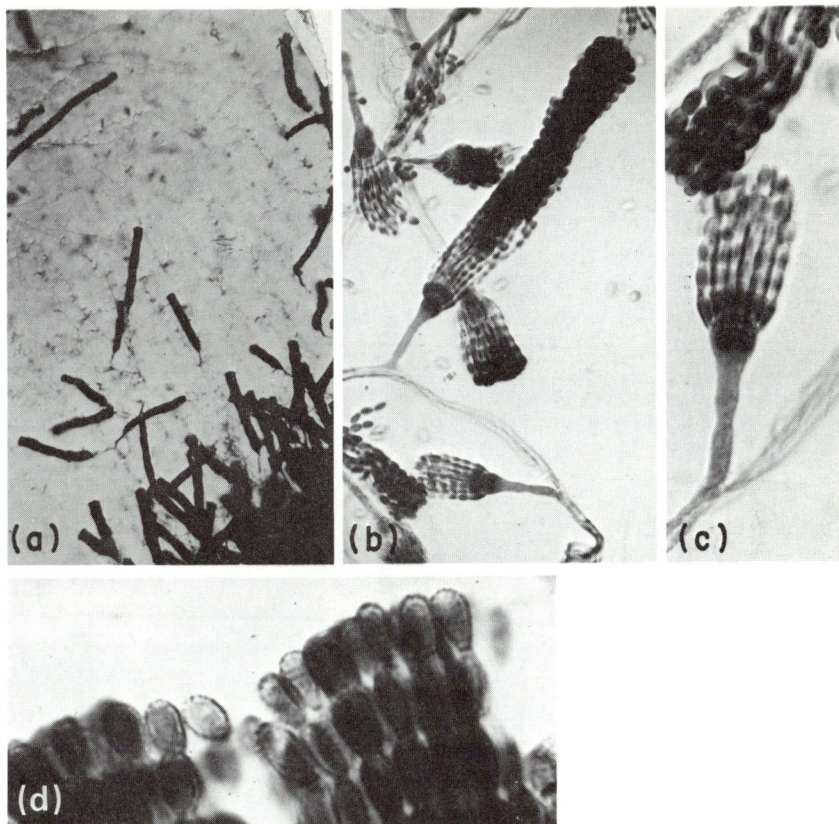

Fig. 10.12 *Aspergillus restrictus* **(a)** Columnar heads as seen in Petri dish, ×50; **(b)** typical conidial heads, ×250; **(c)** typical conidial heads, ×550; **(d)** conidia at apex of conidial column, showing roughening, ×1300.

green; conidial heads columnar of varying length, but about 40 μm broad, when the cultures are jarred the columns may fragment to form masses of green dust; conidiophores smooth, short, often greenish, 2–8 μm diam.; vesicles flask shaped, fertile on upper half or three quarters, often greenish; phialides borne directly on the vesicles, close packed, lower ones deflected upwards, 6–8 × 2–3 μm, often pigmented; conidia small, globose, rough, mostly 2.5–3 μm diam.; cleistothecia white to off-white, producing 8-spored asci; ascospores uncoloured, bivalve, with equatorial crests and ornamented convex surfaces.

A. fumigatus Fresenius is a very common spoilage mould. The heads mounted in lactophenol have a very characteristic appearance. This species is readily distinguished from the last group by its smaller, globose conidia, and by its much more rapid growth.

Several species with a perfect state and with conidial morphology similar to *A. fumigatus* are known, but they are rare. The most common is *Sartorya*

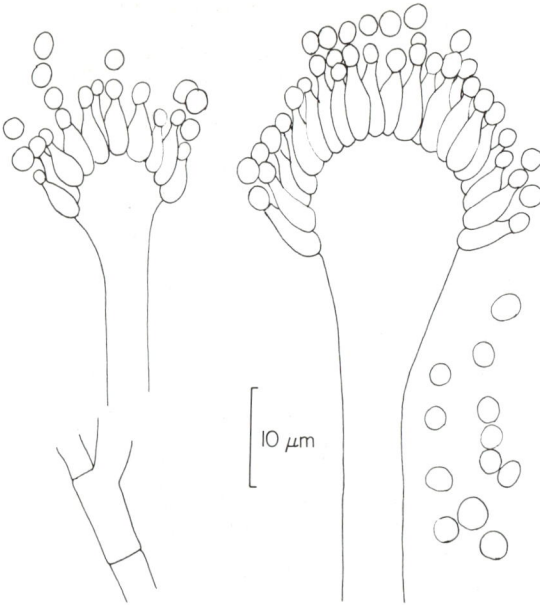

Fig. 10.13 *Aspergillus fumigatus*   Conidial heads, conidia and foot cell.

fumigata Vuill. (=*Aspergillus fischeri* Wehmer), with ascospores with two distinct equatorial crests and convex surfaces bearing anastomosing ridges. Other ascosporic species have differently ornamented ascospores.

    A. fumigatus is the causal agent of a disease of birds, known as Aspergillosis. The disease also occurs in lambs, and can be very serious (Austwick *et al.*,

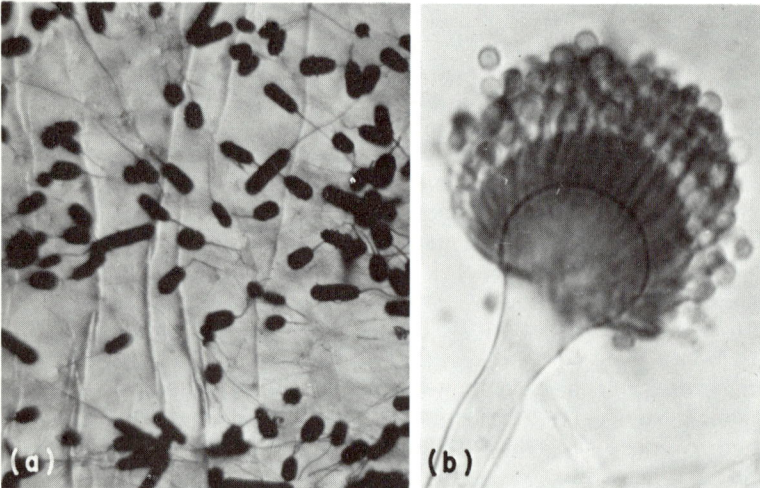

Fig. 10.14 *Aspergillus fumigatus*   **(a)** Columnar heads as seen in Petri dish, ×50; **(b)** typical conidial head, ×1300.

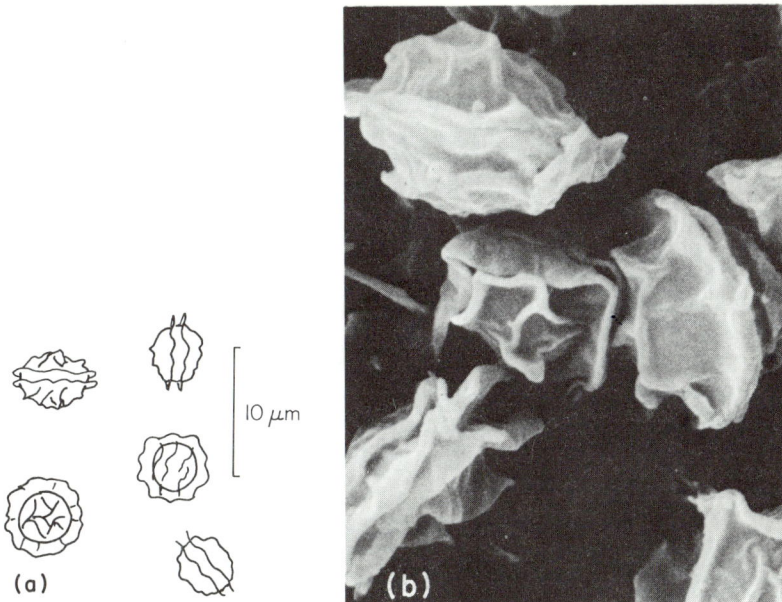

Fig. 10.15 *Aspergillus fischeri* Ascospores. **(a)** Drawing; **(b)** stereoscan photograph, ×7200.

1960). Cases in which human beings have contracted the disease, which has many of the symptoms of tuberculosis, are rare but more common than was originally thought. When handling cultures care should be taken not to distribute conidia in the air.

The species grows well over a wide range of temperatures. It flourishes at 45°C and often at higher temperatures, and is commonly found in decomposing compost. Its presence as a spoilage organism can cause personnel problems when materials become heavily infected due to the clouds of spores of a potentially pathogenic fungus, and care has to be taken during disposal. It is particularly common on hay, but has also been seen on electronic equipment, furs, paint and many other materials, and has been found growing in aircraft fuel.

A. *niger* group

(Lat. *niger,* black) (Figs 10.16–10.19)
Perfect state of *A. japonicus* is *Saitoa* Rajendram and Muthappa.

Growth, colonies spreading rapidly, with mycelium white at first frequently developing areas of bright yellow, producing dark brown to black or purple brown conidial heads; conidial heads globose, radiate or as they grow splitting into several columns of conidial chains; conidiophores arising from the substratum, varying from 200 µm to several millimeters long and 10–20 µm diam., mostly colourless to brown, smooth, splitting when crushed like pieces of cane; vesicles globose, colourless to yellowish brown, fertile over the whole surface; phialides borne directly on the vesicle in some species but metulae usually present, fairly uniform in size, usually brown; metulae varying in length from 10–15 µm in *A. niger* to 120 µm in *A. carbonarius*, usually deep brown; conidia globose to elliptical or horizontally flattened, mostly appearing spiny except at high magnifications when the colour may be seen to be aggregated into bars and nodules between the inner and outer walls, varying in size from 2.5–10 µm diam.; some isolates produce globose sclerotia, white to cream to brown, about 1 mm diam. especially when freshly isolated.

Black Aspergilli are of very common occurrence in all parts of the world, and as is usual in any group of widespread distribution and omnivorous habits, a large number of species and varieties have been described. Successive authors have divided the group in various ways. Raper and Fennell (1965) recognize 10 species and two varieties with metulae in the heads and two species without metulae. The majority of isolates agree well with the diagnoses of *A. niger* but isolates without metulae are quite often received from tropical countries. When large numbers of the latter have been examined they seem to intergrade and can be for most purposes regarded as one species.

A. niger Van Tieghem (Figs 10.16, 10.17) has both metulae and phialides and

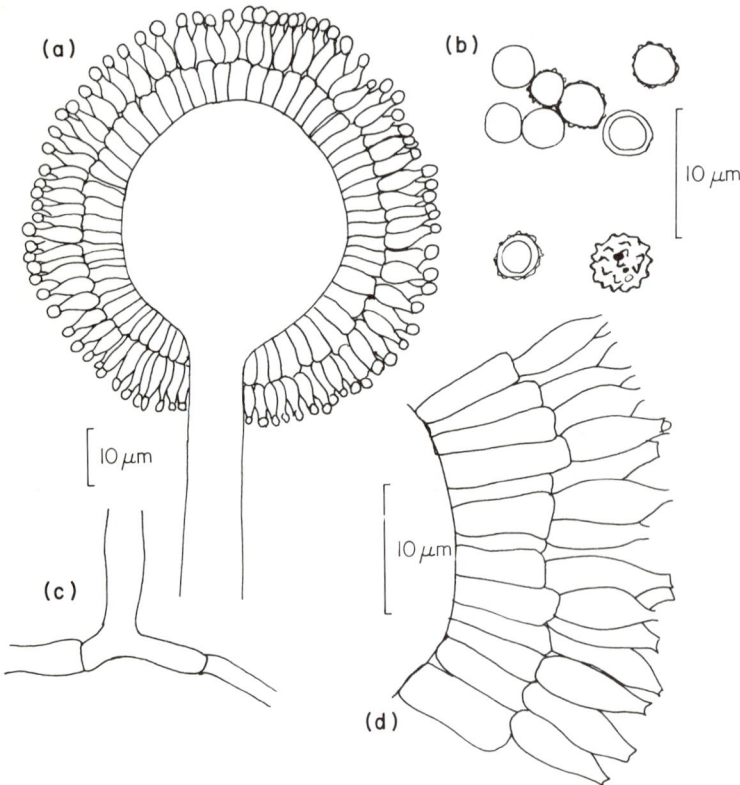

Fig. 10.16 *Aspergillus niger* **(a)** Conidial head; **(b)** conidia; **(c)** foot cell; **(d)** metulae and phialides.

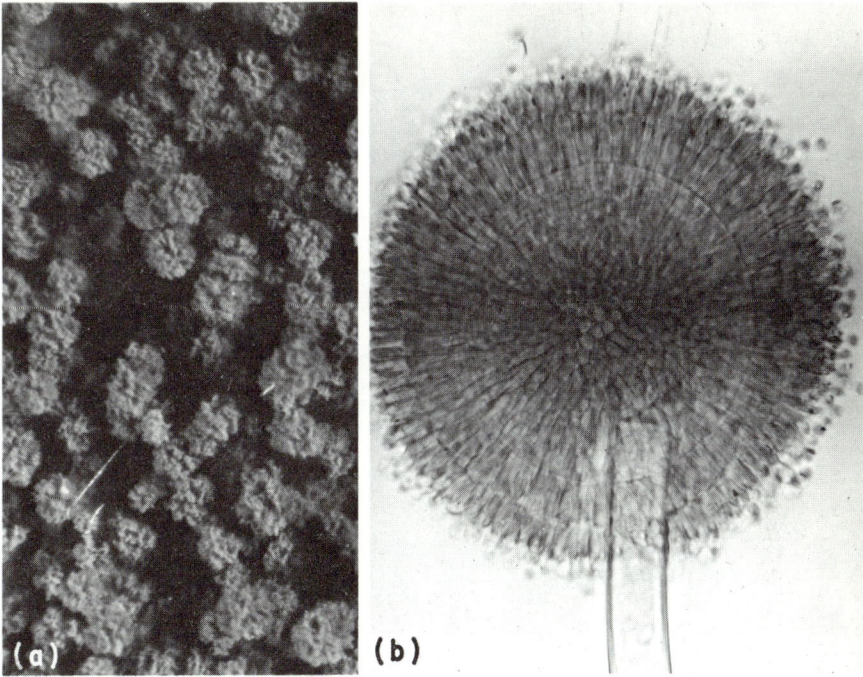

Fig. 10.17 *Aspergillus niger* **(a)** Globose heads as seen in Petri dish, ×20; **(b)** young head not yet fully pigmented, ×500.

small, more or less globose, rough conidia, mostly 4–5 µm diam. but occasionally smaller. Heads vary in colour from dark brown to purplish brown to black, but usually appear black when viewed with the naked eye in daylight.

It is hardly surprising that, with a species of such widespread and common occurrence, mutations have from time to time been obtained. Thom and Church mention several stable mutants obtained experimentally by growing *A. niger* on media containing small amounts of poisonous substances the most distinctive of these being *A. niger* mut. *cinnamomeus* (Schiemann) T. & R., almost colourless, and *A. niger* mut. *schiemannii* (Thom) T. & R., with brownish colonies. Species belonging to this group have interesting biochemical characteristics and workers who are concerned with this side of their activity will need to separate strains on biochemical rather than on morphological grounds.

A. japonicus Saito (Figs 10.18, 10.19) (perfect state *Saitoa japonica*) is characterized by bearing phialides directly on the vesicle. Colonies are definitely purple brown, not very deep, with smallish heads splitting into several columns, vesicles ranging from 20–80 µm or more in diam.; phialides closely packed often appearing short, 5.5–10 × 3–4.5 µm, light brown in colour; conidia globose to elliptical, distinctly spinulose, 3–5 µm diam. but occasionally larger.

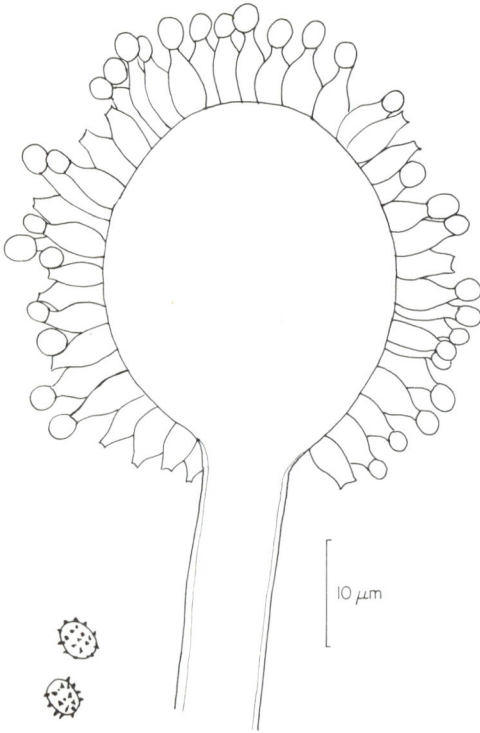

Fig. 10.18   *Aspergillus japonicus*   Conidial head and conidia. Note lack of metulae.

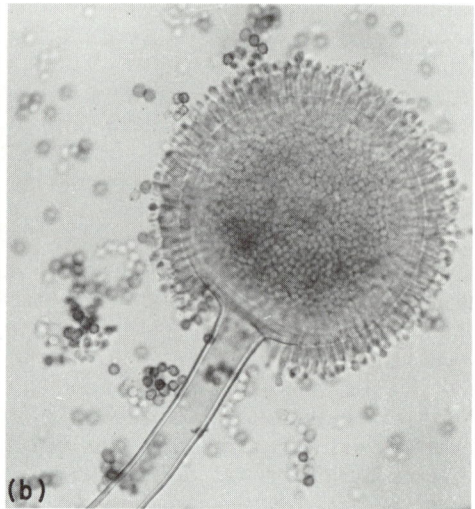

Fig. 10.19   *Aspergillus japonicus*   **(a)** heads splitting into columns as seen in Petri dish (heads in older colonies of *A. niger* also split like this); **(b)** conidial heads with phialides born directly on the vesicle, ×500.

A. candidus Link ex Fries

(Lat. *candidus*, shining white) (Figs 10.20, 10.21)

No perfect state known.

Growth varies from a sparse basal felt with numerous fertile stalks arising from the substrate to floccose with conidiophores arising from both the submerged and aerial mycelium; conidial heads white, globose but often splitting in age as long chains of conidia are formed, varying from less than 100 μm to 250 μm diam., sometimes appearing wet or slimy especially when young; conidiophores erect, colourless or faintly yellowish near the vesicle, smooth, occasionally septate, some short, others up to 1 mm long; vesicles globose, mostly producing metulae over the whole surface, but occasionally only the upper surface is fertile and then more or less columnar heads are produced, some very reduced in size; metulae and phialides are usually both present; metulae enlarged, thick and wedge-shaped, varying from about 5 μm to 15–20 μm or more; phialides 5–8 × 2–3 μm; conidia globose to subglobose, smooth, 2.5–3.5 μm diam.; sclerotia quite numerous in some isolates but usually lacking, cream at first becoming purple to black.

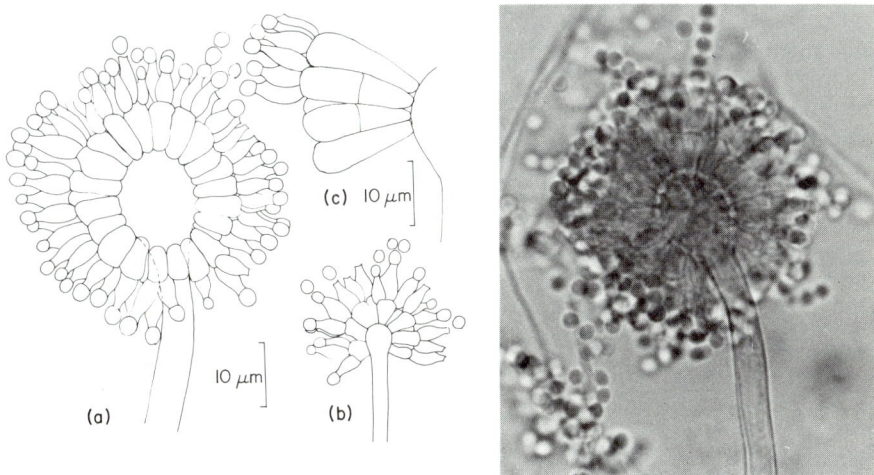

Fig. 10.20 (left) *Aspergillus candidus* **(a)** Conidial head; **(b)** small conidial head; **(c)** enlarged metulae and phialides.

Fig. 10.21 (right) *Aspergillus candidus* Conidial head, ×250.

The name covers a number of somewhat diverse forms with white or creamy white globose heads. Raper and Fennell consider there to be only the one species in the group. It resembles the *A. glaucus* group in its ability to grow on materials of very low water content and in its preference for culture media containing high concentrations of sugar. *A. candidus* is often found in moulding cereals.

In mounted specimens heads are very similar to those *A. niger* except for colour and the enlarged metulae.

A. *ochraceus* group

(Lat. *ochraceus,* ochre-coloured) (Figs 10.22, 10.23)
Perfect state *Syncleistostroma* Subramanian.

Growth, dusty, granular from massed conidial heads, pale yellow, orange-yellow to ochraceous or buff, with submerged mycelium colourless or various degrees of yellow or purplish; conidial heads globose at first, splitting into divergent columns of conidia; conidiophores faintly coloured yellow, usually coarsely roughened, long, 10 μm or more in diam.; vesicles globose, fertile over the entire surface; metulae and phialides usually both present but very tightly packed so that it is difficult to see their outline and only a shadow is seen in mounts where the metulae produce the phialides. Metulae vary in length but are normally 15–30 μm long; phialides 7–10 × 1.5–2.5 μm; conidia globose to elliptical, delicately roughened, faintly yellowish; cleistothecia: in the one ascosporic species described cleistothecia develop within black sclerotia after several months. This is very rare. Sclerotia are frequently produced and often characteristic in shape, size and colour, varying from white to yellow to brown, or pink to vinaceous purple to almost black.

The name covers a series of forms which are of common occurrence and wide distribution and are frequently found as spoilage organisms. Raper and Fennell describe nine species, separating them into three groups according to the conidial colour – pale pure yellow, bright golden yellow and ochraceous. In previous editions of this book the yellow forms were mentioned as *A. sulphureus* (Fresenius) Thom & Church. However, there are several species, of which this is one. They all produce sclerotia and include the species with a perfect state, *Syncleistostroma alliaceum* (Thom & Church em. Fennell & Warcup) Subramanian (=*A. alliaceus* Thom & Church em. Fennell & Warcup). They are all rare.

The majority of the ochraceous coloured isolates match the species *A. ochraceus* Wilhelm. However, one species, *A. melleus,* producing abundant pure yellow to brown sclerotia is quite common.

A. ochraceous Wilhelm produces fairly spreading ochraceous colonies, often developing a reddish purple colour in reverse; conidiophores distinctly rough, and long, vesicles globose with a close range of metulae, 15–30 μm long and phialides 7–10 × 1.5–2.5 μm; conidia small, globose and slightly roughened 2.5–3(3.5) μm diam. Sclerotia are often produced, but may not appear for several weeks, white to pale pink when young, darkening in age to lavender or vinaceous purple.

A. melleus Yukawa (=? *A. quercinus* (Bainier) Thom & Church. Raper and Fennell have some doubts about the actual identity of this species. It has the older name but sclerotia are not mentioned in the original description.) The conidial heads are smaller and slightly paler than in *A. ochraceous* and split into delicate columns. The colonies are usually yellow coloured due to the presence of numerous yellow, small sclerotia, 400–500 μm diam.

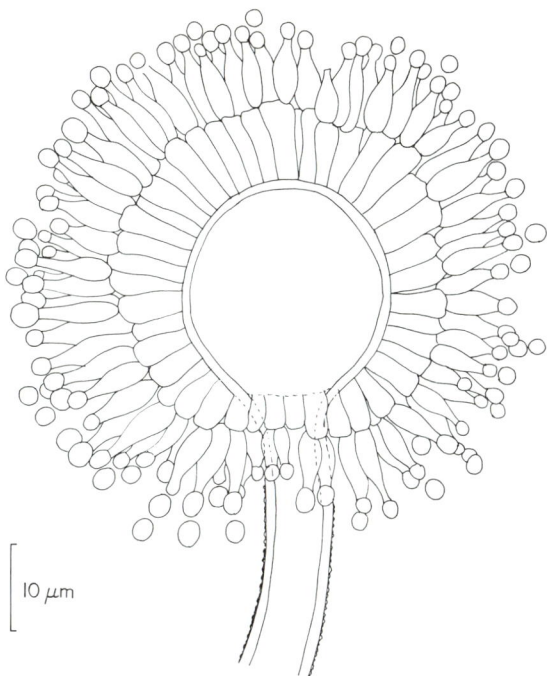

10 μm

Fig. 10.22 *Aspergillus ochraceus* Conidial head.

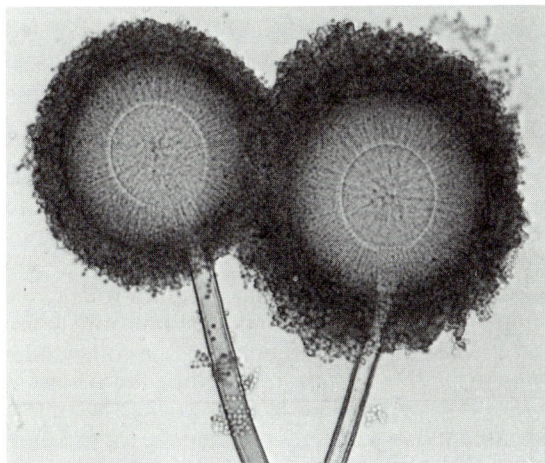

Fig. 10.23 *Aspergillus ochraceus* Typical heads, ×40.

A. *flavus-oryzae* group

(Figs 10.24–10.28)
No perfect state known.

Growth, colonies granular, dusty, light yellow-green to deep yellow-green to yellow-brown to brown; conidial heads globose, radiate or columnar; conidiophores colourless, usually roughened; vesicles globose to subglobose, though small ones remain flask-shaped, fertile over the entire surface; metulae and phialides, phialides may be borne directly on the vesicle or on metulae, both conditions may be seen in one species, isolate or even on one vesicle; conidia globose to subglobose or elliptical, finely roughened to rough, echinulate or tuberculate, often variable in size in one isolate; sclerotia often produced, globose to elongate, red-brown to brown to black.

This common group of species is of considerable economic importance. Some species are widely used in the manufacture of Japanese foods and beverages such as Sake, Miso, Shoyu, Mirin and Amazake. Others are spoilage organisms and when infecting cereals, grain or ground nuts may produce mycotoxins. It is therefore desirable to be able to distinguish one from another. This is not easy as each species and isolate shows considerable variation within the species and isolate, and there is an overlap of characters between species. Identification tends to rest on judging the preponderance and percentage of characters.

Characters of importance in making identifications are colour; the size, shape and marking of conidia and the proportion of heads with phialides only (uniseriate) to heads with phialides and metulae (biseriate). Most isolates show some heads with both conditions. If only a few percent are biseriate then the isolate is regarded as uniseriate. If more then a few percent, often about 20–30% are biseriate the isolate is regarded as biseriate.

Authors differ as to the number of species in the group. Raper and Fennell (1965) included nine species and two varieties. Murakami (1971) made an intensive study of the morphological, physiological and culture characters and submitted the results to component analysis by computer and found two different clusters, the *A. oryzae* group and *A. flavus* group. He placed the Koji moulds used in Japanese food manufacture in the *A. oryzae* group with three species and two varieties and strains other than the Koji moulds in the *A. flavus* group with three species. Kozakiewicz (1978, in the first of a series of proposed papers) is studying the group intensively, basing her work on stereoscan electron microscope studies.

Four species or species clusters are fairly distinct and are described here.

A. *flavus* Link ex Fries (Lat. *flavus,* yellow; a misnomer since the colony is green) (Figs 10.24, 10.25, 10.28). Colonies yellow at first quickly becoming bright to dark yellow green; conidiophores coarsely roughened, up to 1 mm long, and 19–20 μm diam., heads varying in size, loosely radiate or splitting or columnar, biseriate but having some heads with phialides borne directly on the vesicle; phialides 7–10 × 2–2.5 μm; conidia usually globose to subglobose, occasionally elliptical, conspicuously roughened, 3–6 μm diam. but mostly less than 4–5 μm diam. Some strains produce brownish sclerotia.

This is the commonest species in the group, being found on many kinds of organic matter in all parts of the world.

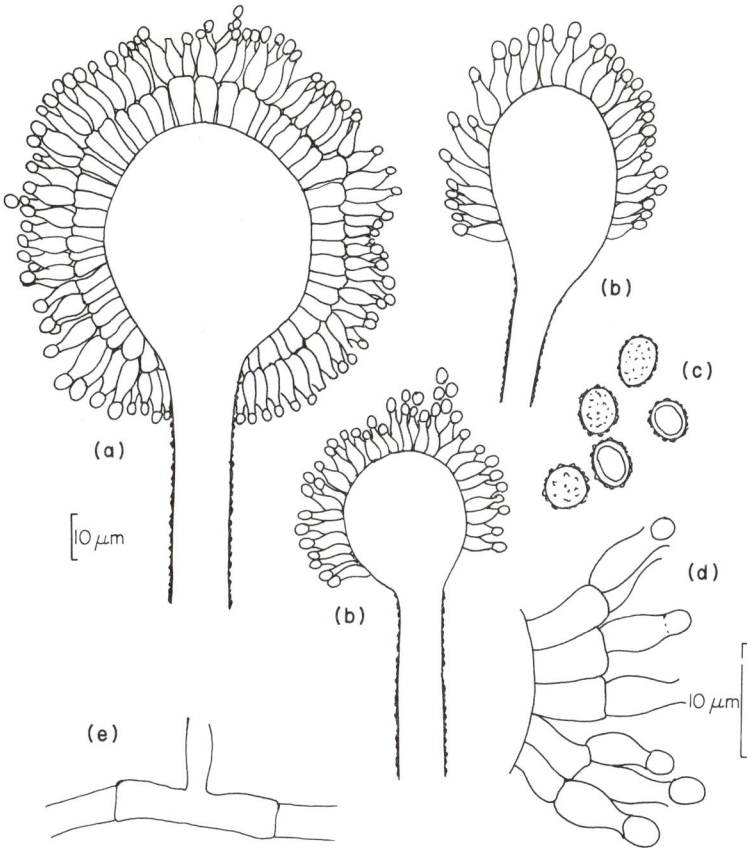

Fig. 10.24 *Aspergillus flavus* **(a)** Conidial head with metulae and phialides; **(b)** conidial head with phialides; **(c)** conidia; **(d)** enlarged metulae and phialides; **(e)** foot cell.

A. parasiticus Speare (Lat. *parasiticus*, parasite). Forms rich green colonies, deeper in shade than the normal *A. flavus*, with comparatively short conidiophores, mostly less than 400 μm, uniseriate, heads radiate, conidia globose, very rough or strongly echinulate, 3.5–5.5 μm diam. Conidia coloured pink when grown on media containing anisaldehyde.

A. oryzae (Ahlb.) Cohn (Lat. oryzae, of rice). (Fig. 10.28). Colonies seldom show true green, but are yellowish becoming yellowish to brownish green; conidiophores long often 2 mm long and up to 4–5 mm long, rather thin walled, usually rough; biseriate; conidia varying much in size but mostly more than 5–6 μm and up to 10 μm diam., globose to elliptical, usually very finely roughened. *A. oryzae* var. *effusus* (Tiraboschi) Ohara (Lat. *effusus*, loose in texture) has deeply floccose colonies, pale dirty greenish yellow, with reverse yellow to rose; conidiophores arise from aerial mycelium.

   A. oryzae and others of the *A. oryzae* group have been isolated frequently from the kojis in Japan, but seem rare elsewhere.

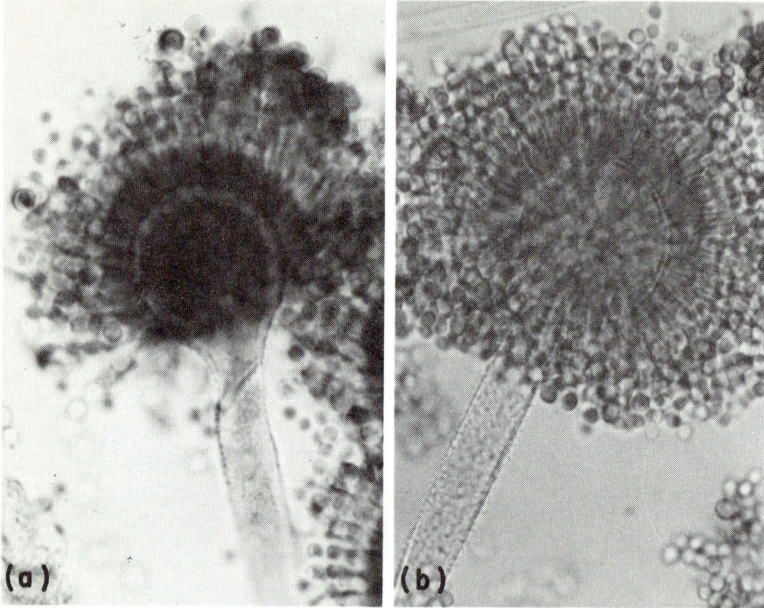

Fig. 10.25 *Aspergillus flavus* **(a)** Conidial head with phialides born directly on the vesicle, ×500; **(b)** conidial head with metulae and phialides, ×500.

A. tamarii Kita (Lat. *tamarii*, of Tamari, a kind of soy sauce) (Figs 10.26, 10.27). Mature colonies brownish green to brown, but often starting as a deep yellow-greenish brown, heads large, radiate, single chains of conidia visible under low magnification; vesicles thin-walled, easily crushed when mounting; metulae normally present but absent in some heads, not unusually mixed with phialides, metulae usually 7–15 μm but occasionally longer, phialides 7–10 × 3–6 μm; conidia dark, cylindrical to pyriform when young becoming globose in age, often adhering in chains in liquid mounts, 5.5–7.5 μm diam., coarsely roughened or tuberculate from nodules of brown colouring matter between the inner and outer walls.

In 1960, 100 000 young turkeys died of a mysterious disease, which was eventually found to be caused by the infection of groundnut feed with strains of the *A. flavus* group. Later tests showed that different strains vary much in toxicity, the ones producing most toxin corresponding well with *A. parasiticus*. The toxic substance has been called 'Aflatoxin'. A report of the Interdepartmental Working Party on Groundnut Toxicity (1962) gives a good account of the early work on the subject. Since then work on aflatoxin has continued and mycotoxins have been found to be produced by many moulds (see Table 10.1, p. 168). However, aflatoxin still seems to be the most important and most written about mycotoxin. The literature on mycotoxins is now enormous and work still continues. Brief surveys listing the main toxins produced and the organisms producing them are given by Austwick (1975) and in the Medical Research Council Memoranda of *Nomenclature of Fungi Pathogenic to Man and Animals* (1977). Recent comprehensive works which

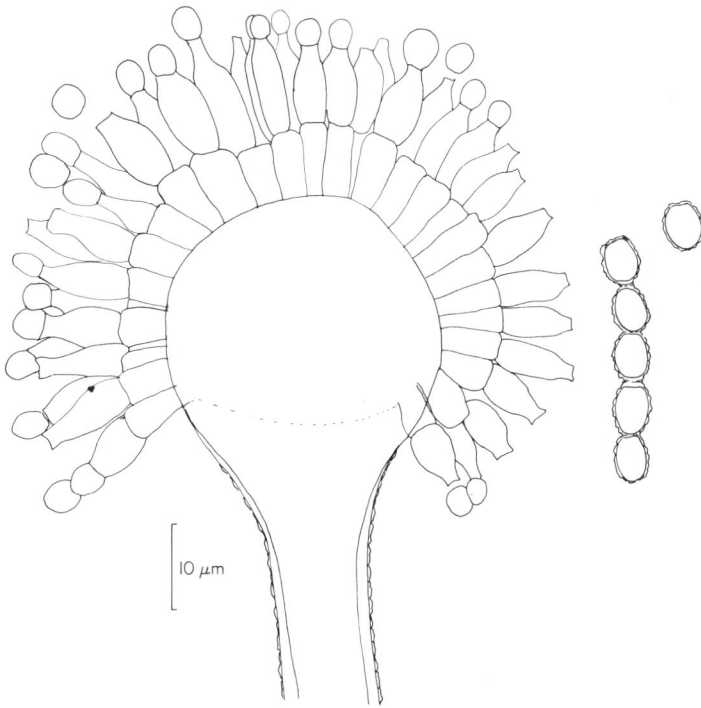

Fig. 10.26 *Aspergillus tamarii* Conidial head and conidia.

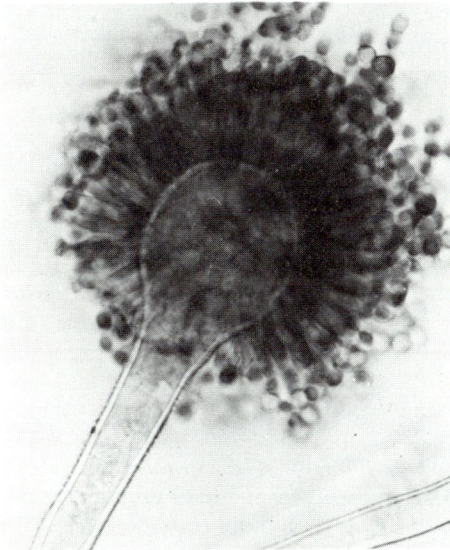

Fig. 10.27 *Aspergillus tamarii* Conidial head, ×470.

give good surveys of work up to the time of publication and good coverage of the literature are the Celanese Chemical Company Manual (1972), Goldblatt (1969), Wyllie and Morehouse (1977) and Ciegler *et al.* (1971).

Apart from the production of mycotoxins the *A. flavus-oryzae-tamarii* group of fungi are of importance in the production of amyolytic, proteolytic and lipolytic enzymes, in antibiotic production and vitamin production. They can cause disease of animals and insects as well as being common spoilage organisms in almost any situation (Raper and Fennell, 1965; Smith and Pateman, 1977).

An interesting chemical reaction of the *A. flavus* group is the production of kojic acid, which in water solution gives a very intense and characteristic blood-red colour with ferric chloride, and this reaction serves as a useful diagnostic test for strains in this group (Birkinshaw *et al.*, 1931; Murakami, 1971). The mould is grown in liquid medium (e.g. Czapek or Raulin-Thom) for several weeks, and a few drops of 10% solution of ferric chloride added to the filtrate to give the bright red colour.

A. wentii Wehmer

(F. A. F. C. Went, Dutch mycologist) (Figs 10.29, 10.30)
No perfect state known.

Growth dense, floccose masses of aerial mycelium are formed, white or tinged with pink or rose when cultures are exposed to light, often completely filling the culture tubes or plates. On some natural substrata it is found as a dense, furry, polychromatic growth; conidial heads large yellow to coffee-coloured, up to 500–800 μm diam., ragged looking or tangled; conidiophores colourless, usually smooth but can be granular, up to several millimetres in length and 10–25 μm diam.; vesicles globose, fertile over the entire surface; metulae and phialides usually both present; metulae varying in length but normally 10–20 μm long; phialides mostly 6–8 × 3 μm; conidia globose or slightly elliptical, usually rough but occasionally almost smooth, about 5 μm diam., yellow to yellow-brown.

The deep floccose growth with rather sparse long-stalked large coffee-coloured heads are characteristic. The species is described as cosmopolitan, occurring in soil and on various organic substrata such as moist grains. However, in the author's (AHSO) experience it is rare. Raper and Fennell (1965) describe four other species and varieties but these appear to be even more rare.

A. versicolor group

(Lat. *versicolor*, multi-coloured or changing colour) (Figs 10.31, 10.32)
No perfect state known.

Growth velvety to floccose, showing a wide range of colour even in one isolate and often patchy, but usually with some area a shade of green; conidial heads variable, radiate to loosely columnar, conidiophores colourless or slightly brown, smooth; vesicles ovate to elliptical, fertile over the upper half or three-quarters; metulae and phialides always present; conidia usually

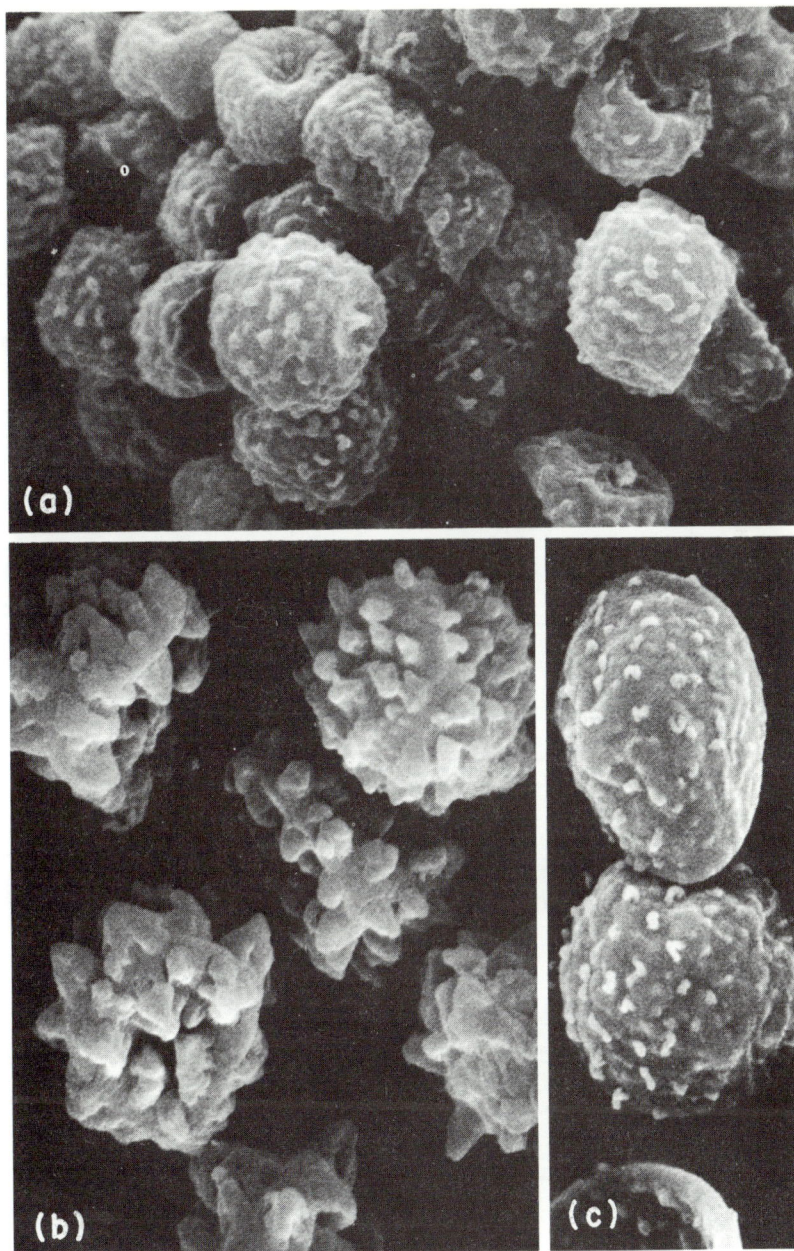

Fig. 10.28 *Aspergillus flavus* group   Stereoscan photographs of conidia, ×10 000.
(a) *A. flavus;* (b) *A. parasiticus;*(c) *A. oryzae.*

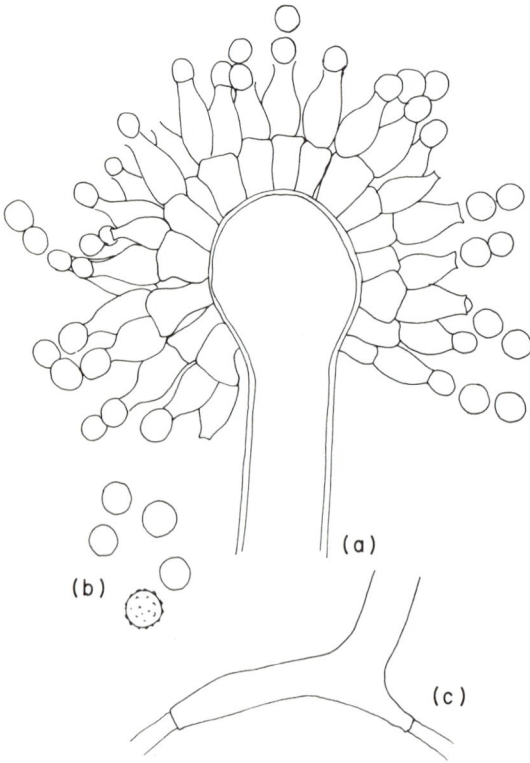

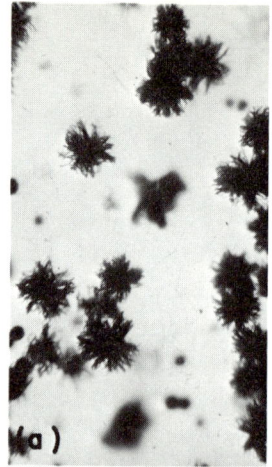

Fig. 10.29 (left)   *Aspergillus wentii*   **(a)** Conidial head; **(b)** conidia; **(c)** foot cell.

Fig. 10.30 (right)   *Aspergillus wentii*   **(a)** Mature heads as seen in Petri dish, ×20; **(b)** typical head, ×250.

globose, and rough; Hülle cells sometimes produced, globose.

Raper and Fennell regard the *A. versicolor* series as containing 17 species and one variety, but a number of the species are uncommon or are represented by a single isolate. However, two species are fairly common.

A. versicolor (Vuill.) Tiraboschi, is of world wide distribution and found in much the same situations as species of the *A. glaucus* group, but requires slightly more moisture for growth. As the name implies cultures show considerable range in colour, from pale green to yellow-green, grey-green, buff or even pink in small areas; reverse deep red or plum colour. Phialides, 5–10 × 2–2.5 μm; conidia globose, delicately roughened, usually 2.5–3 μm but occasionally larger, and turn a beautiful emerald green when mounted in lactophenol.

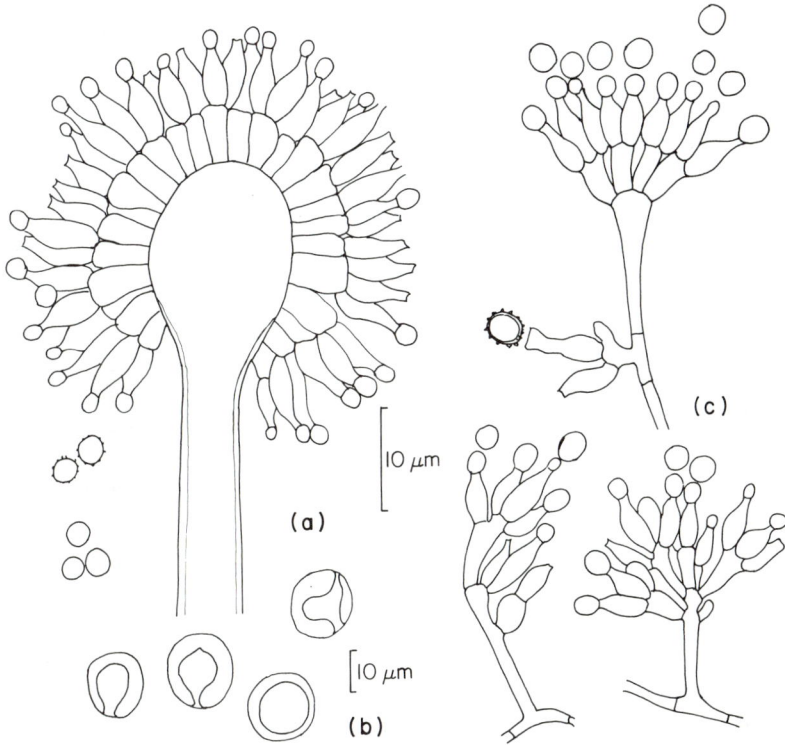

Fig. 10.31 **(a)** *Aspergillus versicolor* Conidia and conidial head. **(b)** *A. versicolor* Hülle cells **(c)** *A. sydowii* Reduced and proliferating conidial heads.

A. sydowii (Bainier & Sartory) Thom & Church (called after P. Sydow, a German mycologist). This species is closely allied morphologically to *A. versicolor,* but forms velvety colonies of a deep bluish green or greenish blue colour with reverse usually very deep red. It is easily mistaken for a *Penicillium* at first sight, until examined under the microscope. Dwarfed heads and even isolated metulae with their clusters of phialides are found in nearly all isolates, and in some isolates are produced almost exclusively.

A. nidulans group

(Lat. *nidulus,* a little nest; refers to the way the cleistothecia are embedded) (Figs 10.33–10.35)
Perfect state *Emericella* Berk. & Br.
    Growth, in most isolates, smooth, velvety, clear green colonies, developing dirty white spots from the centre outwards; reverse deep red to purple.

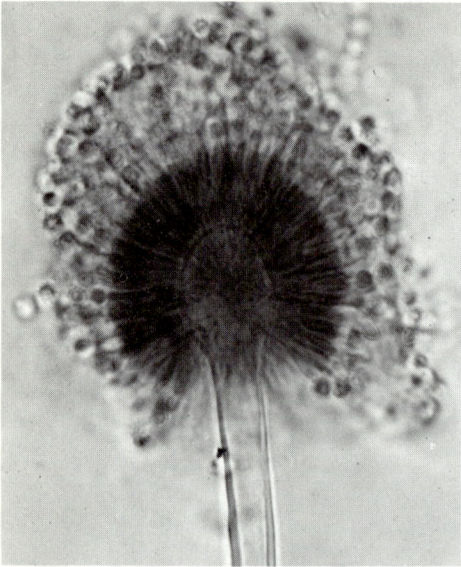

Fig. 10.32 *Aspergillus versicolor* Typical head, ×1000. *A. sydowii* is usually similar.

Sometimes conidial production is reduced and the purple colour of the cleistothecia dominates; conidial heads short columnar; conidiophores brown pigmented, sinuous, smooth with distinct foot cells, usually short; vesicles hemispherical; phialides borne on metulae which are borne on the vesicles, biseriate; conidia globose, rough, about 2.5–4 μm diam.; cleistothecia produced in most species, but may be lacking, globose, dark purple red, typically surrounded by a mass of thick walled cells, called Hülle cells, up to 25 μm diam.; ascospores 8 in an ascus, red to purple, with equatorial crests and convex surfaces smooth or ornamented.

A. nidulans (Eidam) Winter, perfect state *Emericella nidulans* (Eidam) Vuill. is the commonest species; typically light green with cleistothecia buried in off-white masses of Hülle cells. Phialides 5–6 × 2–2.5 μm, conidia 3–3.5 μm; ascospores red or purple with two entire crests and smooth convex surfaces, about 5 × 4 μm. It occurs in soil, as a common mould and has been used in studies on fungal genetics.

The majority of isolates encountered belong to *A. nidulans* proper, but Raper and Fennell (1965) describe 18 species and five varieties while Christensen and Raper (1978) include 30 taxa in their synoptic key to the group, most of these are rare. The following isolates are quite often seen.

A. rugulosus Thom & Raper (=*E. rugulosa*) with two entire crests and rugulose convex surfaces.

A. stellatus Curzi (=*E. variecolor*) with crests dissected and stellate and smooth convex surfaces.

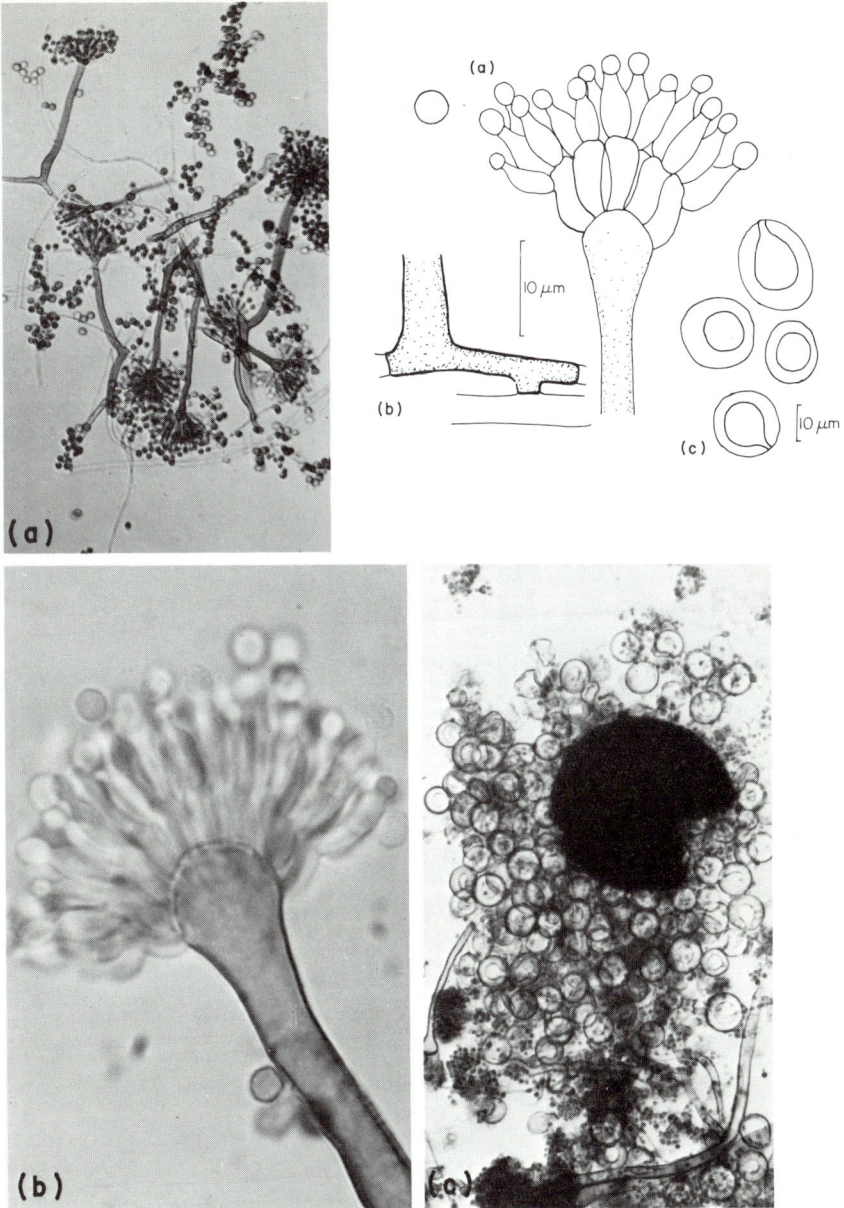

Fig. 10.33 (top right)   *Aspergillus nidulans*   **(a)** Simple conidial head with conidia; **(b)** foot cell; **(c)** Hülle cells.

Fig. 10.34   *Aspergillus nidulans*   **(a)** (top left) Usual appearance of cluster of conidial heads, ×250; **(b)** (bottom left) conidial head, ×1300; **(c)** (bottom right) perithecia with masses of Hülle cells, ×250.

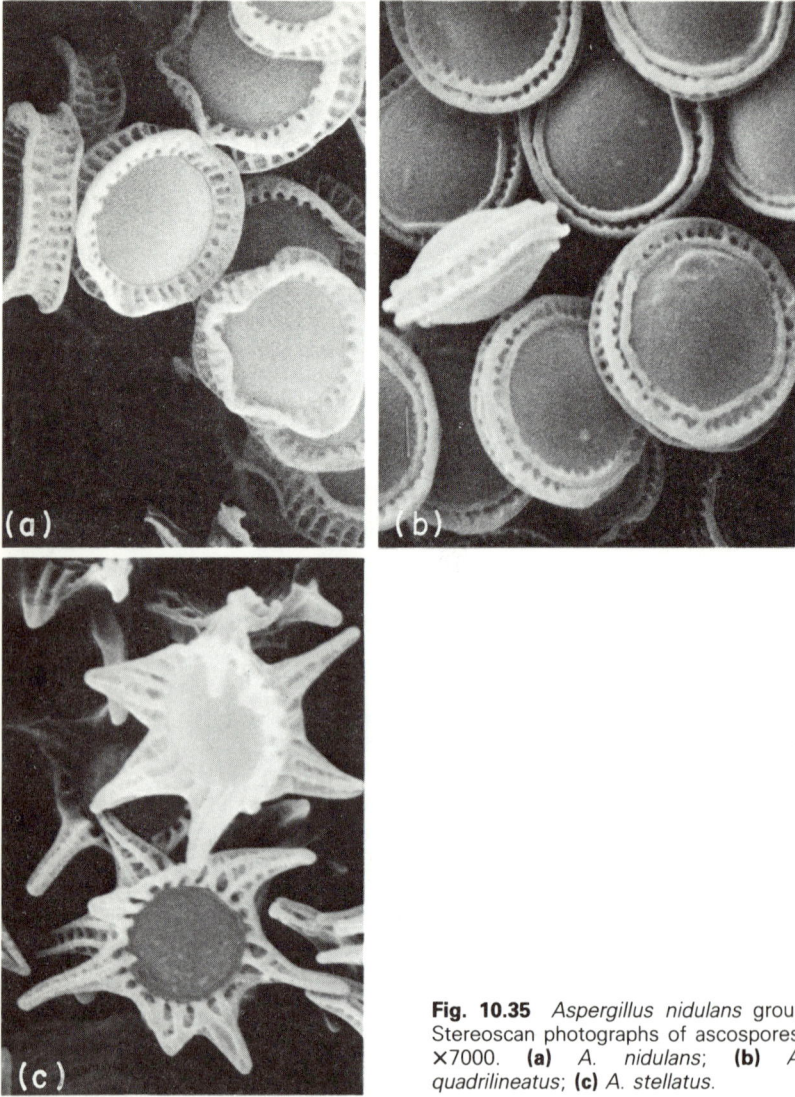

Fig. 10.35 *Aspergillus nidulans* group Stereoscan photographs of ascospores, ×7000. **(a)** *A. nidulans;* **(b)** *A. quadrilineatus;* **(c)** *A. stellatus.*

A. quadrilineatus Thom & Raper (=*E. quadrilineata*) with four very narrow crests. If an apparently crestless isolate is seen it is often found to have four small crests if examined carefully.

A. unguis produces conspicuous coarse encrusted hyphae similar to conidiophores with sterile tips; seldom produces cleistothecia.

A. ustus (Bainier) Thom & Church

(Lat. *ustus*, burnt; from the dark colour) (Figs 10.36, 10.37)
No perfect state known.

Growth floccose, at first brownish yellow, becoming brownish or purplish grey to almost neutral grey, with reverse yellow, dull reddish or purplish; conidial heads radiate to loosely columnar; conidiophores short, smooth, pale brown, 5–10 μm diam. arising from prominent foot cells; vesicles globose, fertile over upper two thirds; metulae and phialides both always present; conidia globose, rough, 3–3.5 μm occasionally up to 5 μm diam.; Hülle cells abundant in some isolates, rarely globose, mostly ovate or sausage shaped, and frequently bent and twisted.

Raper and Fennell recognize five species in the *A. ustus* series. The other four are not of sufficient importance to justify inclusion here.

A. flavipes (Bainier & Sartory) Thom & Church

(Lat. *flavus*, yellow; *pes*, foot) (Figs 10.38, 10.39)
Perfect state *Fennellia flavipes* Wiley & Simmons (1973).

Growth white or silvery white with patches of yellow, pale pink or mauve mycelium, white colour persisting or changing to pale buff or greyish buff; reverse yellow to orange brown or reddish brown; conidial heads becoming definitely columnar in age; conidiophores yellow or brownish yellow, smooth; vesicles subglobose to ovate or elongate, fertile over entire vesicle or only the

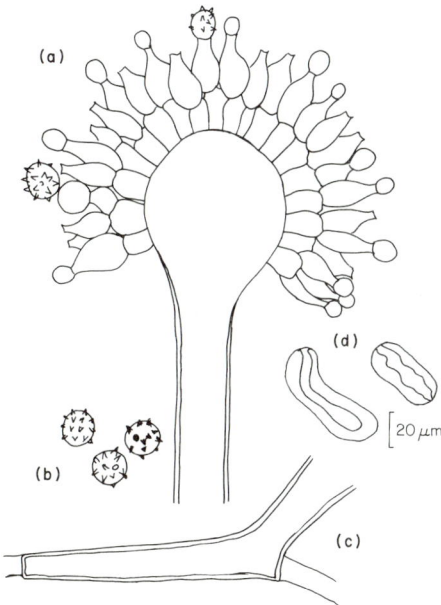

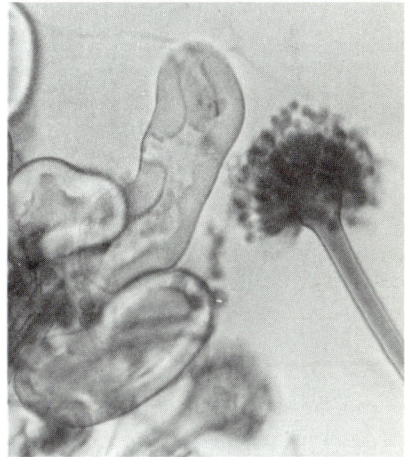

Fig. 10.36 (left)   *Aspergillus ustus*   **(a)** Conidial head; **(b)** conidia; **(c)** foot cell; **(d)** Hülle cells.

Fig. 10.37 (right)   *Aspergillus ustus*   Conidial head and Hülle cells, ×600.

upper part; metulae and phialides both always present, curved upwards; phialides 5–8 × 1.5–2.0 μm; conidia smooth, globose, 2–3 μm diam.; cleisto-thecia formed of yellow thick-walled cells; asci mostly 4 spored; ascospores colourless to pale yellow, subglobose, smooth, with inconspicuous equatorial groove, 6.4–8 × 5.6–6.4 μm. Some isolates produce mycelial knots suggesting sclerotia.

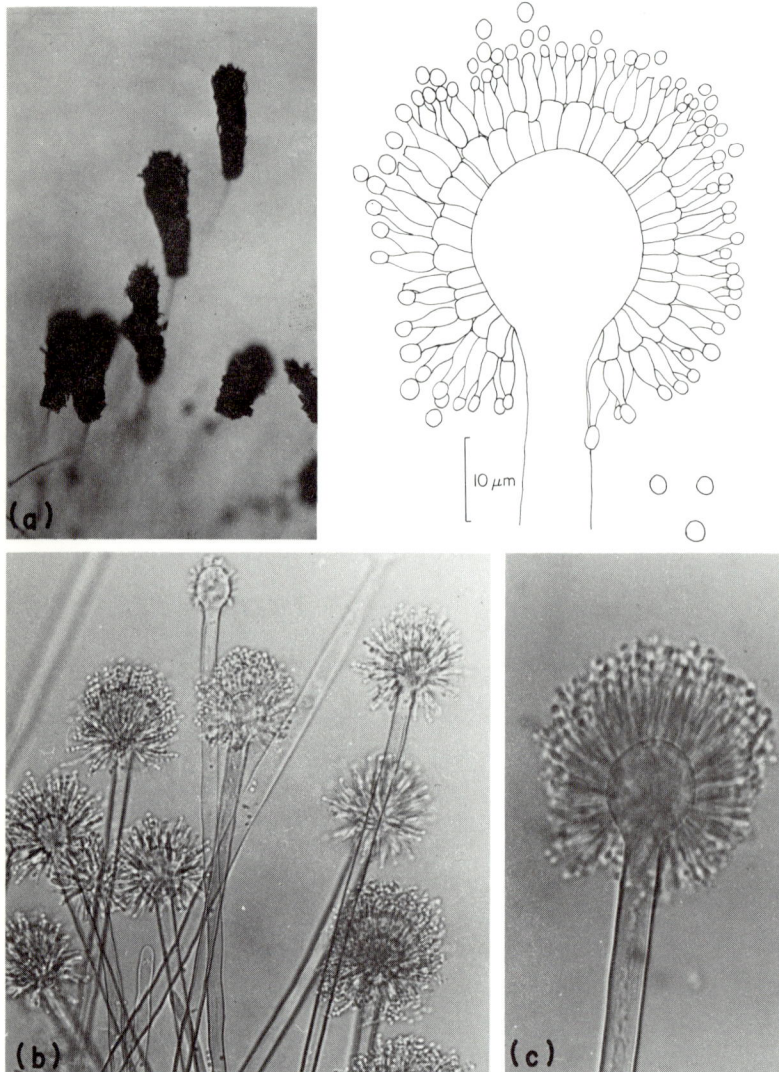

Fig. 10.38 (top right)   *Aspergillus flavipes*   Conidial head.

Fig. 10.39 *Aspergillus flavipes*   **(a)** Columnar heads as seen in old Petri dish culture, ×50; **(b)** cluster of conidial heads, ×250; **(c)** conidial head, ×550.

The buff forms, in old cultures, often approach the appearance of *A. terreus,* but the two are readily distinguished by examination of young vigorously growing colonies. The persistently white isolates are not likely to be confused with *A. candidus* owing to the difference in the shape of the heads.

Raper and Fennell recognize three species belonging to *A. flavipes* group, two of them transferred from the *A. terreus* group as described by Thom and Raper. These are *A. niveus* Blochwitz (with its recently described perfect state, *Emericella nivea* Wiley & Simmons, 1973) easily recognized by its persistent white colonies and loosely columnar heads; and *A. carneus* (van Tieghem) Blochwitz, with heads at first white, then vinaceous fawn, and stalks colourless or nearly so.

A. terreus group

(Lat. *terreus,* of the earth) (Figs 10.40, 10.41)
No perfect state known.

Growth velvety to floccose, buff, cinnamon to sand brown, reverse colourless to bright yellow; conidial heads strongly columnar; conidiophores smooth, colourless; vesicles hemispherical, fertile over upper half to two thirds; metulae and phialides both always present; conidia small, smooth, globose to subglobose; Hülle cells absent but thick hyphae are to be found in the basal mycelium and globose to ovate heavy-walled hyaline cells are produced in some isolates from the submerged vegetative mycelium.

Raper and Fennell accepted one species and two varieties. The majority of isolates belonging to this group are *A. terreus* Thom, but variants from the type occur from time to time (see key).

Key to varieties of *Aspergillus terreus* Thom.

1.	Colonies velvety, cinnamon to orange-brown; conidiophores short	2
1'.	Colonies floccose, golden yellow; heads small; conidiophores up to 500 µm long or even longer	*A. terreus* var. *aureus* Thom & Raper
2.	Sclerotium-like masses produced on malt agar	*A. terreus* var. *africanus* Thom & Raper
2'.	Sclerotium-like masses absent	*A. terreus* Thom

---

A. terreus Thom has colonies cinnamon to sand-brown, velvety; reverse a medium yellow to deep dirty brown, with brown colour more marked in the presence of zinc; heads columnar, up to 500 µm long and 30–50 µm diam.; vesicles dome-shaped, bearing closely packed metulae; phialides 5.5–7.5 ×1.5–2.0 µm; conidia globose or nearly so, about 2 µm diam. Figure 10.41 shows the heads as seen in living culture and Fig. 10.40a, drawn from a mounted specimen, the shape of the vesicle and disposition of phialides, which is very characteristic.

A. terreus is a common soil organism and is found frequently on all kinds of vegetable materials. It has been found on scores of samples of Egyptian cotton, and less frequently on American cottons, and on electronic equipment.

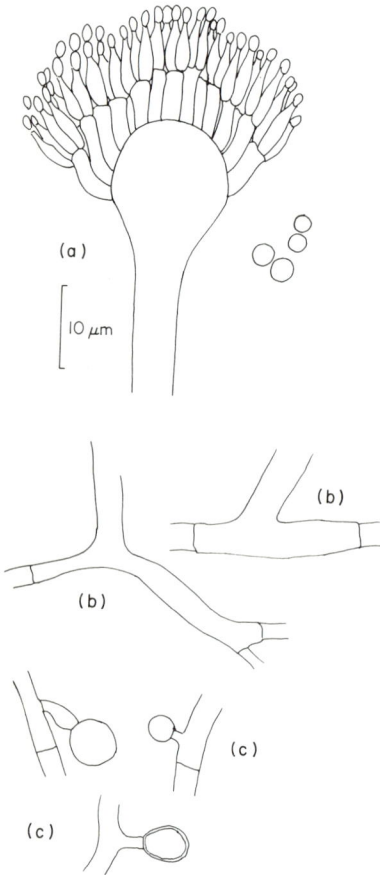

Fig. 10.40 (left) *Aspergillus terreus*
(a) Conidial head and conidia; (b) foot cell;
(c) cells on basal mycelium.

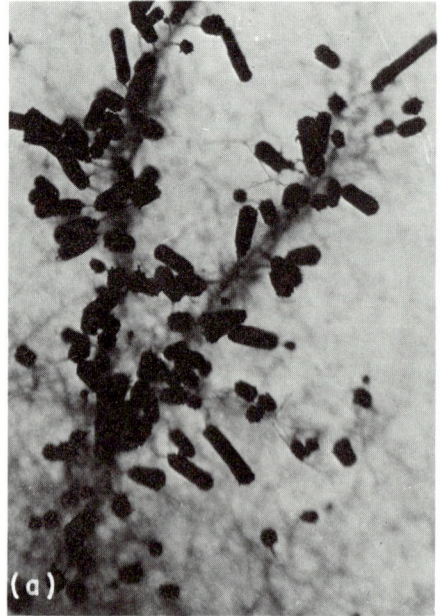

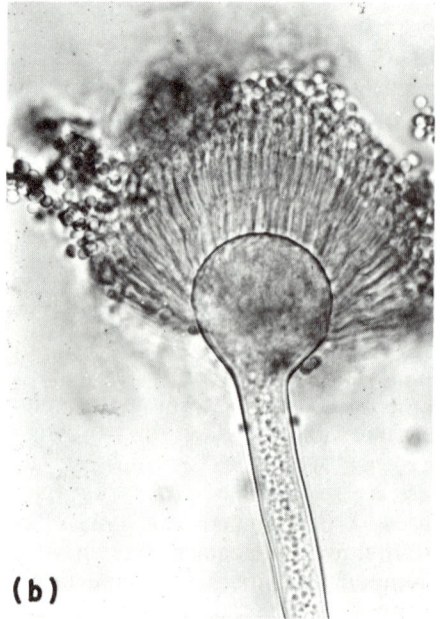

Fig. 10.41 (right) *Aspergillus terreus*
(a) Columnar heads as seen in Petri dish,
×50; (b) typical head, ×800

References

AUSTWICK, P. K. C. (1975). Mycotoxins. *Br. med. Bull.*, **31**, 222–9.

AUSTICK, P. K. C., GITTER, M. and WATKINS, C. V. (1960). Pulmonary aspergillosis in lambs. *Vet. Rec.*, **72**, 19–21.

BAINIER, G. and SARTORY, A. (1911). Étude biologique et morphologique de certains *Aspergillus. Bull. Soc. mycol. Fr.*, **27**, 453–68.

BAINIER, G. and SARTORY, A. (1912). Étude biologique et morphologique de certains *Aspergillus. Bull. Soc. mycol. Fr.*, **28**, 257–69.

BENJAMIN, C. R. (1955). Ascocarps of *Aspergillus* and *Penicillium. Mycologia*, **47**, 669–87.

BIRKINSHAW, J. H., CHARLES, J. H. V., LILLY, G. H. and RAISTRICK, H. (1931). Kojic acid (5-hydroxy-2-hydroxmethyl-y-pyrone). *Phil. Trans.*, Ser. B, **220**, 127–38.

BLASER, P. (1975 (1976)). Taxonomische und physiologische Untersuchungen über die Gattung *Eurotium* Link ex Fries. *Sydowia*, **28**, 1–49.

CELANESE CHEMICAL COMPANY (1972). *Master manual on moulds and mycotoxins.* BERG, G. L., ed. Farm Technology/Agri-Fieldman.

CHRISTENSEN, M. and RAPER, K. B. (1978). Synoptic key to *Aspergillus nidulans* group species and related *Emericella* species. *Trans. Br. mycol. Soc.*, **71**, 177–91.

CIEGLER, A., KADIS, S. and AJL, S. J. (Eds) (1971). *Microbial toxins, VI.* Academic Press, New York.

GOLDBLATT, L. A. (Ed.) (1969) *Aflatoxin, scientific background, control and implications.* Academic Press, New York and London.

INTERDEPARTMENTAL WORKING PARTY ON GROUNDNUT TOXICITY RESEARCH (1962). *Toxicity associated with certain batches of groundnuts.* Report of Agriculture Research Council, Department of Scientific and Industrial Research, Department of Technical Cooperation, Medical Research Council and Ministry of Agriculture, Fisheries and Food. London.

KOZAKIEWICZ, Z. (1978). Phialide and conidium development in the Aspergilli *Trans. Br. mycol. Soc.*, **70**, 175–86.

LOCCI, R. (1972). Scanning electron microscopy of ascosporic Aspergilli. *Riv. pat. veg. Suppl.*, Ser. 4, **8**.

MALLOCH, D. and CAIN, R. F. (1972). The Trichomataceae: Ascomycetes with *Aspergillus, Paecilomyces* and *Penicillium* imperfect states. *Canad. J. Bot.*, **50**, 2613–28.

MANGIN, L. (1909). Quèst ce que *l'Aspergillus glaucus? Ann. Sci. nat.* (Bot.), Ser. 10, **9**, 303–71.

MURAKAMI, H. (1971). Classification of the Koji mould. *J. gen. appl. Microbiol.*, **17**, 281–309.

MEDICAL RESEARCH COUNCIL MEMORANDA (1977). *Nomenclature of fungi pathogenic to man and animals.* (Compiled by AINSWORTH, G. C., AUSTWICK, P. K. C., BARLOW, A. J. E., CRUICKSHANK, C. N. D., GENTLES, J. C., MACKENZIE, D. W. R. and STOCKDALE, P. M.). No. 23: 4th edition. H.M.S.O., London.

RAPER, K. B. and FENNELL, D. I. (1965). *The genus Aspergillus.* Williams & Williams, Baltimore.

SAMSON, R. A. (1979). A compilation of the Aspergilli described since 1965. *Studies in Mycology,* No. 18, CBS. 38pp.

SINGH, R. (1973). The foot cell morphology of genus *Aspergillus*. *Mycopath. et Mycol. appl.,* **49,** 209–15.

SMITH, G. (1933). Some new species of *Penicillium. Trans. Br. mycol. Soc.,* **18,** 88–91.

SMITH, J. E. and PATEMAN, J. A. (Eds) (1977). Genetics and Physiology of *Aspergillus. The British Mycological Society Symposium Series,* No. 1. Academic Press, London, New York, San Francisco.

SUBRAMANIAN, C. V. (1972). The perfect states of *Aspergillus. Current Science,* **41,** 755–61.

THOM, C. and CHURCH, M. B. (1926). *The Aspergilli.* Williams & Wilkins, Baltimore.

THOM, C. and RAPER, K. B. (1941). The *Aspergillus glaucus* group. *Misc. Publ. USDA,* **426,** 1–46.

THOM, C. and RAPER, K. B. (1945). *A manual of the Aspergilli.* Williams & Wilkins, Baltimore.

WILEY, B. J. and SIMMONS, E. G. (1973). New species and a new genus of Plectomycetes with *Aspergillus* states. *Mycologia,* **65,** 934–8.

WYLLIE, T. D. and MOREHOUSE, L. G. (Eds) (1977). *Mycotoxic Fungi, Mycotoxins, Mycotoxicoses. An Encyclopaedic Handbook Vol. 1. Mycotoxic fungi and chemistry of mycotoxins.* Marcel Dekker Inc., New York and Basal.

11

Penicillium and Related Genera

Alive and actively growing, they have individuality as pronounced as their capacities for evil, but the elements of that individuality, color, odor, and habit of growth, are as evanescent as frost designs on a window pane in winter.

C. Thom. *The Penicillia*, 1930

11.1 Introduction

The Penicillia are closely related to the Aspergilli and are just as widespread and omnivorous. Like the Aspergilli, many of them are common and serious agents of destruction, but on the other hand, there are several species which are best known as the means of conducting industrial fermentations. Probably the most important of these are the cheese moulds, the *P. roquefortii* and *P. camembertii* series. Thom's studies (1910, 1930) of the occurrence and mode of action of these moulds has made it possible for production of cheeses of the Gorgonzola, Stilton, Roquefort, and Camembert types to be carried on anywhere with success, instead of being confined to special regions. The first attempts to produce citric acid commercially by mould fermentation used Penicillia, of the group formerly known as *Citromyces,* although modern citric acid manufacture is carried on with the aid of *Aspergillus niger,* owing to the better yields obtained. Other successful large-scale fermentations are the manufacture of gluconic acid, using a member of the *P. purpurogenum* series, and of penicillin, using *P. chrysogenum* or *P. notatum.* The studies of Raistrick and his colleagues have shown that the Penicillia exhibit the most varied metabolic activity, different species being capable of synthesizing from glucose a bewildering variety of substances of complex chemical constitution. Griseofulvin, the only effective antibiotic available for the systemic treatment of fungal infections of skin (ringworm), hair and nails was first produced by *P. griseofulvum.* Amongst many interesting compounds already obtained may be mentioned the production of ergosterol, the parent substance to Vitamin D, and a number of substances closely related to ascorbic acid (Vitamin C). Since the discovery of the importance of mycotoxins numerous toxins have been found to be produced by *Penicillium* species.

As G. Smith says in the previous edition (6th), 'Unfortunately the taxonomy of the genus presents much greater difficulty than that of the related genus *Aspergillus.* In the latter genus there is a wide range of colour to serve as a convenient basis of classification, whereas the great majority of the species of *Penicillium* are some shade of green. Also, whereas the conidial colours of the Aspergilli are comparatively stable throughout the growing period, the greens of the Penicillia change as colonies age and, in addition, vary with changes in cultural conditions. Further, the number of species of *Penicillium* is appreciably greater than the number in *Aspergillus*. The number of specific epithets which have been bestowed runs to many hundreds and, even after many of these have been relegated to synonymy, and others abandoned because of hopelessly inadequate descriptions, there remain, according to the latest taxonomic study of the genus, 137 species and a few varieties.' (Smith referred to Raper and Thom (1949) and since then many species have been added. Kulick (1968) listed descriptions of another 150 new species and still more have been added since his collection.)

In view of the ubiquity and importance of the Penicillia it is not surprising that many taxonomists have attempted to devise a satisfactory scheme of classification. Many of the papers and monographs published in the first half of the century are mainly of historical interest but Thom's monographs of 1910 and especially of 1930 did much to lay a basis for classification, and that of 1930 is still a very useful source of information on early isolates and descriptions. His work culminated in the monograph by Raper and Thom (1949), now out of print, which has for the last 30 years formed the basis of accepted classification. It is still extensively used but many new species have since been described and the genus has become of increasing importance. By virtue of the character of the species they are still difficult to identify and various authors have made attempts to simplify taxonomy of sections of the genus. The ascosporic species have been given names under their perfect states and were the subject of a series of papers, including Stolk and Scott (1967) on the sclerotigenic ascosporic species and Stolk and Samson (1971, 1972) on the ascosporic species with soft-walled cleistothecia. This was followed by studies on the fasciculate asymmetric species by Fassatiova (1977) and Samson, Stolk and Hadlock (1976), on *Penicillium* species from fermented cheese by Samson, Eckardt and Orth (1977) and on the *P. chrysogenum* series by Samson, Hadlok and Stolk (1977).

At present there is no recent monographic treatment of the entire genus but Pitt (1979) is about to produce one. With such a large genus it seems likely, if this work is accepted, that it will stand for some years, but eventually with a genus like *Penicillium*, which shows so much variation, as knowledge accrues and with the use of biochemical techniques and new technology, more revisions of the genus will follow. In the long run the species confines are man made concepts for convenience in pigeon-holing isolates for differentiation and reference purposes.

11.2   Generic description

Penicillium Link ex S. F. Gray

(Lat. penicillus, an artist's brush) (Figs 11.1, 11.2)

The following diagnosis of the genus given by Smith in 1969 (in the previous edition of this book) in accordance with the ideas of Raper and Thom still defines the genus and introduces a few terms which are commonly used in descriptions of species.

Vegetative mycelium colourless or pale or brightly coloured, never dematiaceous, septate, either predominantly submerged or partly submerged and partly aerial, with aerial portion closely matted, loosely floccuse, or partially as ropes of hyphae; fertile branches (conidiophores, stipes) arising from, and more or less perpendicular to, submerged or aerial hyphae, either detached from one another or to some degree aggregated into fascicles or compacted into definite coremia, septate, smooth, or rough, terminating in a broom-like whorl of branches (the penicillus) (Fig. 11.1), the latter consisting of a single whorl of spore-bearing organs (phialides), or twice to several times verticillately branched, with the branching system symmetrical or asymmetrical, the final branches being the phialides; conidia produced in basipetal unbranched chains, globose, ovoid, elliptical, or pyriform, smooth or rough, in most cases green during the growing period but sometimes colourless or in others pale colours; perithecia produced by some species, either sclerotium-like, ripening tardily from the centre outwards, or soft, ripening quickly; sclerotia produced by several species.

In penicilli with more than one stage of branching, branches bearing the phialides are, as in *Aspergillus,* called metulae. The branches supporting the metulae, if comparatively short and obviously part of the penicillus, are known as rami (Fig. 11.2).

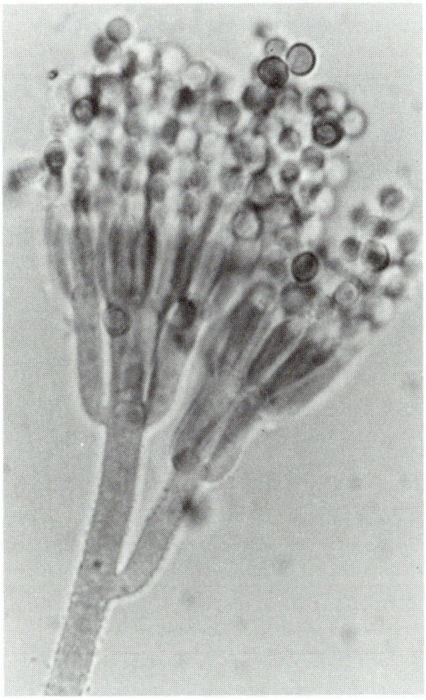

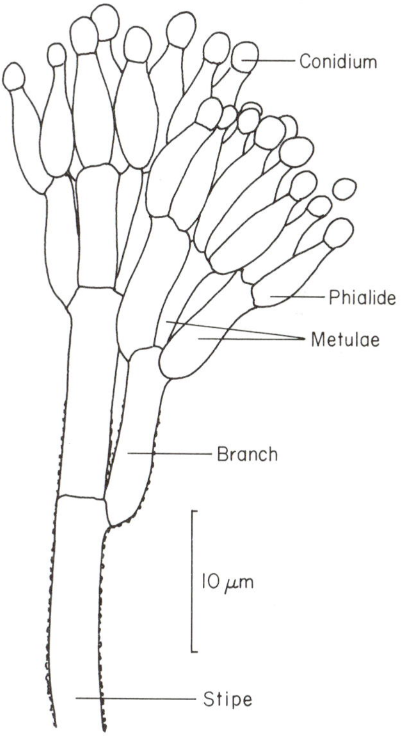

Fig. 11.1 (left) *Penicillium (P. cyclopium)* A typical 'penicillus' with several stages of branching, ×1320.

Fig. 11.2 (right) *Penicillium (P. cyclopium)* Drawing of a typical 'pencillus' showing stages of branching – stipe, branches, metulae, phialides and conidia.

11.3 The perfect states or holomorphs

Langeron (1922) proposed the generic name *Carpenteles* for a species, described by Brefeld in 1874 as *Penicillium glaucum*, producing perithecia which originated as sclerotium-like bodies and gradually formed asci from the centre outwards, and Shear (1934) has revived the name to cover a number of species of similar morphology. Raper and Thom do not recognize the genus but have adopted the name *Carpenteles* for a series of species which produce ascospores. Benjamin (1955) proposed reviving the use of the name *Carpenteles*. In addition, he erected the new genus *Talaromyces*, to include the species of the *P. luteum* series which form soft, mostly yellow, perithecia. Raper (1957) expressed strong disapproval of these suggestions, and argued that the needs of taxonomy are best met by retaining the name *Penicillium* for all the species, whether perithecial or entirely conidial.

It is accepted nomenclatural practice to use the name of the perfect state (teleomorph) where possible. However, it is hoped that ascosporic species with *Penicillium*-like imperfect states (anamorphs) will in future be included in any major works on *Penicillium* as the *Penicillium* state is often the most noticeable and many workers will find it convenient to look for these species in this genus. Several papers relating *Penicillium* species to perfect states have been published and they are to be found in more than one ascosporic genus. Table 11.1 is a list of some of these genera.

Table 11.1 Teleomorphic and anamorphic names of *Penicillium* and allied genera

Teleomorph (perfect)	Anamorph (imperfect)
Talaromyces Benjamin	*Penicillium*
Hamigera Stolk & Samson	*Penicillium*
Byssochlamys Westling	*Paecilomyces*
Thermoascus Miehe	*Paecilomyces*

11.4 Genera closely related to *Penicillium*

Raper and Thom (1949) included three common genera which produce penicillate conidial states. Indeed there are many other genera which do so but most of them are not common. Of these *Paecilomyces* and *Gliocladium* produce conidia by means of phialides and are dealt with here. *Scopulariopsis* has been included in the chapter on Hyphomycetes, as the conidiogenous cells have been shown by Hughes (1953) to be annellophores.

The perfect state (teleomorph) of *Gliocladium roseum* was discovered by Smalley and Hansen (1957). It is a species of *Nectria*, to which they gave the name *N. gliocladioides*. It is interesting that the three genera, originally separated on differences in conidial morphology, should prove to have perfect states differing from the known perfect states of *Penicillium*, and differing widely from each other. The perfect state of *Gliocladium* is *Nectria*, with bright coloured perithecia and two-celled ascospores; that of *Paecilomyces* is *Byssochlamys*, which produces irregular clusters of asci, without any trace of peridium (containing wall) and one-celled ascospores; whilst that of

Scopulariopsis is *Miscroascus*, with dark, carbonaceous perithecia and one-celled ascospores.

Gliocladium Corda
(Gr. *gloios*, mucilage; *klados*, a branch)

This genus differs from *Penicillium* in that a mucilaginous substance is produced by the fruiting organs, and the conidia, instead of standing in detached chains, adhere together, or, in the most typical species, lose entirely the catenary formation and form solid, slimy balls. The line of separation from *Penicillium* is not a sharp one, but in the more common species the distinctive character is well marked and unmistakable.

In addition to producing penicillate branching *Gliocladium* species often produce a verticillate branching system which can easily be confused with *Verticillium* or *Trichoderma*, but some penicillate branching systems are always present.

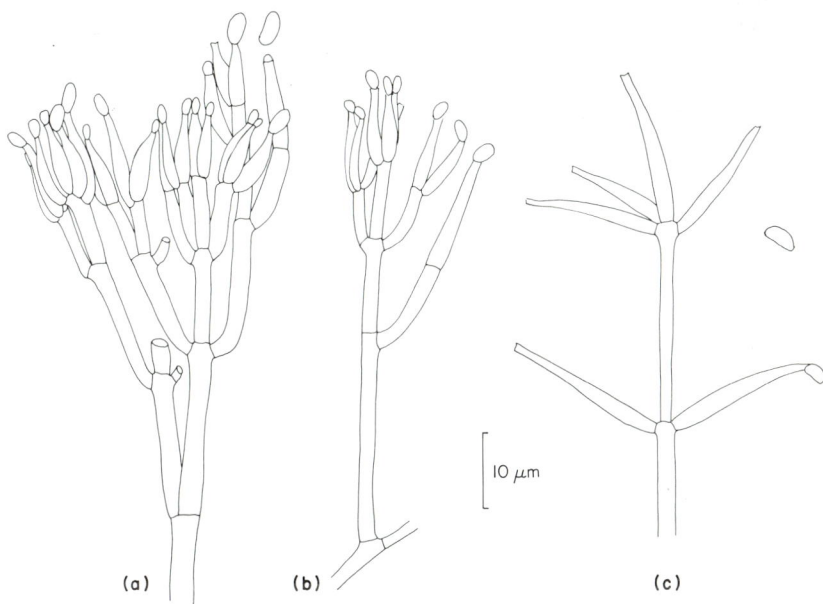

Fig. 11.3 *Gliocladium roseum* **(a)** Complex head with typical 'Gliocladium' type branching; **(b)** simple head; **(c)** simple head with verticillate branching.

G. roseum (Link) Bainier (Lat. *roseus*, pink) (Figs 11.3–11.5). Colonies are loosely floccose in texture, pale pink to salmon becoming green when first isolated, often becoming almost white on continued cultivation, with asymmetrical conidia 5–7 × 3–5 μm, forming slime balls, in some strains quickly, in others tardily. Barnett and Lilly (1962) have shown that this species is a destructive parasite of a number of other fungi.

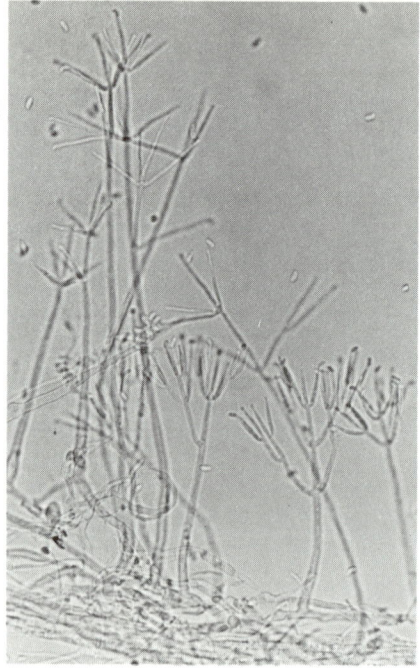

Fig. 11.4 (left)   *Gliocladium roseum*   Typical conidial heads, ×250.

Fig. 11.5 (right)   *Gliocladium roseum*   Verticillate branching, ×420.

G. deliquescens Sopp (Lat. *deliquescens,* melting, dissolving) (Fig. 11.6). Colonies thin, with little aerial mycelium, soon forming dark green fruiting areas and eventually becoming completely slimy; conidia green, 3–3.5 × 2–2.5 μm.

G. catenulatum Gilman & Abbott (Lat. *catenula,* a little chain). Colonies floccose to funiculose, white at first, becoming clear dark green in scattered patches; conidia chains in long slimy columns; conidia greenish, 4–7.5 × 3–4 μm.

In all the species the penicilli are mostly three to four times verticillate, with the branches becoming more slender in successive whorls (see Figs 11.4 and 11.6).

The three species described are all fairly common in soil and have been isolated many times from mouldy fabrics and various items of military equipment.

Paecilomyces Bainier
(Gr. *poikilo-,* varied) (Figs 11.7–11.11)

Colony colour white, pale pink, lilac, yellow-brown, or dirty greyish brown,

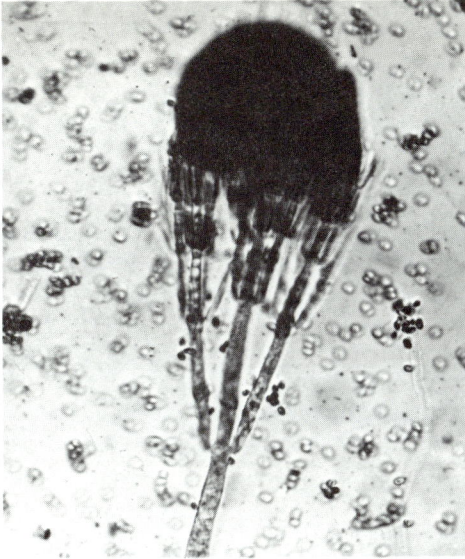

Fig. 11.6 *Gliocladium deliquescens* Conidial head, ×400.

never green; texture closely matted, loosely floccose, or funiculose. Fruiting organs vary considerably in complexity (much as in *Scopulariopsis*), and the conidia are usually ovate and borne in very long tangled chains. The phialides are of very characteristic shape, narrow flask-shaped at the base and terminating in long slender spore-bearing tips, which are usually bent away from the main axis (Fig. 11.7). Some species also produce so-called 'macro-spores', which are comparatively large, globose or ovate, aleuriospores, borne singly or in small clusters, and usually found on the mycelium close to the substrate or even in the submerged mycelium.

Brown and Smith (1957) showed the name *Spicaria* to be a nomen ambiguum and transferred a number of species originally described as *Spicaria* to *Paecilomyces*. Samson (1974) made a comprehensive reappraisal of *Paecilomyces* and some allied Hyphomycetes, placing *P. elegans,* an elegant species with coloured conidiophores and white conidia in chains, in a separate genus, *Mariannea*, and transferring the not uncommon *P. fusisporus* to *Acrophialophora.*

Several species are common spoilage organisms while others occur in soil or as pathogens of insects.

P. variotii Bainier (Dr. Vario, French physician) (Fig. 11.8) is the commonest of all the species especially as a spoilage fungus. Colonies grow well on almost any kind of substrate and are pale dull brown or yellowish brown, loosely floccose and mostly ropy; phialides 15–20 μm long; conidia elliptical, 5–7 × 2.5–3 μm, formed in very long chains. Some isolates produce abundant macrospores, others few or none. Also the complexity of the sporing structures varies considerably.

Byssochlamys fulva (see Chapter 5) has a very similar conidial state but the conidia are cylindrical with truncate base.

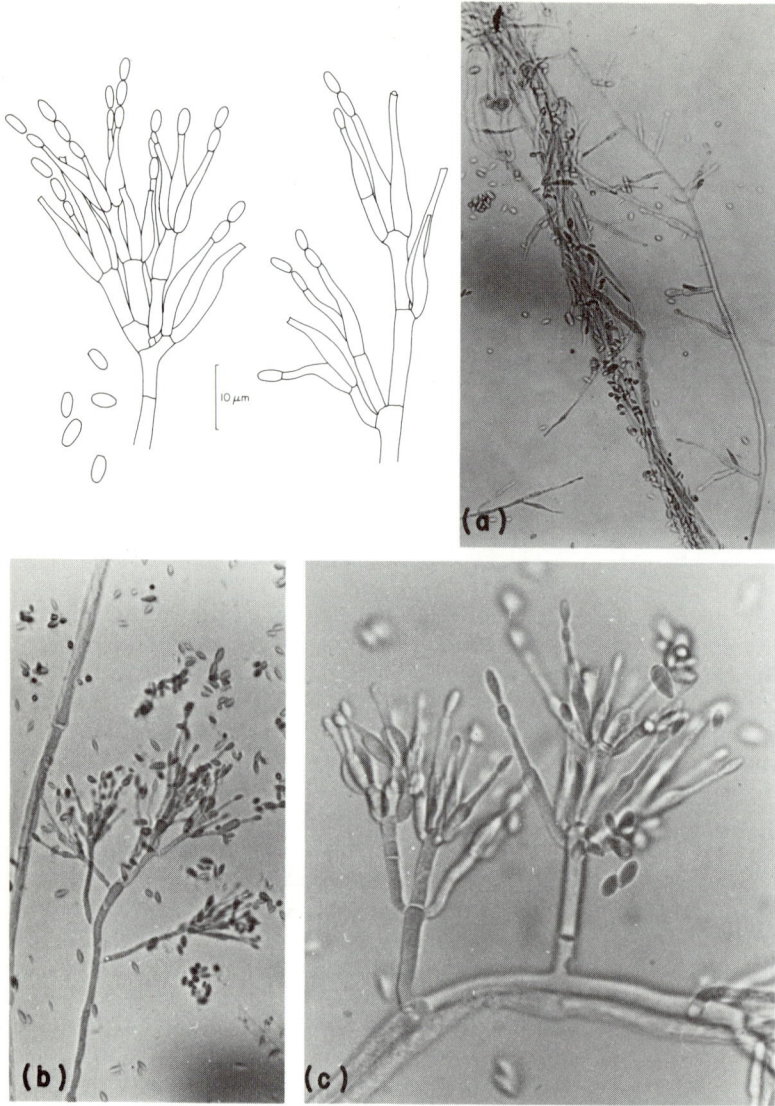

Fig. 11.7 (top left) *Paecilomyces variotii* Conidial heads with irregular branching, tapered phialides and conidia.

Fig. 11.8 *Paecilomyces variotii* **(a)** Ropes of hyphae and various types of spore bearing structures, ×250; **(b)** complex conidial head, ×250; **(c)** conidial heads, with typically tapered phialides, ×550.

P. carneus (Duché and Heim) Brown and Smith (Lat. *carneus*, flesh coloured) (Fig. 11.9). Colonies grow slowly, compact, matted, showing very little tendency to form funicles of hyphae, pure white at first, becoming pale clear pink, especially if exposed to light, with reverse slowly turning deep green; sporing structures simple or fairly complex; phialides mostly 12–16 μm long; conidia elliptical, definitely rough, 3–4 × 2–2.5 μm. Found on a wide variety of substrates, especially soil.

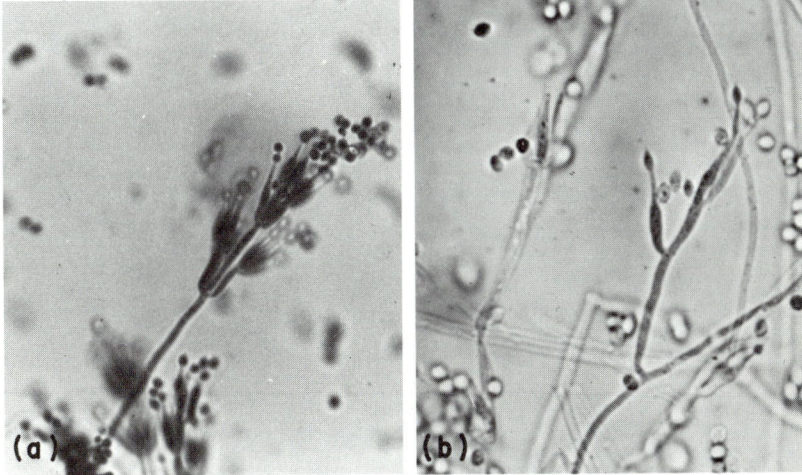

Fig. 11.9   *Paecilomyces carneus*   **(a)** Conidial head, ×550; **(b)** conidial head, ×800.

P. fumosoroseus (Wize) Brown & Smith (Lat. *fumosoroseus*, smoky pink) (Fig. 11.10). Colonies spreading, deeply floccose, white at first then gradually dull pink, with reverse slowly turning pale yellow; sporing structures consisting of solitary phialides or simple verticils; phialides 7–18 μm long; conidia almost cylindrical, 3–4 × 1–2 μm. Best known as a parasite of insects, on which it forms synnema and it has been found on a number of other substrata.

P. lilacinus (Thom) Samson (Lat. *lilacinus*, lilac-coloured) (Fig. 11.11). Originally described by Thom as *Penicillium lilacinum* and was treated as that by Raper and Thom, and Brown and Smith. However, it is a borderline species and was transferred to *Paecilomyces* by Samson (1974). It is a common soil organism and is also found on a variety of organic substrates including plastic contact lenses. Colonies are floccose at first white, then gradually purplish or pale greyish violet; reverse in some strains fairly dark purple, in others only slightly coloured; penicilli varying from complex structures to simple clusters of phialides on short branches from trailing hyphae; phialides 7–8 μm long; conidia elliptical, smooth, 2.5–3 × 2 μm.

P. marquandii (Massee) Hughes is described in Raper and Thom next to *P. lilacinus* as *Spicaria violacea* Abbott. It is distinguished by forming colonies of bluer surface colour, and with reverse bright yellow instead of purple.

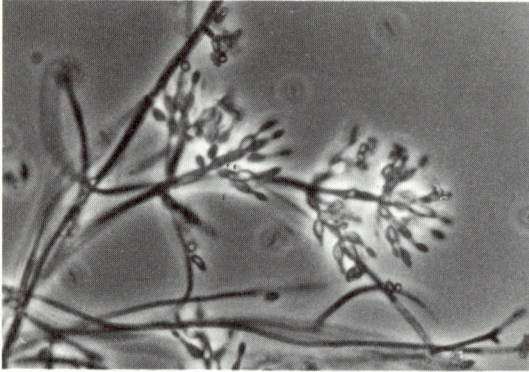

Fig. 11.10 *Paecilomyces fumosoroseus* Conidial head, ×500.

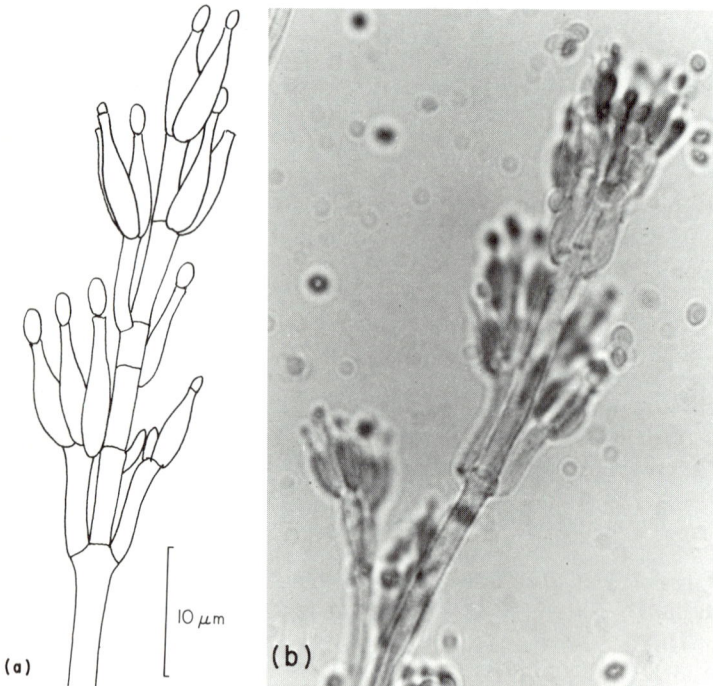

(a)

10 μm

(b)

Fig. 11.11 *Paecilomyces lilacinus* **(a)** Conidial head; **(b)** conidial head, ×1200

P. farinosus (Holm ex S. F. Gray) Brown & Smith (Lat. *farina*, meal, flour). Colonies spreading, deeply floccose becoming mealy and dusty, often developing loose synnema in age, sometimes tinged yellow; reverse becoming yellow; sporing structures simple to complex with verticillate branches bearing whorls of phialides, 5–15 × 1.2–2.5 μm; conidia ellipsoidal, smooth-walled, hyaline, 2–3 × 1–2 μm. Usually occurring on insects, on which definite synnema are formed, but may be isolated from soil. It is a well known insect pathogen.

11.5   The genus *Penicillium*

Raper and Thom (1949) divided the large number of species into three main sections according to the branching systems. In turn these were further subdivided according to branching and texture of the colonies. They were then separated by other characters such as presence or absence of perithecia or sclerotia, conidial shape, size and marking, colony colour, stipe structure and marking. Raper and Thom's keys did not lead to species but to 'series', a series including one to several closely related species.

For many purposes it is sufficient to place a particular isolate in its correct series rather than to attempt a closer identification and in some of the newer treatments of sections of the genus quite a number of the old species are regarded as synonyms. One of the difficulties in identification of *Penicillium* species is that they appear to be to some extent unstable, and some isolates tend to vary slightly in successive cultures, though they are usually approximately stable over periods long enough to permit identification. When trying to identify a *Penicillium* it is frequently impossible to find an exact fit for the isolate in any published description, although it is readily placed 'near to' two or perhaps three species previously described. Thus when relatively few isolates have been seen they may all appear to be new, which may account for the many names in the literature. Thom (1930) showed that when hundreds of isolates are examined, differences become less perceptible and hence his introduction of the series concept.

It is emphasized again and again by Raper and Thom that the accepted species in any series represent little more than convenient centres around which to group the numerous different strains, and it is a great merit to their book that the amount and type of variation to expect is discussed for the great majority of species, and certainly for all the common species.

Raper and Thom's three sections were as follows:

Monoverticillata – with a penicillus with a single whorl of phialides.

Biverticillata-Symmetrica – having a compact whorl of metulae each bearing phialides, the whole penicillus being approximately symmetrical about the axis and in addition with conidia almost always ovate or pear shaped and borne on slender tapering (acuminate, lanceolate) phialides.

Asymmetrica – the largest section included all species in which the penicillus is branched more than once and is assymetrical about the axis, or if approximately symmetrical has not the compact structure and tapering phialides of the Biverticillata-Symmetrica.

Separation on the branching system (Table 11.2 and Fig. 11.12) is slightly modified here to be nearer that suggested by Samson *et al.* (1976) and Pitt (1979). Samson followed Kendrick's (1971) terminology in which the term **conidiophore** refers to the entire conidial apparatus while the term stipe is used for the main stalk which was called a conidiophore by Raper and Thom.

Table 11.2   Division of *Penicillium* according to branching

I Monoverticillate (Aspergilloides)		Lacking branches, bearing a terminal whorl of phialides at the apex of the stipe, e.g. *P. frequentans* (occasional penicilli may have a branch).
II Biverticillata		One stage branching. **a** Biverticillata-Asymmetrica (Furcatum) One stage branching but arranged asymmetrically round the axes, e.g. *P. citrinum*. **b** Biverticillata-Symmetrica (Biverticillium) One stage branching but arranged symmetrically around the axis and with lanceolate phialides, e.g. *P. funiculosum*.
III Terverticillata (Penicillium)		Two or more stage branching, usually asymmetric, e.g. *P. expansum*.
IV Divaricata (Furcatum section Divaricata)		Irregularly branched, e.g. *P. nigricans*.

Colony texture is often used in separation of species. The terminology for this is most easily explained by making reference to Fig. 11.13 (p. 224).

Velvety   All the conidiophores arise from the substrate to give the appearance of short velvet, and there is little aerial mycelium (Fig. 11.13c).

Lanose or floccose   The conidiophores arise from a tangled mass of aerial hyphae (Fig. 11.13a).

Funiculose   Most of the aerial hyphae are aggregated in ropes or bundles and the conidiophores arise from these (Fig. 11.13b).

Fasciculate   Conidiophores arise from the substratum as erect bundles.
  (i)   Small bundles of a few conidiophores with simple conidiophores between the bundles (Fig. 11.13d(i)).
  (ii)   Conidiophores aggregated to form feathery bundles (Fig. 11.13d(ii)).
  (iii)   Conidiophores formed into distinct stalked synnema (Fig. 11.13d(iii)). Fasciculate colonies usually appear granular-velvety to quite

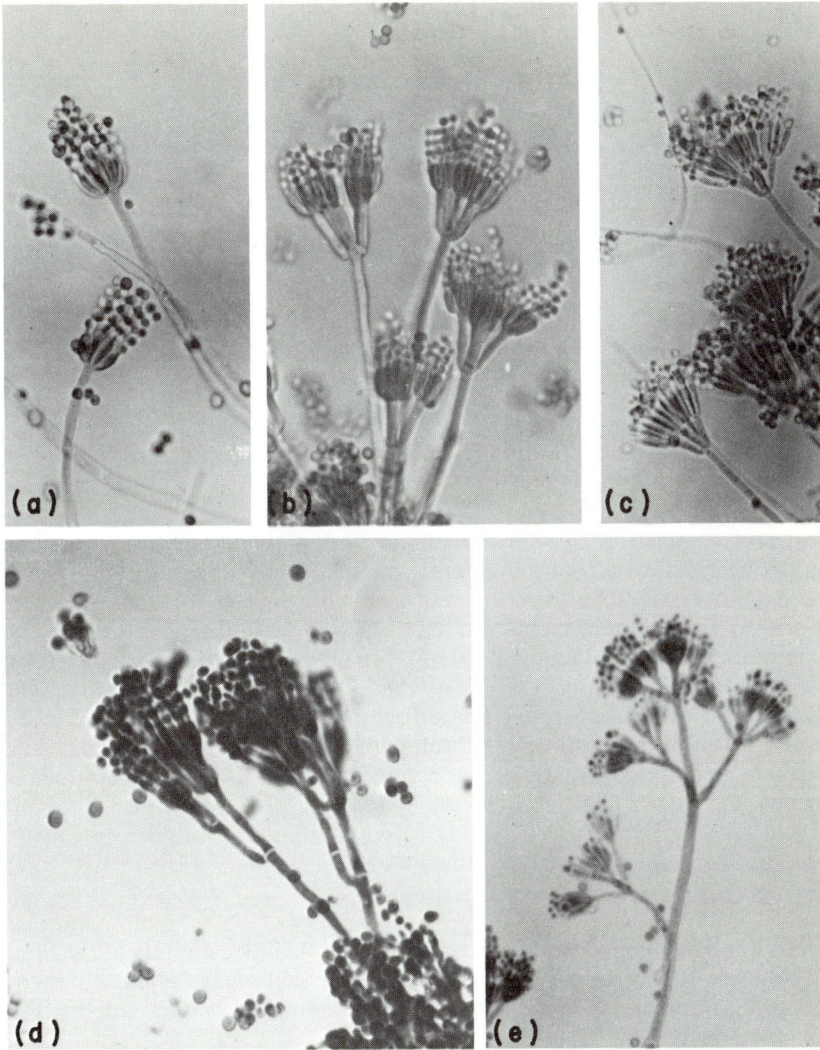

Fig. 11.12   Modes of branching   **(a)** Monoverticillata; **(b)** Biverticillata-Asymmetrica (Furcata); **(c)** Biverticillata-Symmetrica (Biverticillium); **(d)** Tervicillata (Penicillium); **(e)** Divaricata (Furcatum section Divaricata).

granular with definite synnema present. Conidia may be produced in such profusion that crusts are formed.

In attempting to identify any mould it is advisable to get as much information as possible from a study of living colonies before making any microscopic slides. This procedure is absolutely necessary in the case of the Penicillia. Such data as gross characteristics of colonies, origin of conidio-

(a)   **(b)**   **(c)**

(d)

(i)   (ii)   (iii)

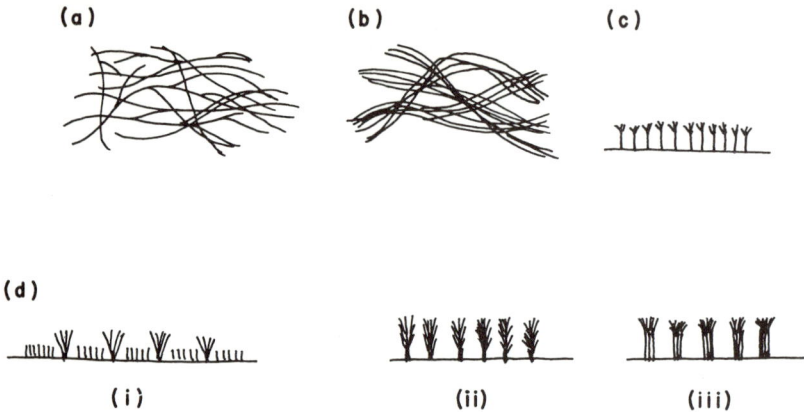

Fig. 11.13   *Penicillium*   Colony textures. **(a)** Floccose or lanose; **(b)** funiculose; **(c)** velvety; **(d)** fasciculate; **(d(i))** small bundles; **(d(ii))** feathery bundles; **(d(iii))** synnema.

phores, and disposition of spore chains are not only essential parts of the diagnoses of species but are used freely in constructing the keys.

The diagnoses given in the following pages are mostly in standard form. They are brief but should be adequate to enable the student to recognize all the commoner species. Details of rami (branches) and metulae have been omitted in order to save space, as these are less important than the characteristics of the whole penicillus and of the phialides and conidia. Unless otherwise stated the descriptions all refer to colonies on Czapek agar.

It has not been possible to give the etymology of all the specific epithets in this genus.

Monoverticillata (Aspergilloides)

Species without branches and bearing a whorl of phialides directly at the apex of the conidiophores (stipes) were referred to the section Monoverticillata by Raper and Thom and will be put in the subgenus *Aspergilloides* by Pitt (1980). (See Fig. 11.14.)

Ascosporic and sclerotial species with monoverticillate conidiophores were dealt with here by Raper and Thom and separately by Pitt.

Pitt regards the presence or absence of a swelling at the apex of the stipe as important and subdivides the group on this character to give the sections Aspergilloides and Exilicaulis. A compromise of these classifications is given in the Key to the Monoverticillata (Aspergilloides).

Ascosporic species (*P. javanicum* series)

These belong to the perfect genus *Eupenicillium* (see Stolk and Scott, 1967; Scott, 1968; Pitt, 1974). The cleistothecia develop as hard sclerotial masses and asci develop slowly from the centre outwards. Raper and Thom placed them in the *P. javanicum* series of the Monoverticillata. *P. brefeldianum*

Key to the Monoverticillata (Aspergilloides)

Cleistothecia or sclerotia present	Cleistothecia present ……………		Ascosporic species *P. javanicum* series (*P. brefeldianum*)
	Sclerotia present	Pink…….	*P. thomii*
		Yellow to orange …	*P. lapidosum*
Cleistothecia and sclerotia lacking	Vesiculate stipes …………………		Aspergilloides
	Non-vesiculate stipes ……………		Exilicaulis

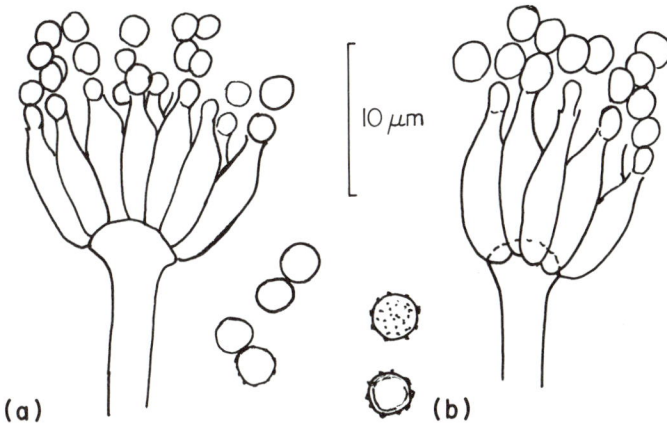

Fig.11.14 Monoverticillata **(a)** *Penicillium frequentans* Conidial head and conidia. **(b)** *P. spinulosum* Conidial head and conidia.

= *Eupenicillium brefedlianum* is the commonest species. They are isolated from soil, not common or of economic importance.

Sclerotial species (*P. thomii* series)

These are characterized by the production of sclerotia. Raper and Thom included five species and two varieties. Pitt (1974) produced a synoptic key to the genus *Eupenicillium* and to sclerotigenic *Penicillium* species. Many of these have monoverticillate penicilli, but there are quite a number with more complex conidial apparatus. Having searched the literature, he accepted 36 species of *Eupenicillium* and 21 species and one variety of sclerotial *Penicillium*. Of the latter five species were the imperfect states of *Eupenicillium* species and three were synonyms.

However, only one species is really common and another worth mention.

P. thomii Maire (named after Charles Thom) (Fig. 11.15). Growth spreading, velvety, grey-green, but usually dominated by pink sclerotia; reverse yellow to brownish; stipes long, slightly rough with vesiculate apex; conidia elliptical, 3–3.5 μm long, to subglobose, smooth in loose columns.

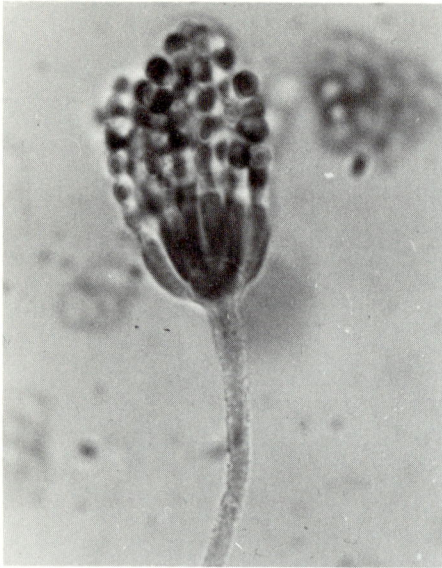

Fig. 11.15 *Pencillium thomii* Conidial head with elliptical conidia and rough stipe swelling at the apex, ×1300.

Fresh isolates usually produce abundant sclerotia but, when kept in cultivation, soon become entirely conidial. This species is fairly common in soil.

P. lapidosum Raper and Fennell (Lat. *lapidus*, stony, full of stones. So named because of the stone-like sclerotia) = *Eupenicillium lapidosum* (Raper and Fennell) Stolk and Scott. Growth yellow to pale orange, granular due to the presence of sclerotia; stipes short, smooth, vesiculate; conidia elliptical, smooth, 2.5–3 × 2–2.5 μm.

The cultures are distinctive due to the brilliant yellow colour in reverse on Czapek's agar, which on malt agar starts as yellow but rapidly turns deep mahogany red. Conidial heads are rare and often branched to form a biverticillate structure. The sclerotia only rarely and then after a very long time ripen to produce ascospores. It is quite common in soil.

Vesiculate species (Aspergilloides)

P. frequentans series

(Lat. *frequentans,* common) (See key to non-ascosporic species) (Figs 11.16, 11.17)

Growth velvety to very slightly floccose, spreading; stipes long, smooth, swelling to a definite vesicle; conidia globose, almost smooth to rough, conidial chains adhering to form columns.

Key to common Monoverticillata (non-ascosporic)

Vesiculate

  quick growth
  {
    long conidiophores and globose conidia
    {
      velvety
      {
        smooth conidia .... P. frequentans
        rough conidia .... P. purpurescens
      }
      floccose, rough conidia .... P. spinulosum
    }
    short conidiophores, elliptical smooth conidia + or − red pigment ...... P. chermesianum
  }

  restricted growth
  {
    long conidiophores — elliptical, rough conidia, strictly monoverticillate .... P. lividum
    moderate conidiophores — elliptical, smooth conidia, strictly monoverticillate ...... P. implicatum
    short conidiophores — elliptical to subglobose, smooth to finely rough conidia, occasionally branched .... P. fellutanum
  }

Non-vesiculate
{
  slow growth, short conidiophores, globose conidia, and floccose
  {
    conidia echinulate + no pigment ...... P. restrictum
    conidia finely rough + vinaceous pigment .... P. vinaceum
    conidia smooth + bright yellow pigment .... P. citreoviride
  }
}

P. frequentans Westling = *P. glabrum* (Wehmer) Westling. Pitt shows the latter to be the older name, though the name *P. frequentans* used by Raper and Thom is very well known. It is the commonest of the monoverticillate species.

It grows rapidly on all culture media, with colonies normally a rich green, becoming darker and browner in age, and reverse at first yellow then deep

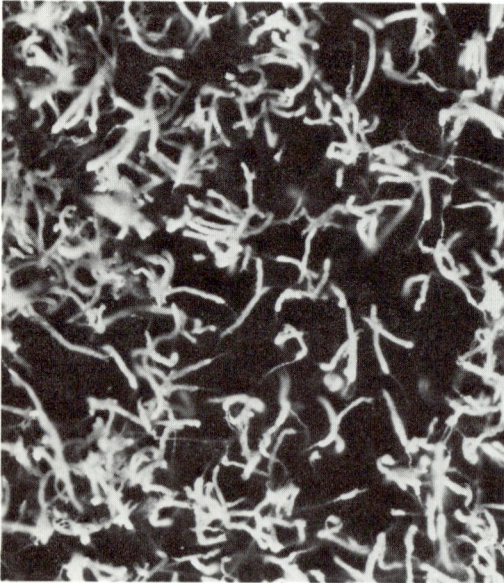

Fig. 11.16 *Penicillium frequentans* Columns of spore chains as seen in Petri dish, ×50.

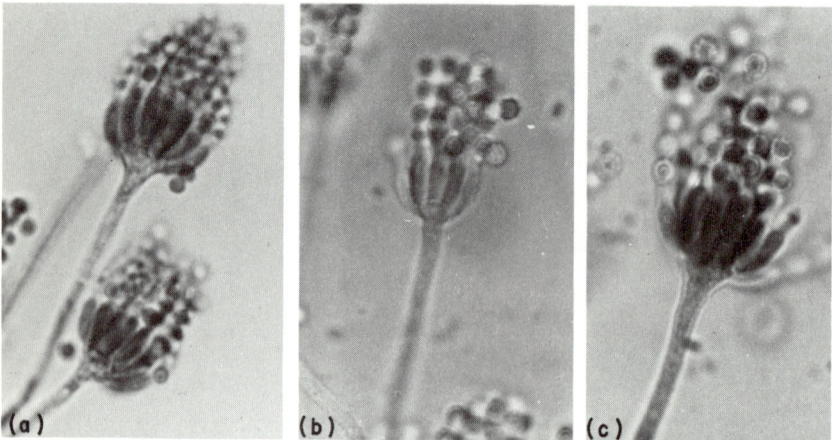

Fig. 11.17 **(a)** *Penicillium frequentans* Note distinct swelling at the apex of the stipe and almost smooth globose conidia, ×1040. **(b)** *P. purpurescens* Note very rough globose conidia and roughening almost arranged in bars, ×1040. **(c)** *P. spinulosum* Rough globose conidia, ×1040.

brown, some strains differ somewhat from the usual type, the conidial colour being pale blue-green and reverse colourless or nearly so. Conidiophores usually smooth, appreciably swollen at the tip; phialides mostly 9–10×3–$3.5\ \mu$m; conidia globose or almost so, smooth or slightly rough, about $3\ \mu$m diam., packed into solid columns (Fig. 11.17a).

P. spinulosum Thom (Lat. *spinulosus,* with little spines) (name accepted by Pitt). Colonies are broadly spreading, and consist of a somewhat loose felt, blue-green or grey-green, becoming definitely greyish in old cultures, with broad white edge during the growing period; reverse white to cream or faintly pink; conidiophores arising from either submerged or aerial hyphae, smooth or slightly rough, with apices definitely swollen; phialides not very numerous, 6–9×2.5–$3\ \mu$m; conidia globose or nearly so, spinulose, 3–$3.5\ \mu$m diam., borne in loose columns (Fig. 11.17c).

P. purpurescens (Sopp) Raper and Thom (Fig. 11.17b) (Lat. *purpureus,* purple coloured, hence becoming purple) (named accepted by Pitt) is another closely related species and seen quite often. It has spreading blue-green velvety colonies, producing reddish-purple to brown shades in reverse; long conidiophores swelling at the apex bearing crowded phialides, 7–10×2.5–$3.5\ \mu$m; conidia distinctly rough, globose, large, 3.5–$4.5\ \mu$m in diam. or even larger.

P. chermesianum Biourge (name accepted by Pitt) (Fig. 11.18a). Growth spreading, appearing velvety but bearing short stipes on matted floccose aerial hyphae, occasionally producing a red pigment in colony reverse; stipes short, smooth, swelling at the apex, strictly monoverticillate; phialides crowded, incurved, up to 15 in a verticil, 6–8×2–$2.5\ \mu$m; conidia elliptical, smooth, 2–2.5×1.5–$2\ \mu$m.

P. chermesianum occurs from time to time in soil or organic substrata subject to contact with soil.

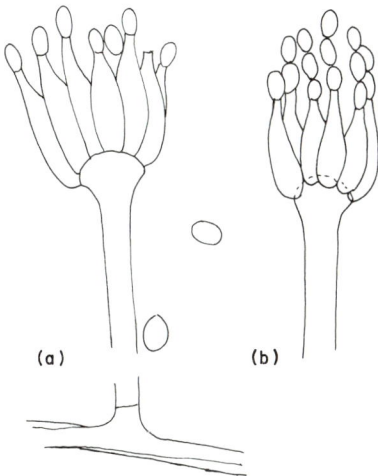

Fig. 11.18 (a) *Penicillium chermesianum* Conidial head. **(b)** *P. implicatum* Conidial head.

P. lividium series

Raper and Thom included three species in their *P. lividum* series but only *P. lividum* is worth mention and even this is not common.

P. lividium Westling (Lat. *lividus,* dark blue) (name accepted by Pitt). Growth velvety to lanose, somewhat restricted, a vivid distinctive blue-green in colour; stipes long, rough, swelling at the apex; conidia elliptical, rough, 3–4 × 2.5–3 μm.

P. implicatum series

Both Raper and Thom and Pitt use this name for a series of monoverticillate species with restricted colonies, though Pitt gives it a somewhat wider application.

P. implicatum Biourge (Lat. *implicatus,* tangled) (Fig. 11.18b) is of quite common occurrence in soil and in soil contaminated materials. Growth restricted, rich blue-green, often on a background of yellow to orange, reverse bright orange to red; stipes smooth, swelling at the apex, not very long but arising from the substrate, strictly monoverticillate; conidia elliptical to subglobose, more or less smooth, 2.5–3 × 2–2.5 μm; often produced so profusely that they form crusts which break when the petri dish is tapped.

P. fellutanum Biourge (Fellut, old name of Feluy, Belgian village) (Figs 11.19, 11.20). Growth slow, bluish-green to sage green or fairly dark with centre often white, growth matted with submerged outer zone; stipes short arising from matted aerial hyphae, distinctly swelling at the apex, often with one to several branches terminating in monoverticillate heads; phialides closely packed, 6–8 × 1.5–2 μm; conidia subglobose to elliptical, smooth to finely roughened, adhering in loose columns, 2.5–3 × 2–2.2 μm.

Non-Vesiculate species (Exilicaulis)

Only two series of species with stipes which do not swell at the apex are of interest to the industrial mycologist.

P. restrictum series

There are several species in this series but only one is common.

P. restrictum Gilman & Abbott (Lat. *restrictus,* restricted, close). Growth very restricted, floccose colonies bearing sparse dark green conidial heads on aerial hyphae, no reverse colour; stipes very short arising from trailing hyphae; phialides very few in a verticilli, loosely arranged, very ampuliform, with swollen base tapering to a narrow neck, 5 × 1.5 μm; conidia globose and distinctly roughened or echinulate, about 2.5 μm in diameter (Fig. 11.21a).

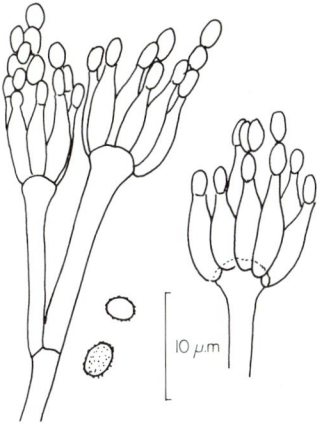

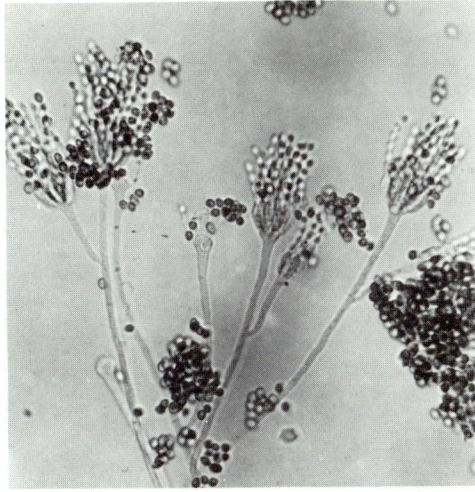

Fig. 11.19 (left)   *Penicillium fellutanum*   Conidial heads showing branching, swelling at apex of stipe, and elliptical rough conidia.

Fig. 11.20 (right)   *Penicillium fellutanum*   Conidial heads with branched stipes, ×500.

The conidial apparatus appears not unlike the fragmenting heads often produced by *Aspergillus sydowii*.

P. vinaceum Gilman & Abbott is a very striking species with somewhat similar conidial apparatus but with smooth conidia (Fig. 11.21b). It is also characterized by large vinaceous drops produced on the white to grey mycelium and by a dark vinaceous pigment in the colony reverse and diffusing into the medium.

P. citreonigrum Dierckx (Lat. *citreus*, lemon yellow; *niger*, black, dark coloured). Raper and Thom preferred the name *P. citreoviride* Biourge which is a more recent name as the only culture of *P. citreonigrum* they could obtain came from Biourge and was not necessarily the type. Raper and Thom record the species as quite common in soil and on military equipment, though I (AHSO) have seldom seen it. Knowledge of its synonymy is of importance as under the name *P. citreoviride* it is recorded as producing the mycotoxin citreoviridin on rice, which in Japan was the cause of the notorious cardiac beriberi.

Growth is very restricted, strongly wrinkled, velvety to slightly floccose with pale yellow to bright yellow mycelium, tardily tinged pale grey-green by conidia; reverse and agar bright yellow becoming darker or brown in age; stipes arising from trailing hyphae, only moderately long, 50–100 μm, smooth, not developing a vesicle; phialides 8–12 × 2.5 μm; conidia globose in loosely parallel chains, loose columns or slightly divergent, thin walled, nearly smooth, 2–3 μm in diameter.

P. fellutanum was keyed next to *P. citreoviride* by Raper and Thom but it has vesiculate conidiophores and lacks yellow pigment.

10 μm

(a)   (b)   (c)

(d)

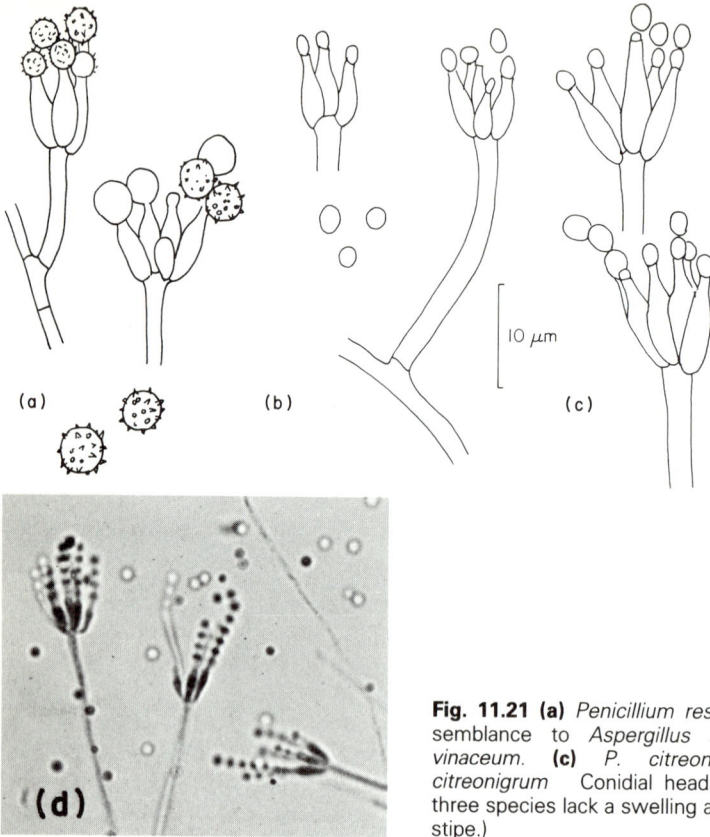

Fig. 11.21 (a) *Penicillium restrictum* Note resemblance to *Aspergillus sydowii*. **(b)** *P. vinaceum.* **(c)** *P. citreonigrum.* **(d)** *P. citreonigrum* Conidial heads, ×550. (These three species lack a swelling at the apex of the stipe.)

Biverticillata-Asymmetrica (Furcatum)

Raper and Thom placed together all species which had asymmetric branching whether they were once or twice branched and then separated these species into main groups by colony texture. Thus species with a terminal whorl of metulae each bearing phialides appear in several sub-sections. This is a distinctive character and shown by several additional species described since 1949. Pitt (1979) uses it as the basic character for his section Furcatum, which he subdivides into three series – Megaspora with large globose spores, Oxalica with spreading colonies and Citrina with restricted colonies (see key to Biverticillata-asymmetrica species). The section is taxonomically interesting, but only a few species are common or of economic significance.

P. simplicissimum is described under the section Divaricata (p. 266), but the majority of conidial heads have an apical whorl of metulae.

Key to common species of Biverticillata-Asymmetrica (Furcatum)

Large globose conidia		Megaspora series (rare)
Conidia otherwise ... {	spreading colonies	Oxalica series
	restricted colonies	Citrina series

Oxalica series

Conidia rough .... {	colonies floccose, stipes rough, conidia slightly elliptical, phialides abruptly tapered	*P. simplicissimum* (in Divaricata)

Conidia smooth ...... {	spherical conidia		*P. novae-zeelandiae* and others
	large elliptical conidia, colonies velvety	good growth on Czapek	*P. oxalicum*
		poor growth on Czapek	*P. digitatum*

Citrina series

Colonies restricted and velvety .... {	0–3 unequal metulae often dark green, black reverse	*P. corylophilum*
	2–4 equal metulae often bright yellow reverse	*P. citrinum*
	5 or more metulae, colourless reverse	*P. paxilli*
	3–5 or more compressed metulae, bright colours in reverse	*P. herquei* and others

P. citrinum Thom (Lat. *citrinus,* lemon yellow) (Figs 11.22, 11.23). The most common species in this section. Growth restricted, dull grey-green with narrow white margin, velvety with pale dull yellow drops; reverse yellow to dull orange with agar coloured similarly or pinkish; penicilli with smooth long stipes bearing a cluster of several metulae of more or less equal length,

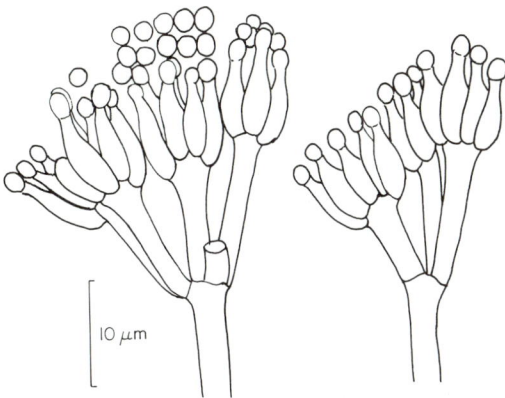

Fig. 11.22 *Penicillium citrinum*   Conidial heads with few to several branches.

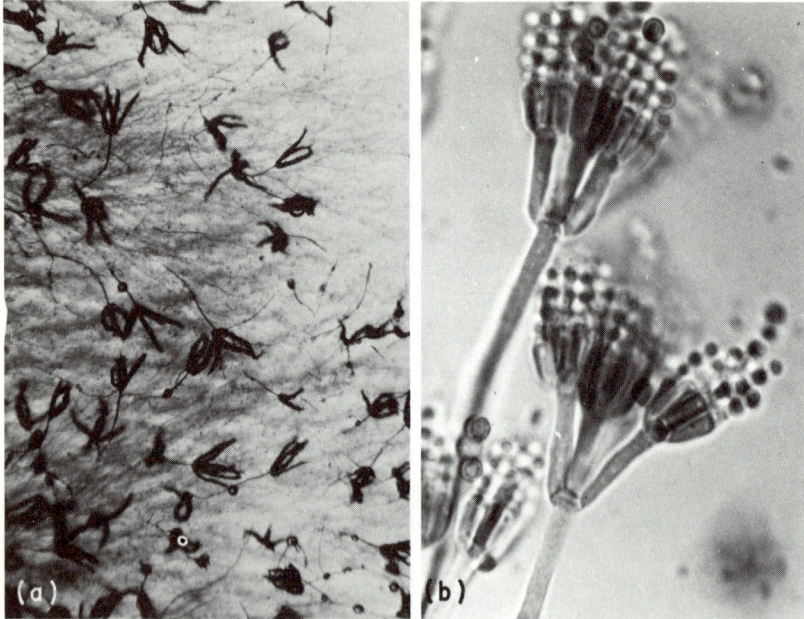

Fig. 11.23   (a) *Penicillium raistrickii*   Divergent columns of conidial chains as seen in Petri dish culture, ×90. *P. citrinum* is similar in appearance. **(b)** *P. citrinum*   Asymmetric biverticillate penicillus, ×1200.

2.5–3 μm diam.; conidia globose to subglobose, smooth, 2.5–3 μm diam. packed in solid divergent columns one to each metula.

It produces citrinin which is now regarded as an important mycotoxin, though when first isolated by Hetherington and Raistrick (1931) it was examined for its antibiotic activity.

Some isolates which produce more floccose growth, less pigment and less distinct columns of conidia used to be referred to *P. steckii* Zaleski but it is now considered as a synonym of *P. citrinum*.

P. corylophilum Dierckx (Gr. *korylos*, hazel tree; *phileo*, to love). Growth somewhat restricted, blue-green to grey-green and finally olive brown, with reverse pale dirty brown, on malt agar remaining green, with reverse very dark, almost black; penicilli bearing 2 to 3 unequal metulae; stipes smooth, 2–2.5 μm diam.; conidia globose to elliptical, 2.5–3 μm in long axis, smooth, in roughly parallel to tangled chains, but not in true columns.

P. paxilli Bainier (of *Paxillus*, a genus of Agarics) is often found on decaying basidiomycetes. It forms greyish colonies with large tangled heads, with short somewhat swollen compressed metulae in groups of 5 or more on the apex of the stipe.

P. oxalicum Currie & Thom (name refers to production of oxalic acid) (Figs 11.24, 11.25). Growth broadly spreading, dull green with margin shading through pale blue-green to white, velvety, with dense masses of conidia which tend to break off in crusts when the culture is jarred, and give the cultures a silky appearance when viewed under low magnifications; reverse colourless,

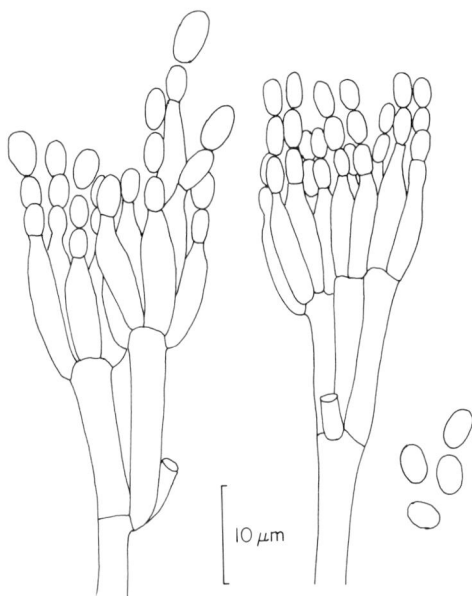

Fig. 11.24 *Penicillium oxalicum* Conidial heads with asymmetric biverticillate penicillus and large elliptical conidia.

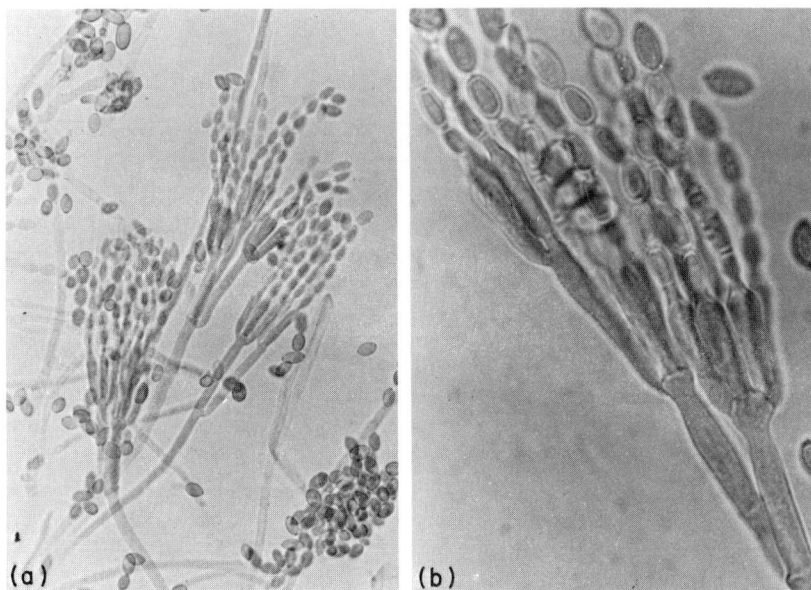

Fig. 11.25 *Penicillium oxalicum* **(a)** Conidial heads, ×110; **(b)** conidial head, ×440.

pale yellow or pinkish; penicilli irregularly biverticillate; stipes smooth, 3.5–4.5 μm diam.; phialides usually about 10 μm, but sometimes up to 15 μm by 3–3.5 μm; conidia elliptical, smooth, mostly 5–5.5 × 3–3.5 with chains massed into columns.

Easily recognized by its heavily sporing, dark green colonies and large elliptical conidia. When kept in cultivation it tends to become floccose, with conidium production much reduced. It is common in soil especially from the tropics and occurs occasionally as a parasite of corn seedlings.

P. digitatum Saccardo (Lat. *digitus,* a finger) (Figs 11.26, 11.27). Growth very poor on Czapek's agar but rapidly spreading on malt and potato dextrose agar, smooth, velvety, dull yellowish green; penicilli simple but with all parts large; stipes short, mostly 4–5 μm diam., bearing metulae and phialides few in the verticil, 15–30 μm long; conidia ovate, smooth, large, mostly 6–8 × 4–7 μm but sometimes up to 12 μm or even more in long axis.

The penicilli are very fragile and it is exceedingly difficult to prepare good slides. When unbroken penicilli are found, the aptness of the specific epithet is apparent, for they have a distinct resemblance to skeleton hands (Fig. 11.27).

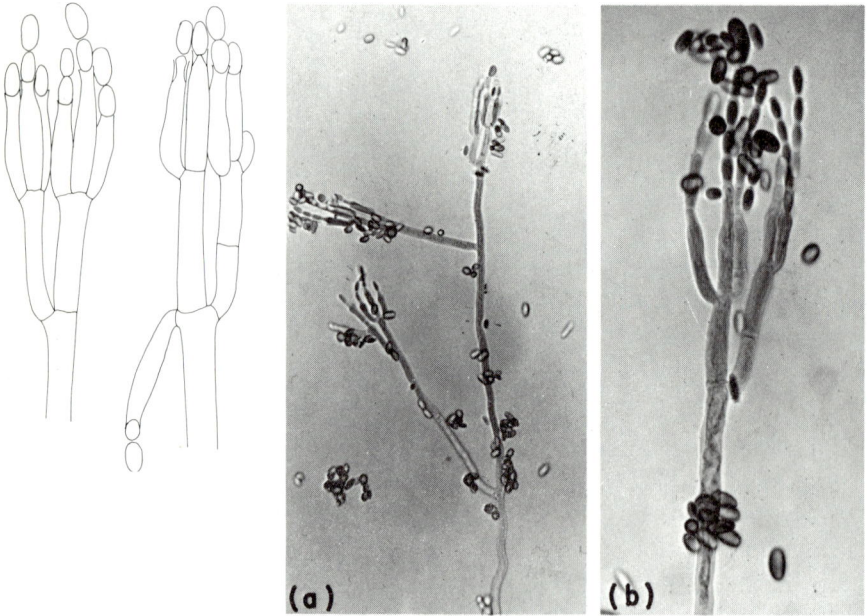

(a)   (b)

Fig. 11.26 (left)   *Penicillium digitatum*   Conidial heads.

Fig. 11.27 (right)   *Penicillium digitatum*   **(a)** Conidiophores, ×250; **(b)** conidiophore, ×500.

This occurs chiefly on citrus fruits of all kinds and is the cause of serious financial loss in the industry. It can readily be obtained from mouldy oranges, lemons or grapefruit, on which infected areas are olive to yellowish green, not blue green, by which it can be distinguished from *P. italicum* which causes soft rot of citrus fruits.

Other species showing a terminal whorl of metulae are **P. raistrickii** (Fig. 11.23a), **P. novae-zeelandii** and **P. herquei**. However, none of these is common, though the latter two are distinctive due to the production of vivid green to almost black pigments in the colony reverse.

Biverticillata-Symmetrica (Biverticillium)

This section can be divided into species which produce a perfect (teleomorphic) state and those which fail to produce ascospores (see Key to Biverticillata-Symmetrica species).

Ascosporic species (*P. luteum* series or *Talaromyces*)

Penicillium species showing penicilli typical of the Biverticillata-Symmetrica, that is with symmetrical biverticillate penicilli and lanceolate phialides, which produce an ascosporic state with soft or rudimentary walls, yellow, white or buff in colour were classified as the *P. luteum* series by Raper and Thom. Benjamin (1955) introduced *Talaromyces* as the name for this perfect state. Udagawa followed this in his paper in 1966. Stolk and Samson, in a series of papers (Stolk, 1965; Stolk and Samson, 1971, 1972) reviewed the whole situation. In 1971 they showed that *Talaromyces* could be separated into 3 genera, true *Talaromyces* in which asci are produced in chains, and *Byssochlamys* and *Hamigera* in which the asci are produced terminally by means of croziers. In 1972 they made a comprehensive study of *Talaromyces* and Pitt (1979) will follow their work.

They divided the genus into four sections as follows:

Emersonii – weakly thermotolerant, brown to fawn in colour, with *Paecilomyces* or *P. cylindrosporum* conidial states.
Thermophila – strongly thermophilic, grey-green to brown, and divaricate conidial state.
Purpurea – predominantly monoverticillate, with *P. restrictum*-like conidial state.
Talaromyces – mesophilic, with biverticillate-symmetrical conidial state, but with occasional reduced conidial apparatus, sometimes consisting of single phialides.

Talaromyces section

As with the *Aspergillus glaucus* series, shape and markings of the ascospores form the main bases of classification. However, most of the species tend, on continued cultivation, to become entirely conidial or even sterile. Emmons (1935) showed the species to be homothallic. Most isolates retain their characteristics best on corn-meal agar and preservation in silica gel is frequently satisfactory, though selective cultivation is often best.

Key to common species of Biverticillata-Symmetrica

Talaromyces (Ascosporic)

elliptical spinulose ascospores
- spreading growth
  - club shaped perithecial initials .... *T. flavus*
  - irregular perithecial initials .... *T. wortmannii*
- restricted growth
  - helically coiled initials, smallish ascospores .... *T. helicus*
  - *T. luteus*
  - *T. trachyspermus*
  - *T. thermophilus*

Yellow cleistothecia ....

elliptical ascospores with transverse bands ....

White to cream cleistothecia and elliptical spinulose coinida ....

White to pinkish brown growth, thermotolerant ....

Biverticillium (Non-ascosporic)
Producing coremia ....
- Funnel shaped heads, lanceolate phialides .... *P. duclauxii*
- .... other Bivert.-Symm. sp.

Not producing coremia ......
- Asymmetrically arranged or compact .... Furcatum series

Non-coremial species

Short conidiophores mostly from trailing hyphae or ropes of hyphae

- globose conidia .... rough conidia, broad, compact penicilli .... *P. verruculosum* and others
- elliptical conidia ....
  - very ropy colonies, slightly rough conidia .... *P. funiculosum*
  - conidia in conical mass .... *P. piceum*
  - smooth large heavy conidia, yellow orange mycelium .... *P. islandicum* / *P. aculeatum*

Long conidiophores arising mostly from the substrate

- rough conidia
  - globose .... purple red pigment, narrow funnel shaped heads .... *P. purpurogenum*
  - elliptical ....
    - no purple red pigment, yellow or orange and restricted growth .... *P. rugulosum*
    - red colours in reverse .... *P. purpurogenum*
- smooth conidia (elliptical) .... yellow to brown colours in reverse with yellow aerial mycelium .... *P. variabile*

Raper and Thom described 11 species, while Stolk and Samson included 12 species in their section *Talaromyces* and six species in their other sections. Several additional species have been added. However, most of these are seen infrequently and only two species are fairly common in soil and hence appear on materials that come in contact with or are contaminated with soil. A few others are worth mentioning.

Talaromyces flavus (Klöcker) Stolk & Samson = *P. vermiculatum* Dangeard and ***T. wortmannii*** (Klöcker) C. R. Benjamin = *P. wortmannii* Klöcker.
The two species are very similar and are best considered together. Thom (1930) regarded them as synonymous. They both form soft yellow perithecia which are without a true wall, being bounded by a soft hyphal web. The ascospores are almost identical, being ovate, without furrow, finally spinulose all over, 3.5–5 × 2.5–3.5 μm. The points by which they may be distinguished are: colonies of *T. flavus* are spreading and of soft texture, those of *T. wortmanii* restricted and consisting of tough felts; the perithecial initials are quite different, those of *T. flavus* consisting of long, thick, deeply staining, club-shaped hyphae, around which thinner hyphae are tightly coiled, those of *T. wortmannii* being much smaller (Fig. 11.28c, d), and, owing to the texture of the colonies, more difficult to find, and consisting of irregular knots of somewhat thickened hyphae. Figure 11.28a, b shows two of the many different patterns which can be observed.

Some of the other ascosporic species have been reported in this country and are always to be found occasionally. Their distinguishing characteristics are given below.

T. trachyspermus (Shear) Stolk & Samson = *P. spiculisporum* Lehman (Lat. *spiculus*, a little spine). Cleistothecia white to cream, with definite walls, ascospores elliptical, finely spinulose all over, 3–3.5 ×2.2–2.8 μm.

T. luteus (Zukal) Benjamin (Lat. *luteus*, golden yellow) = *P. luteum* Zukal. Cleistothecia yellow, without definite walls; ascospores elliptical with 3–4 transverse bands, approximately 4.5 × 2.2–2.8 μm.

T. helicus (Raper & Fennell) Benjamin (name refers to cleistothecial initials) = *P. helicum* Raper & Fennell (Fig. 11.28e, f). Cleistothecia yellow, soft without definite walls, arising from helically coiled initials; ascospores elliptical, finely spinulose, 2.5–3 × 1.4–1.8 μm.

T. thermophilus Stolk = *P. dupontii* Griffon & Maublanc is a thermophillic species and quite frequently isolated from warm spoilage situations provided the isolation is made at a reasonably high temperature as it produces little growth below 30°C. Growth is good at 45°C. Colonies floccose to funiculose, mycelium white, pinkish to brown, conidial areas grey-green, reverse brownish. Cleistothecia seldom produced, occasionally seen on oat grains. Ascospores when seen 3.5–4.5 × 2.2–3.5 μm. Penicillus divaricate and delicate with small, smooth, ellipsoidal conidia, 2.5–4.5 × 1.5–2.5 μm. Chlamydospores, 4.5–6.5 μm often seen and swollen vegetative hyphae are typical of the species.

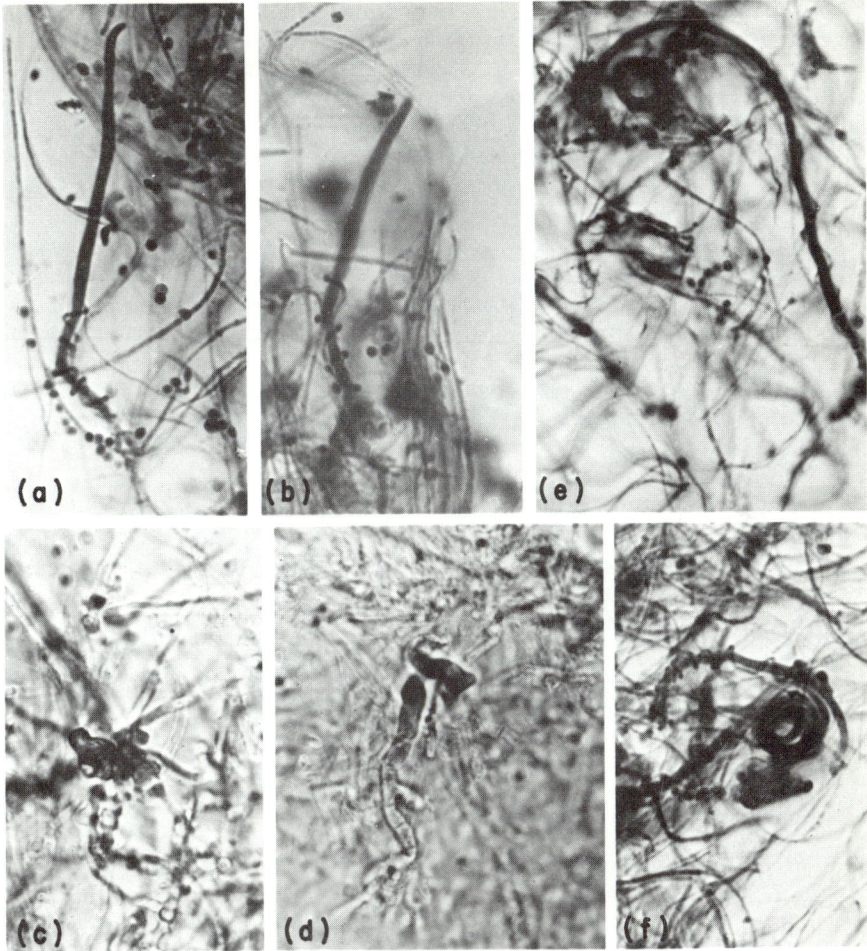

Fig. 11.28 Cleistothecial initials **(a)**, **(b)** *T. flavus*, ×550; **(c)**, **(d)** *T. wortmannii*, ×1000 (note these are almost twice the magnification of the others); **(e)** and **(f)** P. helicum, ×550.

Non-Ascosporic species (Biverticillium)

Raper and Thom in their key to the Biverticillata-Symmetrica first separated the ascosporic and sclerotial species and the remaining species into series according to the colony texture, with coremial, funiculose, velvety, restricted and lanose colonies. Pitt (1979) treats the ascosporic species as *Talaromyces*, the coremial species as section *Coremigenum* and the remaining species as the section *Biverticillium*.

Coremial species or Coremigenum

Pitt included five species in the series in his key of which Samson, Stolk and Hadlock (1976) included two in their fasciculate species. However, only one of these is seen often enough to mention.

P. duclauxii Delacroix (E. Duclaux, French microbiologist). Different strains vary in appearance, some growing as a forest of slender erect coremia, 4–5 mm high, others developing irregular spiky clumps of coremia fertile along the entire length. Surface colour grey-green; reverse yellow then purplish red, with colour diffusing somewhat; penicilli typically biverticillate with acuminate phialides; conidia elliptical to subglobose, rough. A fairly abundant species, reported from many sources.

Non-Coremial species or Biverticillium

This is quite a large section of the genus *Penicillium* and Raper and Thom included 17 species several of which are now included in the Biverticillata-Asymmetrica or Furcatum. Pitt also includes 17 species in his key but several of these are new.

One of the problems of identifying these is the variability of the species. It is proposed to describe only a few of the species which occur frequently as the others are seldom seen.

P. funiculosum Thom (Lat. *funiculus,* a thin rope) (Figs 11.29, 11.30). Colonies spreading, with tough basal felt of mycelium and aerial growth as ropes or tufts of hyphae, sporing in irregular patches; reverse usually pink to reddish, in some isolates becoming very deep red, in others colourless or almost so; penicilli typical; conidiophores short, mostly arising from funicles (Fig. 11.30), smooth, about 3 μm diam.; phialides 10–12 × 2–2.5 μm; conidia elliptical, with thick walls, smooth or slightly rough, 2.5–3.5 × 2–2.5 μm, borne in tangled chains or distinct columns. Many strains deteriorate in

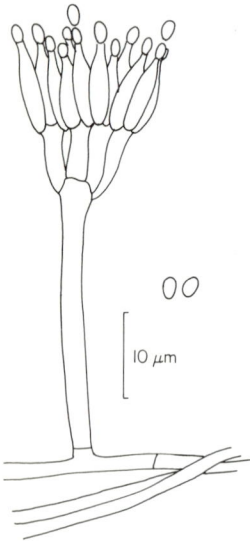

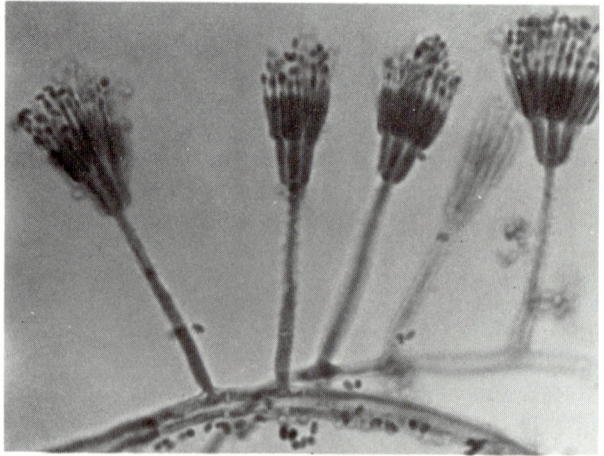

10 μm

Fig. 11.29 (left)   *Penicillium funiculosum*   Conidiophore arising from hyphal rope.

Fig. 11.30 (right)   *Penicillium funiculosum*   Conidiophores arising from hyphal rope, ×550.

culture, growing as a sodden felt, sporing tardily and sparsely. Very common in soil and on decaying vegetation. It has also been reported as producing a core-rot of *Gladiolus* corms (Jackson, 1962).

P. verruculosum Peyronel (Lat. *verruculosus,* warted) (Fig. 11.31). Colonies floccose, funiculose, rather loose in texture, a mixture of yellow-green and yellow; reverse greenish or pale dull brown; penicilli rather short and broad; conidiophores smooth, 2.5–3 μm.; phialides 8–10 × 2.5 μm, more abruptly tapered than is usual in this section; conidia globose or nearly so, very rough, 2.8–3.5 μm diam.

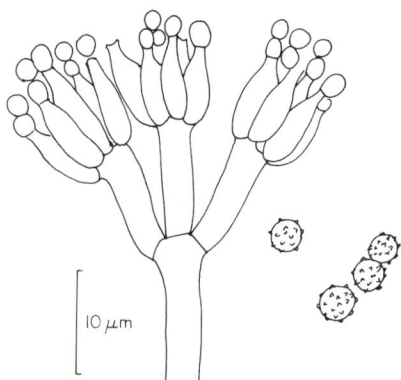

Fig. 11.31 *Penicillium verruculosum* With heavy metulae and phialides and rough, globose conidia.

This is very similar to *P. aculeatum* Raper & Fennell which differs in producing more restricted velvety colonies, lacking yellow mycelium and producing red pigment in the colony reverse . Both species with their short, broad, heavy penicilli, somewhat ampuliform phialides and rough globose conidia appear similar to some species in the Biverticillata-Asymmetrica such as *P. herquei* and *P. novae-zeelandii.*

P. islandicum Sopp (Lat. *islandicus,* Icelandic) (Figs 11.32, 11.33). Colonies very striking in appearance, restricted, fairly thick and tufted, a mixture of orange, red, and dark green; reverse dull orange to red, becoming dull red-brown; penicilli typical but rather short and conidiophores often short arising from aerial trailing ropes of hyphae, phialides, some lanceolate but many somewhat abruptly tapered for Biverticillata-Symmetrica, heavy walled; conidia elliptical, smooth, heavy walled, 3–4.5 × 2–3 μm, borne in tangled chains.

It has been found to produce important mycotoxins, luteoskyrin and islanditoxin on mouldy rice.

P. piceum Raper & Fennell (Lat. *Picea,* the spruce. Name suggested by the resemblance of the conidial heads to miniature spruce trees, *Picea excelsa.* The name is unfortunate since Lat. *piceus* = pitch black) (Figs 11.34, 11.35). Colonies grow fairly rapidly, fairly thick matted, yellow at first then producing irregular patches of dull yellowish green, with reverse dull orange to

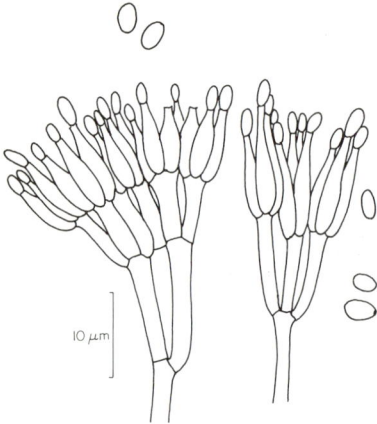

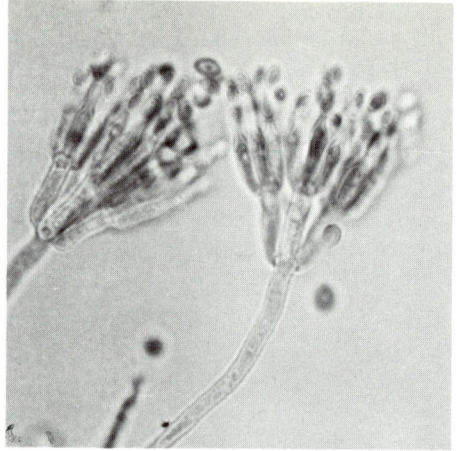

Fig. 11.32 (left)   *Penicillium islandicum*   Conidial heads.

Fig. 11.33 (right)   *Penicillium islandicum*   Conidial head, ×1300.

brown. The mature heads are very characteristic, the chains forming solid masses which are more or less conical (Fig. 11.35a). Penicilli are typically biverticillate; conidiophores short, phialides incurved, conidia subglobose to elliptical, rough, 2.5–3 × 2.2–2.8 μm.

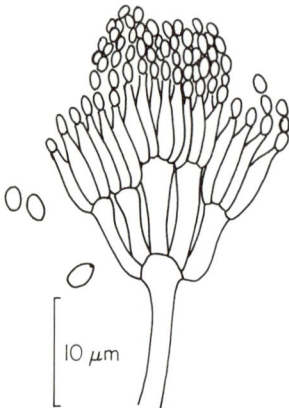

Fig. 11.34   *Penicillium piceum*   Conidial head.

P. purpurogenum Stoll (Lat. *geno*, to cause; hence producing purple colour) (Figs 11.36, 11.37). Colonies usually velvety, more rarely somewhat floccose, sporing heavily, deep green on a yellow mycelium, becoming very dark green; reverse and agar intense blood red or purplish red (on malt agar almost colourless); penicilli appressed, relatively long with narrow closely parallel phialides giving the heads a narrow look; conidiophores fairly short, smooth, arising from the substrate, about 3 μm diam.; phialides 10–12 × 2–2.5 μm; conidia elliptical to subglobose, with thick walls, in most strains rough, occasionally smooth, 3–3.5 × 2.5–3 μm. Common in soil and reported as

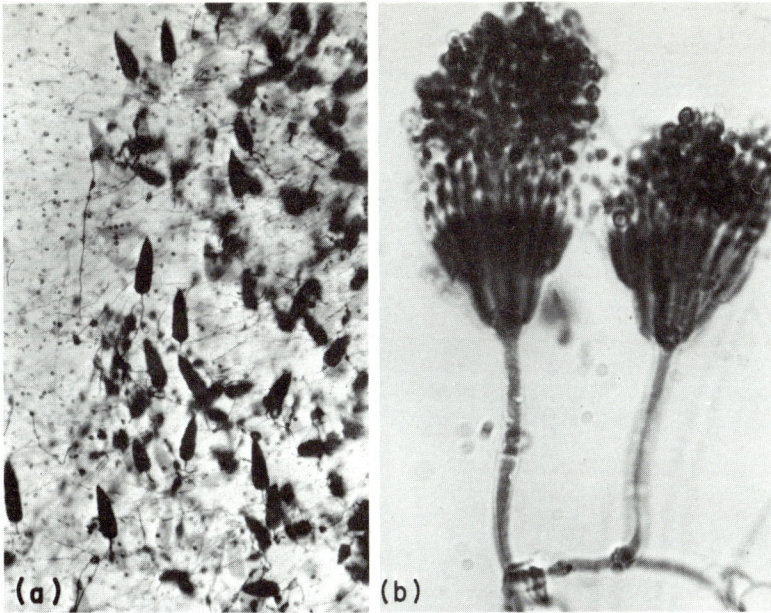

Fig. 11.35 *Penicillium piceum* **(a)** Characteristic shape of conidial heads, ×50; **(b)** typical conidial heads, ×975.

occurring on a variety of substrates. Some isolates with light grey-green colour, more restricted colonies, less diffusing pigment and smooth conidia were referred to *P. rubrum* Stoll, which is now considered (Pitt) a synonym of *P. purpurogenum*.

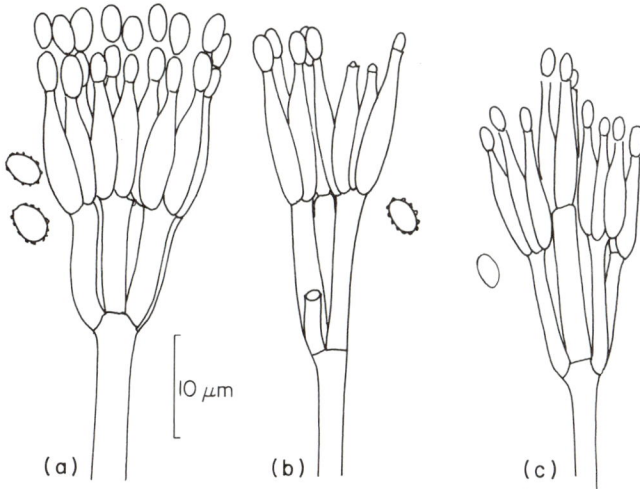

Fig. 11.36 *Penicillium purpurogenum* Appressed conidial heads with narrow parallel phialides, especially in (c). **(a)** and **(b)** Typical isolates with rough conidia; **(c)** isolate previously referred to *P. rubrum* with smooth conidia.

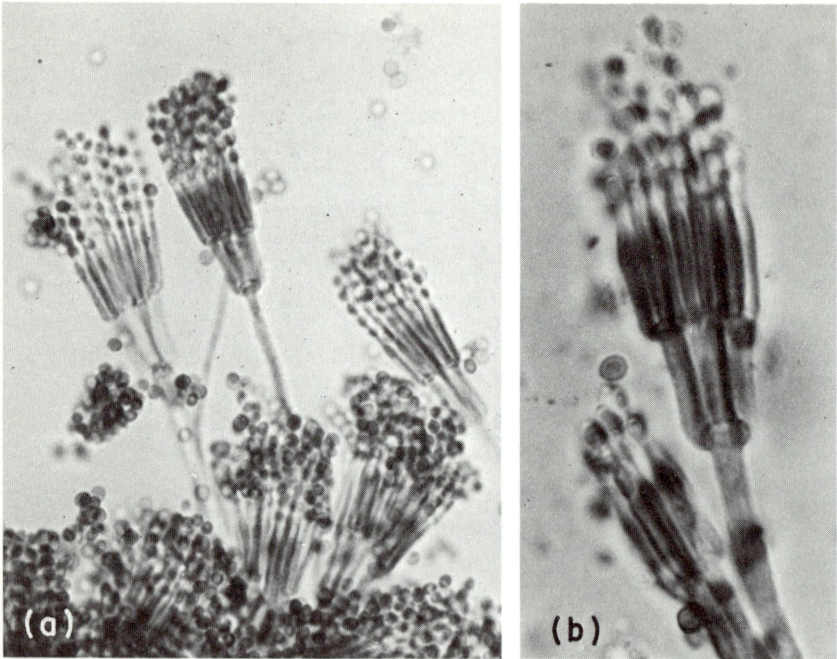

Fig. 11.37 *Penicillium purpurogenum*   Conidial heads. **(a)** ×550; **(b)** ×1300.

P. variabile Sopp (Lat. *variabilis,* variable). Colonies growing moderately well, velvety or almost so, sage green, grey-green or almost grey, often sporing in patches with sterile areas white, yellow, or orange, and with margin white, cream, or yellow; reverse yellow, brownish or greenish; penicilli typical, standing below a superficial growth of yellow hyphae; conidiophores arising from the basal felt, varying in length, smooth 2.5–3 μm diam.; phialides 10–12 × 2 μm approximately; conidia elliptical, often with some-what pointed ends, smooth to finely roughened, 3–3.5 × 2–2.5 μm. This species includes most strains which, before the publication of the Manual, would have been identified as belonging to Thom's 'P. *luteum* series, non-ascosporic'. As the name implies, different isolates vary much in appearance, especially in the amount of pigmentation of the mycelium.

P. rugulosum Thom (Lat. *rugulosus,* wrinkled) (Fig. 11.38). Colonies res-tricted, almost velvety to definitely floccose, rich green becoming greyer; reverse at first colourless, slowly becoming deep yellow to orange to reddish; penicilli mostly typical but not infrequently irregular, with metulae of different lengths; conidiophores smooth, 2.5–3 μm diam.; phialides 10–12 × 1.8–2 μm; conidia elliptical, markedly roughened, 3–3.5 × 2.5–3 μm, borne in tangled chains. An interesting occurrence of this species, and, less frequently, of some of the other biverticillate species, is as parasites of the black Aspergilli. Hyphae of the *Penicillium* twine up the stalks of the *Aspergillus* and smother the black heads with fruiting masses of a dark olive-green colour. Such parasitic species causes trouble from time to time in

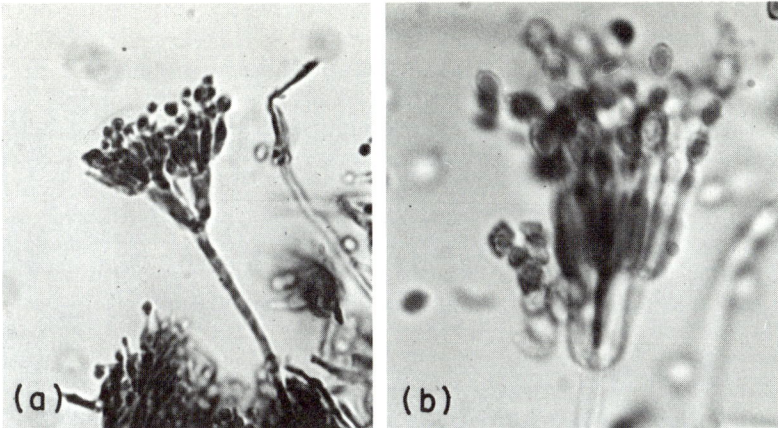

Fig. 11.38 *Penicillium rugulosum* Conidial head with rough elliptical conidia. **(a)** ×550;
(b) ×1300.

factories where citric acid is manufactured by the mould fermentation
process, attacking the mycelial mats of *A. niger* and causing portions to sink
in the solution.

P. herquei, P. tardum and similar species (Fig. 11.39). Some restricted lanose
species with biverticillate but compressed rather broad and heavy penicilli
and more or less ampuliform phialides were keyed here by Raper and Thom.
They are now regarded as being nearer *P. citrinum* in the section Biverticillata-Asymmetrica and are uncommon.

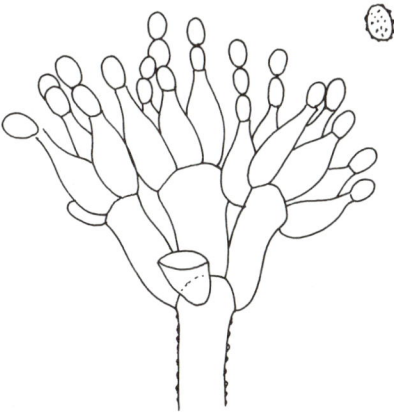

Fig. 11.39 *Penicillium herquei* Conidial head
(note the heavy and swollen metulae and
phialides more like those seen in *P. citrinum*).

Terverticillata (Subgenus Penicillium)

Penicillium species with two stages of branching, that is a terminal verticil of
metulae, each bearing phialides, with one or two branches below are very
common and many are of considerable economic importance. They include

species used in cheese ripening, production of penicillin and other antibiotics and are the cause of spoilage of household materials, fruit, cereals and food stuffs, often with the production of toxins harmful to man and animals.

They are quite numerous and show great similarities one with another. As would be expected with widespread and common species of complex genetic make-up there is great variation within the species. It is difficult to separate one species from another, and identification often presents difficulties. Being at the same time of economic importance it is not surprising that they have been the subject of many taxonomic studies. The green cheese moulds inspired Thom's first studies resulting in his monograph of 1910. Work continued with a multiplication of species and synonyms until Raper and Thom's monograph of 1949. Since then still more species of *Penicillium* have been described. The recent realization of the importance of mycotoxins in foodstuffs and many of them being produced by *Penicillium* species has inspired yet further work.

There are some rare and obscure species in the group but those commonly occurring tend to fall into a number of fairly constant species or species aggregates. Different authors have made varying interpretations of separation and naming for these and the literature is full of a multiplicity of names referring to similar organisms. Smith, in the previous edition, followed Raper and Thom (1949) and included most of the common species. It is proposed to follow his descriptions which are still accurate (they are, after all, still the same organisms) but at the same time to cross reference them with the recent work of Fassatiova (1977), Samson *et al.* (several papers) and Pitt (1979). It would be impracticable here to give a comprehensive list of synonyms, but a simple list cross referencing names of common species as used by the above authors and likely to be met in the literature is given in Table 11.3.

Raper and Thom made a primary separation using colony characters, into series with velvety, lanose, funiculose and fasciculate colonies. Samson dealt with sections which he regarded as interesting, for example the *P. chrysogenum* group, the cheese moulds and the Fascicilate species, while Fassatiova examined the Fasciculate species. Pitt provides a complete monograph of the genus, in which he refers the twice branched species to his subgenus *Penicillium*. Having separated the cylindrical spored species he divides the remaining species on surface markings of the stipes, colony texture and rate of growth under standard conditions. Separation here is a modification of this to include only the common species or species aggregates (see key to Terverticillata species, p. 250).

Species with smooth stipes
Species with globose conidia

P. chrysogenum series

This series was revised by Samson, Hadlok and Stolk in 1977, who came to the conclusion that there was only one very variable species, with spreading colonies, and a compact verticil of metulae and phialides with one or more divergent branches from lower nodes of the stipe, bearing phialides or metulae and phialides, each metula producing a fairly definite column of spore chains.

Table 11.3 Table of different names frequently used for common Terverticillate *Penicillium* species

Pitt (1979)	Raper & Thom (1949)	Fassatiova (1977)	Samson (various papers)
P. digitatum	*P. digitatum*	–	–
P. italicum	*P. italicum*	–	–
P. camembertii	*P. camembertii*	–	*P. camembertii* (syn. of *P. caseicolum*)
P. chrysogenum	*P. chrysogenum*	–	*P. chrysogenum*
P. chrysogenum	*P. notatum*	–	*P. chrysogenum*
P. hirsutum	*P. corymbiferum*	–	*P. verrucosum* var. *corymbiferum*
P. expansum	*P. expansum*	*P. expansum*	*P. expansum*
P. aurantiogriseum	*P. cyclopium*	*P. cyclopium* + *P. cyclopium* var. *aurantiovirens*	*P. verrucosum* var. *cyclopium*
P. crustosum	*P. crustosum*	–	*P. verrucosum* var. *cyclopium*
P. echinulatum	*P. cyclopium* var. *echinulatum*	*P. echinulatum*	*P. echinulatum*
P. viridicatum	*P. viridicatum*	*P. olivinoviride*	*P. verrucosum* var. *verrucosum*
P. olivicolor	*P. ochraceum*	–	*P. verrucosum* var. *ochraceum*
P. griseofulvum	*P. urticae* (Smith used *P. patulum*)	–	*P. griseofulvum*
P. brevicompactum	*P. brevicompactum* + *P. stoloniferum*	–	–
P. puberulum	*P. puberulum*	*P. cyclopium*	?*P. verrucosum* var. *cyclopium*
P. verrucosum	?*P. viridicatum* (name not accepted by Raper & Thom)	–	*P. verrucosum* (perhaps different interpretation)
P. granulatum	*P. granulatum*	–	*P. granulatum*
P. claviforme (but places in Bivert-symmetrica)	*P. claviforme*	–	*P. claviforme*
Geosmithia putterilii	*P. pallidum*	–	–

– not discussed by this author

Key to common species of Terverticillata (*Penicillium*)

Species with smooth stipes

globose conidia
- spreading colonies loose penicilli .... *P. chrysogenum*
- restricted colonies compact penicilli .... *P. brevi-compactum*

cylindrical conidia
- poor growth on Czapek agar .... *P. digitatum*
- grey rots of Citrus (seldom elsewhere) .... *P. italicum*
- blue rots of apples (in soil and elsewhere) .... *P. expansum*
- grey restricted growth, spreading heads, very short phialides .... *P. griseofulvum*

Species with rough stipes

floccose colonies / funiculose colonies
- white or very pale green .... *P. camembertii*
- not green .... *P. pallidum* (= *Geosmithia* series)

velvety colonies
- spreading velvety poor growth on Czapek, almost black green reverse .... *P. roquefortii*

fasciculate colonies .... Fasciculata

Fasciculata

conidia rough
- producing pink sclerotia at 20°C .... *P. echinulatum*
- coremia with pink stalks .... *P. gladioli*

conidia smooth or nearly so
- not as above, with fasciculate colonies, smooth .... *P. claviforme*
- globose conidia and rough stipes .... *P. viridicatum* series

P. viridicatum series

- restricted yellow green colonies .... *P. viridicatum*
- more spreading grey green and blue green .... *P. cyclopium*
- blue green colonies with deep wine red drops on surface, causing rot of lilaceous bulbs ... *P. corymbiferum* (= *P. hirsutum*)
- conidia forming crusts on malt agar .... *P. crustosum*

P. chrysogenum Thom (Lat. *geno*, to cause; Lat. ex Gr. *chrysos,* golden yellow) (Figs 11.40–11.42). Colonies broadly spreading, blue-green to bright green, with broad white margin during the growing period, smooth velvety, usually becoming greyish or purplish brown in age with overgrowth of white or rosy hyphae; reverse yellow, with colour diffusing somewhat; drops usually numerous, colourless to bright yellow; penicilli fairly complex with divergent branches (Fig. 11.42) with all parts smooth; stipes 2.5–4 μm diam.; phialides ampuliform with a reduced neck, 7–10 × 2–2.5 μm; conidia elliptical to subglobose, 3–4 μm in long axis, smooth; some penicillin produced by most strains. Some strains are stable in culture, especially if kept on organic media, but others tend to produce more and more floccose mycelium in successive cultures and eventually become quite atypical.

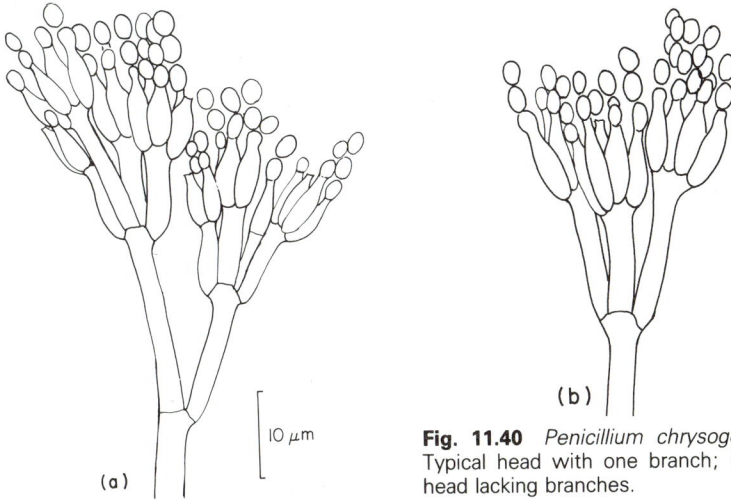

10 μm

(a)

(b)

Fig. 11.40 *Penicillium chrysogenum* **(a)** Typical head with one branch; **(b)** simple head lacking branches.

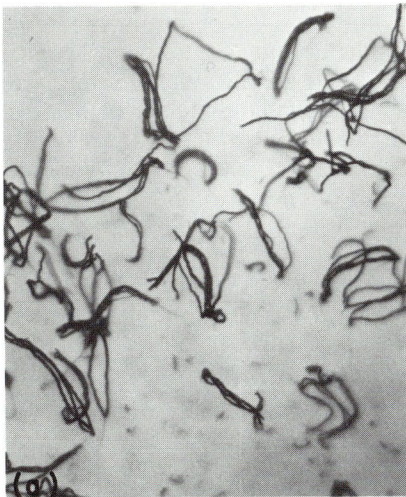

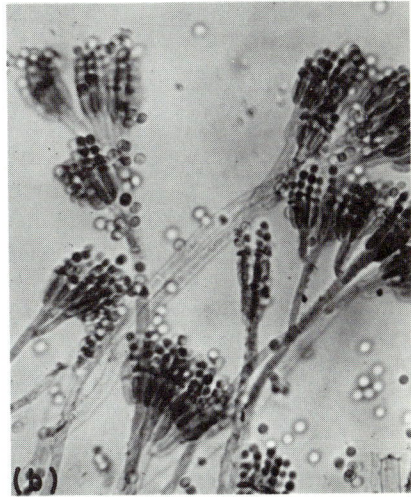

Fig. 11.41 *Penicillium chrysogenum* **(a)** Columns of conidia as seen at margin of Petri dish culture, ×20; **(b)** conidial heads (from Fleming's strain of *P. notatum*), ×500.

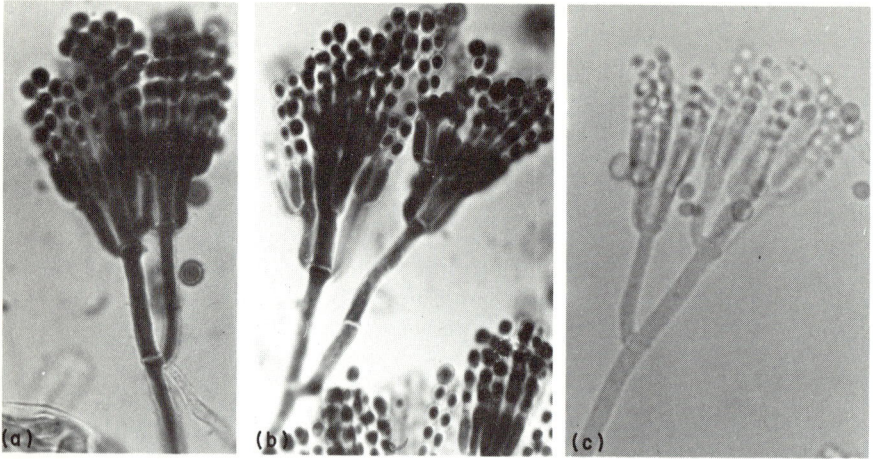

Fig. 11.42   *Penicillium chrysogenum*   Various branching patterns. **(a)** ×850; **(b)** ×850; **(c)** the most typical pattern, ×900.

Isolates previously referred to *P. notatum* Westling by Raper and Thom tend to have simple penicilli, seldom branched, and with globose conidia, about 3 μm diam., but this difference is not now regarded as significant.

P. brevi-compactum series

The name very aptly describes the penicillus as there are frequently three or four stages of branching and yet the total length of the penicillus is only 35 μm to 45 μm at the most (see Fig. 11.43). Invariably the metulae and rami are closely appressed to the main axis, giving a very distinctive appearance, even to penicilli which show comparatively little branching. Colonies usually appear velvety, but the conidiophores do not arise directly from submerged hyphae but from a thin surface mycelial felt. In young cultures the penicilli may be suggestive of *P. citrinum,* but the metulae are always more compacted together and are shorter in proportion to the phialides. Older cultures are unmistakable, since the spore chains are no longer parallel but are twisted and tangled in a very characteristic way (see Fig. 11.44a). Under very moist conditions, marginal extension of colonies is often by stolon-like hyphae.

Raper and Thom included three species in this series. However, one of these *P. paxilli* Bainier produces penicilli rarely branched below the metulae and therefore seems to belong with *P. citrinum* in the Biverticillata-Asymmetrica. Pitt regards *P. brevicompactum* and *P. stoloniferum* as synonyms.

P. brevi-compactum Dierckx (Lat. *brevis,* short; *compactus,* compact) (Fig. 11.43, 11.44). Colonies restricted in growth, grey-green with narrow edge shading through pale blue-green to white, becoming greyer and often brownish in age; drops dull yellow to brownish; reverse dull yellow or greenish brown; penicilli mostly complex, particularly in freshly isolated cultures; stipes coarse, 4–5 μm diam., smooth or slightly rough, often with

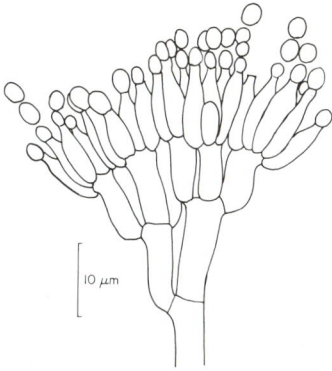

Fig. 11.43 (left) *Penicillium brevicompactum* Typical conidial head with appressed short branches and metulae.

Fig. 11.44 (below) *Penicillium brevicompactum* **(a)** Heads in living culture showing the very characteristic tangled conidial chains, ×100; **(b)** typical head, ×1200.

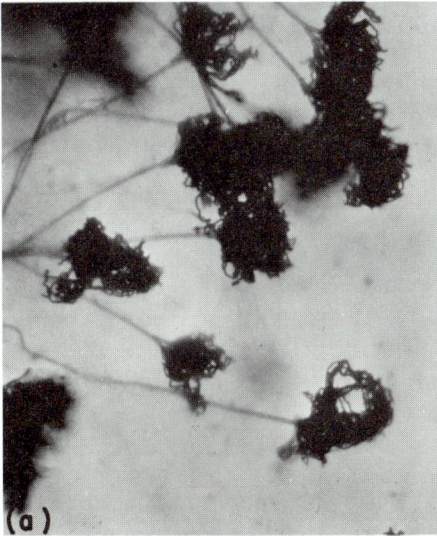

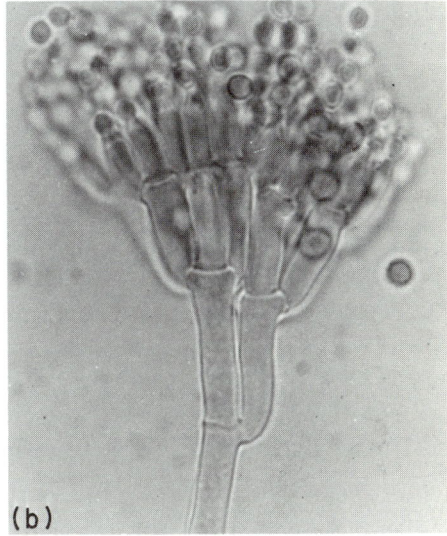

apex enlarged; branches and metulae enlarging upwards, more or less wedge-shaped, short, mostly 4–5 μm diam. but often inflated and reaching 6–7 μm; phialides 7–10 × 3–3.5 μm; occasionally broader; conidia globose or subglobose, slightly rough, 3.5–4 μm diam. Occasionally a secondary growth appears on top of an established culture, pale blue-green with long stiff stipes and enormous conidial heads, appearing almost like an *Aspergillus*. In some seasons it occurs abundantly as a parasite of agarics. It is found growing on textiles and a variety of other manufactured products, and is a common infection of stored maize, the consumption of infected grains being supposed at one time to be the cause of Pellagra.

Isolates producing thinner, more yellowish-green and more wrinkled colonies, with stipes thinner and sinuous, and elements of the penicillus not inflated were referred by Raper and Thom to *P. stoloniferum* Thom. However, the compact nature of the penicillus remains and the two species grade so completely that it appears to be one variable species.

Species with cylindrical conidia

P. digitatum Saccardo, the common spoilage mould of citrus fruit could be classified here but has been included in the Biverticillata-Asymmetrica.

P. italicum series

This series includes only the one species.

P. italicum Wehmer (Lat. *italicus*, Italinan) (Figs 11.45, 11.46). Colonies somewhat restricted, pale grey-green, usually showing obvious fasciculation and often producing curious prostrate coremia at the edge of established colonies; odour fragrant; reverse pale tan to yellowish brown; penicilli normally with three stages of branching but sometimes more complex; stipes smooth, 4–5 µm diam.; phialides few in number, mostly 8–15 × 2.5 µm; conidia at first cylindrical, almost *Oidium*-like, becoming elliptical, smooth, 4–5 × 2.5–3.5 µm. The chains of cylindrical conidia, which often make it difficult to decide where the phialides end, are highly characteristic (Fig. 11.46). Most strains deteriorate on continued cultivation, becoming floccose and non-fasciculate. Found frequently on all kinds of citrus fruits, and can always be obtained when required from such fruits which show blue-green patches of mould. It is distinguished from *P. digitatum* not only by colour and texture of the colonies on fruit, but also by the type of rot produced. Fruits attacked by *P. digitatum* shrivel and dry up, whereas *P. italicum* produces a soft rot which rapidly reduces the fruit to a slimy pulp. Even a very small colony of *P. italicum* will render the whole fruit nauseous. This species is found occasionally on other substrates but is of little importance outside the citrus fruit industry.

P. expansum series

The colony colour is normally grey-green, but may tend towards yellow-green or blue-green. Zonation is usually evident and may be marked. Raper and Thom included two species here, *P. expansum* and *P. crustosum*. However, although *P. crustosum* produces slight rot of apples it has very rough conidiophores and seems nearer to the *P. viridicatum* series.

P. expansum Link ex S. F. Gray (Lat. *expansus*, spread out) (Figs 11.47, 11.48). Different strains vary considerably in the freedom with which they produce fascicles or coremia, and in the degree of zonation of colonies when grown in Petri dishes. Colonies may be granular from the first, or may be almost velvety and show fasciculation only in age.

Colonies spreading rapidly, dull green with white margin, becoming eventually brownish, reverse in some strains colourless, in others becoming deep brown, with the colour often patchy; penicilli long and compact with branches appressed against the stipe (Fig. 11.48); stipes normally smooth, but occasionally slightly rough, 3–3.5 µm diam.; phialides cylindrical with a short neck, 8–12 µm sometimes up to 15 µm by 2–3 µm; conidia elliptical at first,

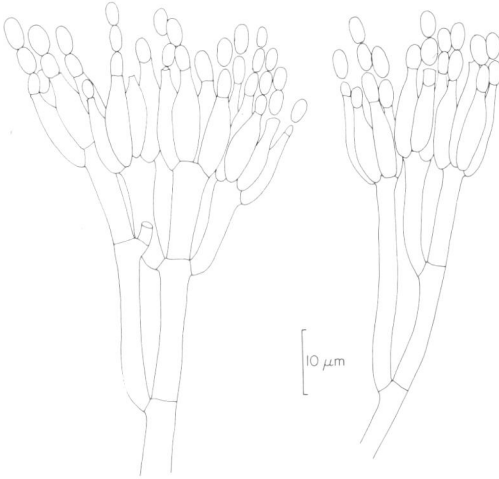

Fig.   11.45   (left)   *Penicillium italicum* Typical elongate conidial heads with elliptical conidia.

Fig.   11.46   (below)   *Penicillium italicum* Characteristic penicilli, with chains of bead like conidia **(a)** ×550; **(b)** ×500; **(c)** ×1300.

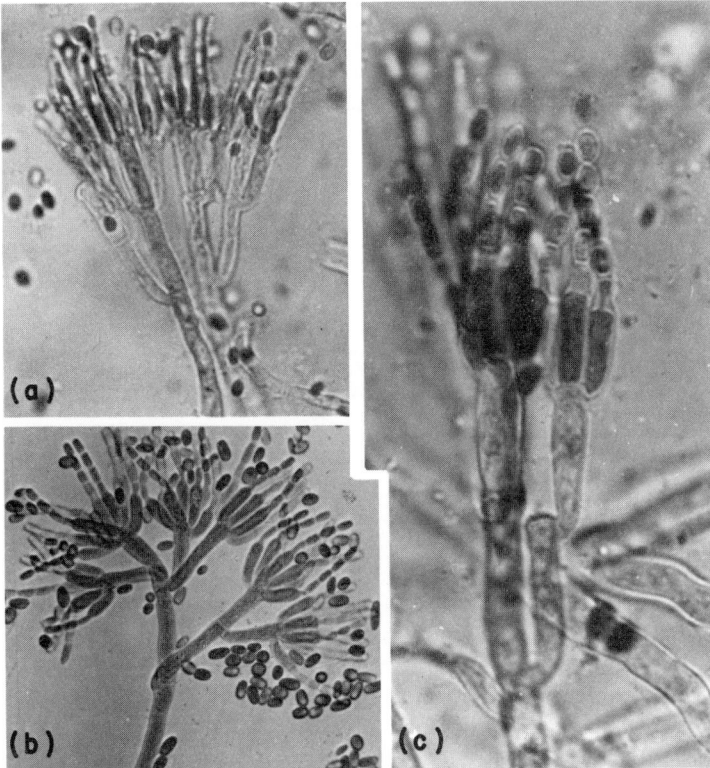

remaining so or becoming subglobose, smooth, 3–3.5 μm in long axis, borne in loose columns or in tangled chains; conidial chains often persistent in fluid mounts. Occurs on a variety of substrates, including textiles and paper, but is best known as the cause of a very distinctive brown rot of apples in storage,

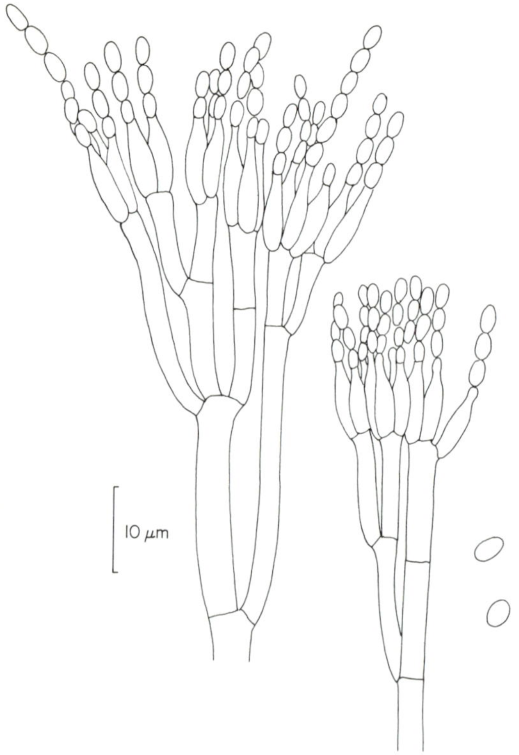

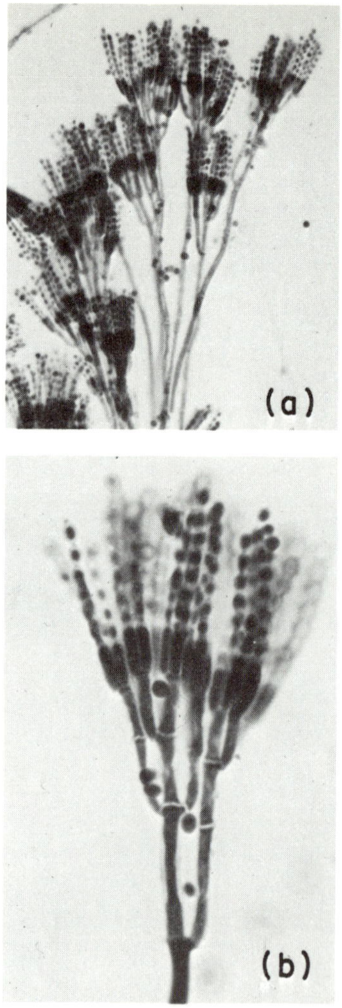

Fig. 11.47 (above)   *Penicillium expansum*
Narrow elongate penicilli.

Fig. 11.48 (right)   *Penicillium expansum*
Typical penicilli; **(a)** ×250; **(b)** ×500.

and also, to a lesser extent, of a number of soft fruits. In advanced stages of the disease conidial areas appear on the brown sodden patches in the form of blue-green coremia.

Some strains of this common organism are distinguished from *P. cyclopium* with difficulty. The main points of difference are the conidiophores of *P. cyclopium* are usually rough and of *P. expansum* smooth; the penicilli of *P. expansum* have a long drawn-out appearance (Fig. 11.48) whereas those of *P. cyclopium* tend to be shorter; the spores of *P. cyclopium* are mostly globose and those of *P. expansum* elliptical; *P. expansum*, when inoculated into sound apples, produces large areas of soft brown rot in 7–10 days, whereas *P. cyclopium* has little or no effect; and *P. expansum* produces a sweetish, rotten apple odour and *P. cyclopium* a sharp earthy odour.

P. griseofulvum series

There is only one species in this series, though it has several synonyms, of which *P. patulum* Bainier and *P. urticae* Bainier appear frequently in the literature.

P. griseofulvum Dierckx (Figs 11.49, 11.50). Colonies restricted with abrupt margins, pale grey-green to almost pure light grey, thick with prominent fascicles; drops colourless, large; reverse pale dull yellow to brownish; penicilli loosely divergent with three to four stages of branching; conidiophores smooth, sinuate, 3–4 μm diam.; phialides short, 4.5–6 μm long, crowded; conidia elliptical to subglobose, smooth, 2.5–3 μm in long axis, borne in divergent chains. Common in soil and fairly frequently isolated from other sources. It produces the substance variously known as 'patulin', 'expansin', 'claviformin' etc. which was originally thought of as an antibiotic but now regarded as an important mycotoxin.

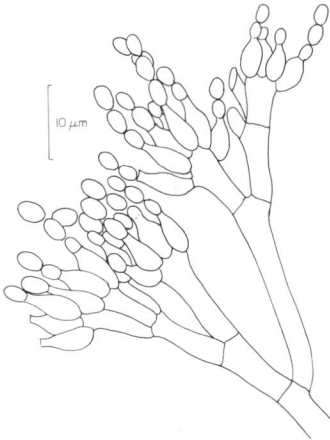

Fig. 11.49 (left)   *Penicillium griseofulvum*
Typical conidial head.

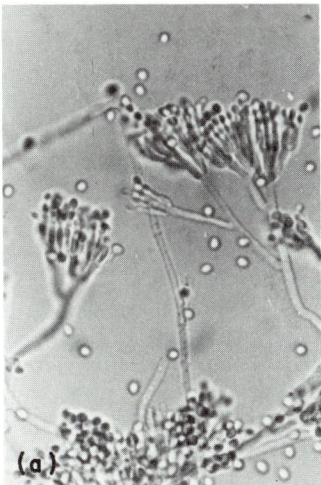

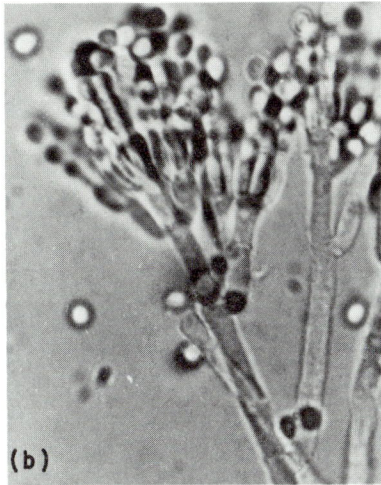

Fig. 11.50 (below)   *Penicillium griseofulvum*   Conidial heads. **(a)** ×550; **(b)** ×1300.

Species with rough stipes

Floccosa (with floccose colonies)

P. camembertii series

Raper and Thom included two species in this series, one being green, *P. camembertii*, and the other white, *P. caseicolum*. Samson *et al.* (1977) on examination of the considerable literature on these fungi and study of the type strains and additional isolates concluded that separation on colour is not sufficient. Both are used in the cheese industry and produce cheeses of somewhat different flavours. The white isolates appear more common.

P. camembertii Thom (of Camembert cheese) = *P. caseicolum* Bainier (Fig. 11.51). Colonies floccose, white remaining so or slowly turning pale greyish green from the centre outwards; reverse uncoloured or cream; penicilli irregular, with rami and metulae at the same level and with few branches at each stage; stipes slightly rough, 2.5–3.5 μm diam.; phialides 9–14 μm by about 2.5 μm conidia elliptical at first then subglobose, up to 5 × 4.5 μm, smooth, borne in tangled chains.

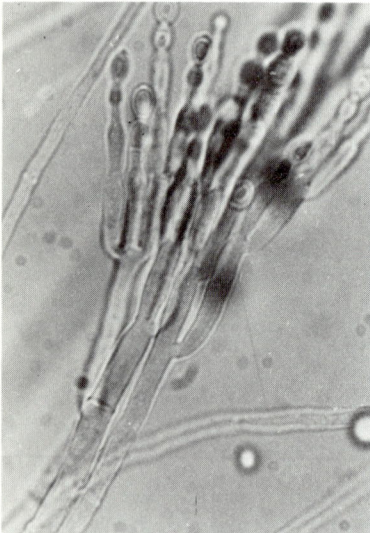

Fig. **11.51** *Penicillium camembertii* Conidial head, ×1300.

P. commune series

Raper and Thom described seven species in this series of their lanata subsection. However, they are not common and Samson *et al.* (1976) included *P. commune* with the fasciculate species. Pitt regards *P. commune* and *P. lanosum* as synonyms of *P. puberlum*. It seems probable that many of the species previously assigned here may have been isolates of other species which had gone fluffy with cultivation.

Funiculosa (with funiculose colonies)

Raper and Thom used this section to classify isolates showing ropy growth and divided it into two series. One series was green and the other did not show any green colour. The green species were not all distinct and it seems likely that ropy isolates are often deteriorated forms from other sections. Pitt regards most of them as synonyms of other species or of doubtful designation. The ropy forms which do not show any green colour, Raper and Thom's *P. pallidum* series, have a distinct type of penicillus and Pitt has transferred them to a new genus, *Geosmithia* (see Table 11.3). They are not common.

Velutina (with velvety colonies)

P. roquefortii series

Raper and Thom included two species in this series, *P. roquefortii* and *P. casei*. However, Samson, Eckardt and Orth (1977) consider *P. casei* as a synonym of *P. verrucosum* var. *cyclopium*. Pitt regards the isolates, described by Raper and Thom as *P. casei*, as *P. roquefortii*. The species is usually very distinct, though it has sometimes been confused with *P. cyclopium*, but the stipes are more warted and coarser and the conidia larger.

P. roquefortii Thom (of Roquefort, a variety of blue-veined cheese). Colonies usually broadly spreading, blue-green then duller dark green, smooth, velvety with irregular margin consisting of radiating lines of conidiophores, usually termed 'arachnoid' from its similarity to a spider's web; reverse sometimes almost colourless but more usually very dark green, almost black; penicilli distinctly asymmetric, commonly with three stages of branching; stipes rough, 4–6 μm diam.; metulae often rough; phialides 8–12 × 3–3.5 μm; conidia globose or nearly so, mostly 4–6 μm diam. but occasionally larger, smooth, borne in loose columns or tangled chains. Most isolates grow normally on almost any culture medium but occasional strains, especially those from French Roquefort, grow very poorly on Czapek agar. All strains grow well on malt agar and many of them produce smooth conidiophores. This species is used for ripening cheeses of the Roquefort, Stilton and Gorgonzola types. Many local varieties of the blue-veined cheese also owe their characteristic appearance and flavour to *P. roquefortii* although it cannot be said that the mould is consciously used. During recent years a tan-coloured mutant of the mould has appeared, and has caused a good deal of trouble in the industry, because it detracts from the appearance of the cheese. The remedy is to pasteurize the milk, since this appears to be the main source of infection, clean the air of the factory, and work with pure cultures of the green mould. It is also common in silage and compost heaps.

Samson, Eckardt and Orth (1977) made a comprehensive study of *Penicillium* species from cheese. *P. roquefortii* has recently been reported to produce toxic substances (Wei and Liu, 1978).

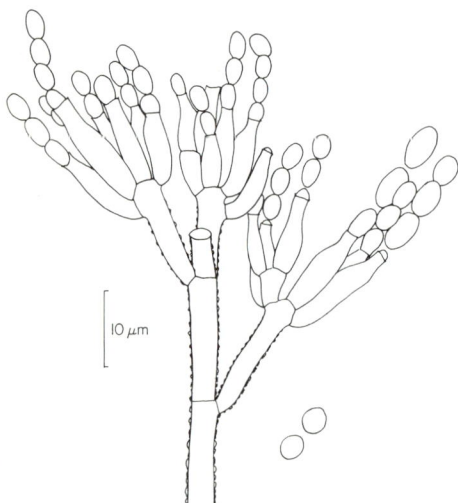

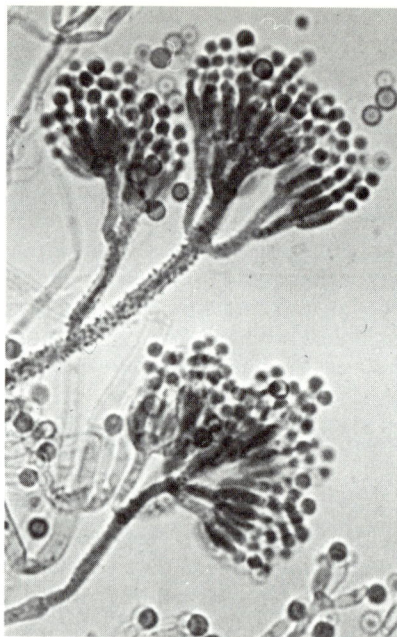

Fig. 11.52 (left)   *Penicillium roquefortii*   Spreading conidial head with very rough stipe.

Fig. 11.53 (right)   *Penicillium roquefortii*   Conidial heads showing rough stipe, ×550.

Fasciculata (with fasciculate colonies) or series Viridicata

Classified in this subsection are some of the most interesting species of the genus, species which are of widespread occurrence and of considerable economic importance. The distinguishing feature of the group is the production of erect bundles (fascicles) of conidiophores (stipes) either small and simple bundles or true coremia consisting of tightly compacted bundles. Species which regularly produce coremia are readily recognized as belonging here, and the same may be said of certain others whose colonies present a rough, granular appearance. There are, however, a number of species with a more or less velvety appearance when grown on ordinary laboratory media and which show obvious fasciculation only at the edges of colonies, or only when grown under special conditions or on special media. Most such species have a suggestion of mealiness on the surface of colonies, and examination of the edges of growing colonies will usually reveal small tufts of stipes. Another way of demonstrating fasiculation in cases of doubt is to pick off a small portion of a colony, from the edge or just inside it, immerse in a drop of alcohol on a slide, then mount in lactophenol without teasing out. Small bundles of conidiophores are thus readily observed (see Fig. 11.54a).

This is the area which has attracted most interest from modern authors as many workers have experienced difficulty in separating the species, especially as some of the commonest species are mainly separated on the basis of colour. Since Raper and Thom's monograph major revisions have been undertaken by Fassatiova (1977). Samson *et al.* (1976) and Pitt (1979) with attempts at chemical separation by Kulik and Vincent (1973), Lee *et al.* (1976) and Dart *et al.* (1976).

Samson *et al.* (1976) in their paper on fasciculate Penicillia describe 13 species and six varieties. Of these the species *P. verrucosum* Dierckx had five varieties, which included species regarded by Raper and Thom as *P. viridicatum*, *P. corymbiferum* and *P. cyclopium*. Fassatiova modified some of the descriptions of these species and gave them different names. Differences based largely on colour, shades of green, and rate of growth, with little difference in structure of the penicillus merge into each other when many isolates are examined. However, broadly defined more or less distinct types of growth tend to appear regularly. This is not the place to go into the problems of nomenclature and priorities, and in the immediate future, until some compromise is reached, the same taxa will tend to be designated by different names according to opinion. So to help with interpretation of the literature the commonest names likely to be met with will be given for each taxa.

Samson, Hadlock and Stolk (1977) included all the fasciculate terverticillate species in their study. For convenience in separation several of the fasciculate species with smooth stipes have been discussed already. Only the commoner species will be dealt with here. Fassatiova included *P. expansum*, which usually has smooth stipes, in her study and called her section after this fungus. It is certainly closely related, see p. 254.

P. viridicatum Westling (Lat. *viridicatus*, made green; is the name used by Raper and Thom and by Pitt (=*P. verrucosum* Dierckx in Samson; = *P. olivinoviride* Biourge emend. Fassatiova in Fassatiova). Pitt also uses the name *P. verrucosum* for isolates which are more restricted in growth.

Colonies growing restrictedly, 2.5–3.5 cm in 14 days, bright yellow-green, sometimes with a narrow somewhat bluish-green zone just inside the white margin, fairly thick, usually distinctly granular; reverse pale dull yellow to dull brown; odour pronounced earthy; penicilli with normally two to three stages of branching, often with branches and metulae borne at the same level, stipes rough, long, 100–400 × 2.5–4 µm; phialides flask-shaped, 7–10 × 2–3 µm; conidia globose to subglobose, sometimes elliptical when first formed, smooth to very finely roughened, 3–4 µm diam. Not very common. Reported as producing mycotoxins.

P. cyclopium Westling (Gr, *kyklops*, round-eye; probably refers to the clean circular outline of colonies). This is the name used by Raper and Thom (=*P. verrucosum* var. *cyclopium* (Westling) Samson, Stolk and Hadlock in Samson *et al.* = *P. cyclopium* Westling emend. Fassatiova in Fassatiova; = *P. aurantiogriseum* Dierckx in Pitt). (Fig. 11.54, also Figs 11.1, 11.2.)

Colonies rather dull blue-green, with brighter zone inside the white margin, almost velvety but showing distinct fasciculation in the younger areas; reverse usually pale peach but occasionally fairly bright yellow or purplish brown; the penicilli as for *P. viridicatum*. This is probably the commonest of the Penicillia, at least in this country. It is found on almost every conceivable type

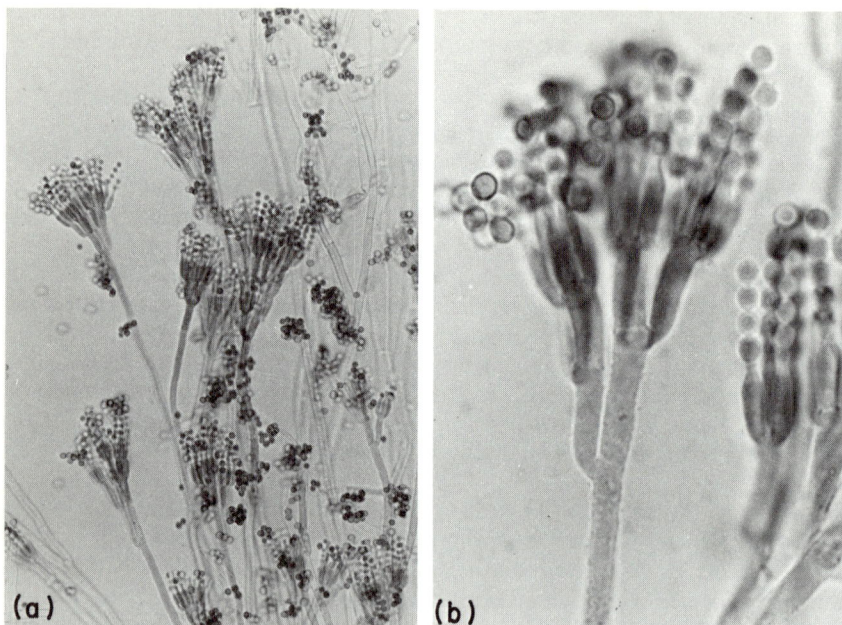

Fig. 11.54 *Penicillium cyclopium* **(a)** Mounted specimen showing fasicle, ×275; **(b)** typical penicillius, with rough stipe and appressed branch, ×1100.

of organic substrate; it occurs, along with *P. brevicompactum*, on various species of agarics, and it is a parasite of bulbous plants. Tulips, in particular, are liable to be attacked at the growing point, producing, as a result, distorted growth with blooms mis-shapen or lacking. It is also reported as producing mycotoxins.

P. martensii Biourge was distinguished from *P. cyclopium* by Raper and Thom as having a bluer colour but is now generally regarded as a synonym of *P. cyclopium*.

Isolates similar to *P. viridicatum* and *P. cyclopium* lacking green colour, either white or ochraceous, are seen from time to time and have been described as varieties of these species or as separate species. However, they are relatively rare.

P. crustosum Thom (Lat. *crustosus*, crusty) was classified in the same series as *P. expansum* by Raper and Thom and regarded as a synonym of *P. verrucosum* var. *cyclopium* by Samson, Hadlock and Stolk. However, isolates are frequently seen which are dull yellow to grey-green, fairly spreading, almost velvety, and forming crusts of conidia which break off in irregular pieces when the dish is tapped. The penicilli are very similar to those of *P. viridicatum* and *P. cyclopium* usually with pronounced roughening of the stipes. It is found fairly frequently on various materials and produces a brown rot of pomaceous fruit.

P. echinulatum Raper and Thom ex Fassatiova (Gr. *ekhinos*, hedgehog, sea-urchin; referring to the spines on these animals). Raper and Thom described a variety of *P. cyclopium* with spiny conidia. Fassatiova regarded it as a separate species and this has been accepted by Samson *et al.* and Pitt. It is very similar to *P. cyclopium* except for its slightly darker colour and rough walled conidia.

P. corymbiferum Westling (Lat. ex Gr. *corymbus*, a cluster of flowers; *fero*, to bear) (=*P. verrucosum* var. *corymbiferum* (Westling) Samson, Stolk and Hadlok in Samson *et al.*; Pitt recognizes the name *P. hirsutum* Dierckx).

Colonies spreading, yellow-green to olive-green, with broad white margin, mycelium yellow, markedly fasciculate; drops yellow-brown to blood red (usually on Czapek and not on malt agar); reverse deep reddish brown with the colour diffusing somewhat; penicilli compact, with three stages; stipes rough, 3.5–4.5 μm diam.; phialides 9–12 × 2–2.5 μm; conidia globose, smooth, variable in size but mostly about 3 diam., borne in tangled chains. Not uncommon, especially in soil. It is an active parasite of lilaceous bulbs and root stocks and frequently isolated from this source.

P. claviforme Bainier (Lat. *claviformis*, club-shaped). Some species have stipes mostly aggregated in definite coremia. One of these, *P. claviforme*, is occasionally seen and its prominent coremia with pinkish stalks and large green heads are unmistakable. It is common on dung, where the coremia are readily visible to the naked eye, and it is sometimes isolated from soil.

P. gladioli Machacek (of *Gladiolus*; from its parasitic habit) (= *Eupenicillium crustaceus* Ludwig) was included in the Fasciculata by Raper and Thom. It is parasitic on gladiolus corms and produces pinkish sclerotia at 20°C which occasionally ripen to form ascospores. It is seldom seen other than on *Gladiolus*.

Divaricata (Furcatum section Divaricata)

Most species of *Penicillium* have definite distinctive branching systems. However, some species do not follow this pattern and produce branches irregularly, almost verticillately or in a mixture of branching systems. Raper and Thom included these under their section Asymmetrica subsection Divaricata, while Pitt places them in his subgenus Furcatum as a section Divaricata. The species included by both authors vary somewhat. Raper and Thom include some which seem truely biverticillate, for example *P. raistrickii* while Pitt includes species which were regarded by Raper and Thom as branched Monoverticillata, for example *P. waksmanii*. However, these marginal species are rare and both authors classify the common species with irregular branching together as Divaricata. Here it is proposed to treat them as three main series. This is a simplification but should assist separation of the common species. (See key to common species of Divaricata.)

Key to common species of Divaricata

No green colour ....

- growth spreading, elliptical conidia, abruptly tapered phialides with long tube
  - delicate conidiophores almost smooth ...... *P. janthinellum*
  - (fragmentary heads .... *P. ochrochloron*)
  - robust rough conidiophores .... *P. simplicissimum*

Green colour

- growth restricted, globose conidia
  - smooth conidiophores rough conidia .... *P. nigricans* (= *P. janczewskii*)
  - rough conidiophores smooth conidia .... *P. canescens*

Paecilomyces

P. janthinellum series or series Janthinella

Only two species are common and one other worth mention. They all produce spreading floccose colonies, grey-green conidia, penicilli borne irregularly from trailing aerial hyphae; phialides which are characteristic of the series are abruptly tapered to a narrow conidium bearing tube, and conidia globose to elliptical.

P. janthinellum Biourge (Lat. *janthinellus,* somewhat violet) (Figs 11.55, 11.56) is an extremely common organism. It is widely distributed in soil and, like most common soil Penicillia, is found on a wide variety of substrates. For the beginner it is a puzzling species because different isolates may show very considerable differences in appearance of colonies. Colonies usually form a tough felt, with surface slightly floccose or ropy, radially furrowed, pale grey-green or greenish grey; reverse in many freshly isolated strains reddish to purple, in others, and most old strains, colourless to pale dirty yellow; respiration drops pale brownish or lacking; penicilli irregular, sometimes monoverticillate, sometimes forming fairly compact heads with metulae and phialides, but with the metulae often of different lengths, sometimes with loose irregular branching; stipes delicate, smooth or slightly rough, 3–3.5 μm diam.; phialides divergent, 8–10 × 2–2.2 μm, tapering abruptly, but with comparatively long slender tips (this species forms one of the transitions between *Penicillium* and *Paecilomyces*); conidia mostly elliptical but some-

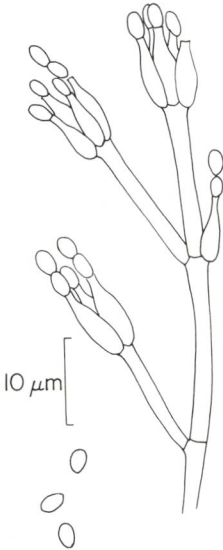

Fig. 11.55 (above) *Penicillium janthinellum* Divaricate penicillus, and tapered phialides.

Fig. 11.56 (right) *Penicillium janthinellum* **(a)** Divaricate penicillus as seen in living culture, ×100; **(b)** conidial head, ×1100.

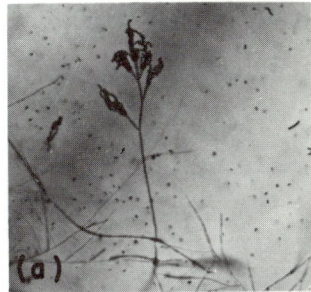

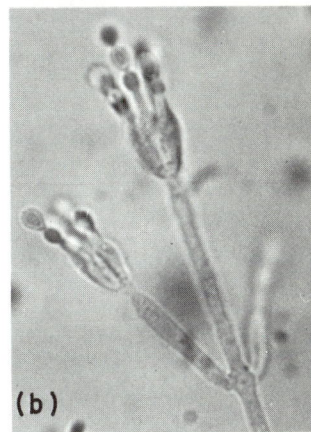

times nearly globose, delicately roughened, 3–3.5 μm in long axis, borne in divergent or tangled chains.

P. simplicissimum (Oudemans) Thom (Lat. *simplicissimus*, very simple) (Figs 11.57, 11.58). Colonies pale blue-green, thick velvety with a tough basal felt; reverse colourless to pale dull yellow; penicilli usually consisting of a group of 2–4 metulae and appearing as a bunch of monoverticillate heads, and could be included in the Biverticillata-asymmetrica group, or occasionally as several short branches from trailing hyphae, each bearing a cluster of phialides; long stipes, heavy, rough, about 2.5 μm diam.; phialides tapering abruptly at tip, 8–10 × 2–2.5 μm; conidia elliptical to subglobose, 2.5–3 μm in long axis, delicately roughened, in divergent to tangled chains. Differs from *P. janthin-ellum* in its brighter conidial colour, lack of red colour in reverse, frequent biverticillate pencilli and long, rough robust stipes.

P. ochrochloron Biorge produces very sparse fragmentary conidial heads and is comparatively rare, but was found to attack military equipment and is

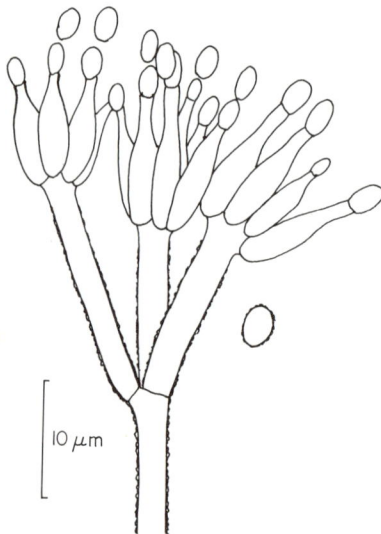

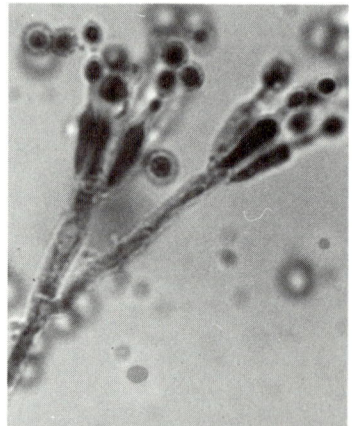

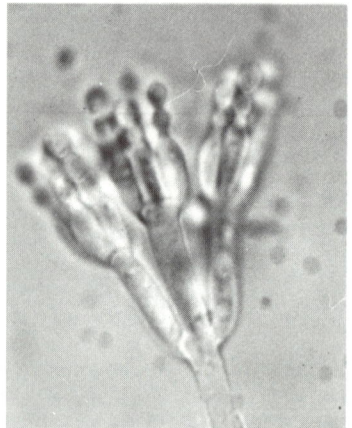

Fig. 11.57 (above) *Penicillium simplicissimum* Penicillus with terminal whorl of metulae, rough heavy stipe, rough elliptical conidia and tapered phialides.

Fig. 11.58 (right) *Penicillium simplicissimum* Conidial heads, ×1300.

particularly resistant to copper based fungicides. It is used as a test organism for testing the resistance of materials to fungal attack. Its structure is very fragmentary and the phialides are typical of this series.

P. lilacinum series

Raper and Thom included species lacking green conidia and with divaricate branching as the *P. lilacinum* series of the Divaricata. Samson (1974) transferred *P. lilacinum* to *Paecilomyces*. It is definitely a marginal species and as it lacks green colour it is described under *Paecilomyces* (p. 219).

P. canescens and *P. nigricans* series (*Series Canescentia*)

Pitt treats these two series as one, and indeed the species are indistinct and difficult to distinguish as they tend to merge. All species produce restricted growth, divaricate branching, stipes or conidia or both rough, phialides ampuliform with a distinct, short, broad neck, and globose conidia. Of the several species in the series two are sufficiently common to be described here.

P. janczewskii Zaleski (= *P. nigricans* Bainier (Lat. *nigricans,* becoming black)). Pitt shows that the name *P. janczewskii* has priority though this is rather disappointing as the name *P. nigricans* is well known and approriate. A soil organism of wide distribution.

Colonies thick velvety, in fresh isolates almost pure grey becoming very dark in age, in old strains usually a definite greenish shade at first, becoming slowly dark grey; reverse deep orange-red to brownish orange; penicilli very divaricate, consisting either of a fairly definite cluster of metulae, or metulae

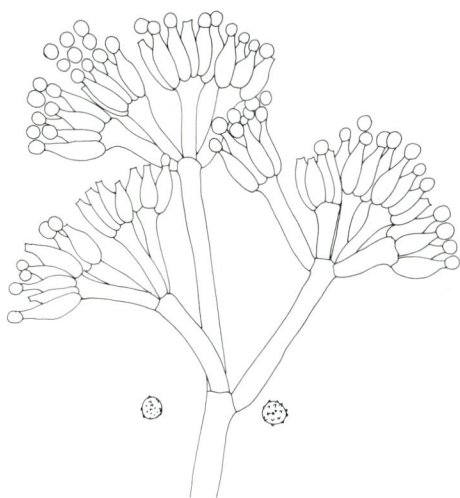

Fig. 11.59　*Penicillium janczewskii*　Divaricate penicillus, rough, globose conidia.

Fig. 11.60　*Penicillium janczewskii*　Conidial head showing divaricate penicilli, ×500.

scattered along the terminal portion of the stipe, or of two or more small clusters of metulae; stipes smooth, 2.5–3 µm diam.; phialides 7–8 × 2 µm approximately; conidia globose, very rough, spiny, 3–3.5 µm diam., in chains at first roughly parallel then tangled.

P. canescens Sopp (Lat. *canescens*, becoming white) is also a soil organism, but not so common. However, it is an interesting spoilage organism as being one of the dominant fungi found on the paint of box girder bridges. It differs from *P. janczewskii* in producing grey-green colonies with conidia produced in columns, with stipes rough and globose conidia smooth or almost so (Fig. 11.61).

Fig. 11.61 *Penicillium canescens* Penicillus, ×1000.

Table 11.4 Species of Penicillium most commonly recorded as producing mycotoxins

Species	Mycotoxin
Penicillium cyclopium	Cyclopiazonic acid, penicillic acid
P. viridicatum	Ochratoxins and citrinin
P. citreo-viride	Citreoviridin
P. islandicum	Luteoskyrin, islanditoxin, cyclochlorotine
P. rubrum and P. purpurogenum	Rubratoxins
P. rugulosum	Rugulosin
P. patulum = P. urticae = P. griseofulvum	Patulin, griseofulvin
P. expansum	Patulin, citrinin
P. citrinum	Citrinin
P. crustosum	Penitrem A

11.6   References

BARNETT, H. L. and LILLY, V. G. (1962). A destructive mycoparasite, *Gliocladium roseum. Mycologia,* **54,** 72–7.

BENJAMIN, C. R. (1955). Ascocarps of *Aspergillus* and *Penicillium. Mycologia,* **47,** 669–87.

BREFELD, O. (1874). Botannische Untersuchungen uber Schimmelpilze. Heft 2: Die Entwicklungsgeschichte von Penicillium Leipzig.

BROWN, A. H. S. and SMITH, G. (1957). The genus *Paecilomyces* Bainier and its perfect stage *Byssochlayms* Westling. *Trans. Br. mycol. Soc.,* **40,** 17–89.

DART, R. K., STRETTON, R. J. and LEE, J. D. (1976). Relationships of *Penicillium* species based on their long-chain fatty acids. *Trans. Br. mycol. Soc.,* **66,** 525–9.

EMMONS, C. W. (1935). The ascospores in species of *Penicillium. Mycologia,* **27,** 128–50.

FASSATIOVA, O. (1977). A taxonomic study of *Penicillium* series *expansa* Thom emend. Fassatiova. *Acta Universitatis Carolinae-Biologica 1974,* **12,** 283–335.

HETHERINGTON, A. C. and RAISTRICK, H. (1931). On the production and chemical constitution of a new yellow colouring matter, citrinin, produced from glucose by *Penicillium citrinum* Thom. *Phil. Trans., Ser. B.,* **220,** 269–95.

HUGHES, S. J. (1953). Conidiophores, conidia and classification. *Can. J. Bot.,* **31,** 577–659.

JACKSON, C. R. (1962). *Penicillium* core-rot of *Gladiolus. Phytopathology,* **52,** 794–7.

KENDRICK, B. (ed.) (1971). *Taxonomy of Fungi Imperfecti.* University of Toronto Press, Toronto and Buffalo.

KULIK, M. M. (1968). A compilation of descriptions of new *Penicillium* species. Agricultural Handbook, 351, Agricultural Research Service, USDA.

KULIK, M. M. and VINCENT, P. G. (1973). Pyrolysis-gas-liquid chromatography of fungi: observations on variability among nine *Penicillium* species of the section Asymmetrica, subsection Fasciculata. *Mycopath. Mycol. appl.,* **51,** 1–18.

LANGERON, M. (1922). Utilité de deux nouvelles coupures génériques dans les Perisporiacés Diplostephanus n.g. et Carpenteles n.g. *C.R. Soc. Biol., Paris,* **87,** 343–5.

LEE, J. D., STRETTON, R. J. and DART, R. K. (1976). Classification of *Aspergillus* and *Penicillium* species. *Microbios Letters,* **2,** 163–7.

MOORE, W. C. (1941). New and interesting plant diseases. *Trans. Br. mycol. Soc.,* **25,** 206–10.

PITT, J. I. (1974). A synoptic key to the genus *Eupenicillium* and to sclerotigenic *Penicillium* species. *Can. J. Bot.,* **52,** 2231–6.

PITT, J. I. (1979). The genus *Penicillium* and its telemorphic states, *Eupenicillium* and *Talaromyces.* Academic Press, London.

RAPER, K. B. (1957). Nomenclature in *Aspergillus* and *Penicillium. Mycologia,* **49,** 644–62.

RAPER, K. B. and THOM, C. (1949). *A manual of the Penicillia.* Williams and Wilkins, Baltimore.

SAMSON, R. A. (1974). *Paecilomyces* and some allied Hyphomycetes. *Studies in Mycology,* **6.** Centraalbureau voor Schimmelcultures, Baarn.

SAMSON, R. A., ECKARDT, C. and ORTH, R. (1977). The taxonomy of *Penicillium* species from fermented cheeses. *Antonie van Leeuwenhoek,* **43,** 341–50.

SAMSON, R. A., HADLOK, R. and STOLK, A. C. (1977). A taxonomic study of the *Penicillium chrysogenum* series. *Antonie van Leeuwenhoek,* **43,** 169–75.

SAMSON, R. A., STOLK, A. C. and HADLOK, R. (1976). Revision of the subsection Fasciculata of *Penicillium* and some allied species. *Studies in Mycology,* **11.** Centraalbureau voor Schimmelcultures, Baarn.

SCOTT, DE B. (1968). *The genus* Eupenicillium *Ludwig.* Council for Scientific and Industrial Research Report No. 272, South African CSIR, Pretoria.

SCOTT, DE B. (1968). Studies on the genus *Eupenicillium* Ludwig. IV. New species from soil. *Mycopath. Mycol. appl.,* **36,** 1–27.

SHEAR, C. L. (1934). *Penicillium glaucum* of Brefeld (Carpenteles of Langeron) refound. *Mycologia,* **26,** 104.

SMALLEY, F. B. and HANSEN, H. N. (1957). The perfect stage of *Gliocladium roseum. Mycologia,* **49,** 529–33.

STOLK, A. C. (1965). Thermophilic species of *Talaromyces* Benjamin and *Thermoascus* Miehe. *Antonie van Leeuwenhoek,* **31,** 262–76.

STOLK, A. C. and SAMSON, R. A. (1971). Studies on *Talaromyces* and related genera. I. *Hamigera* gen. nov. and *Byssochlamys. Persoonia,* **6,** 341–57.

STOLK, A. C. and SAMSON, R. A. (1972). The genus *Talaromyces.* Studies on *Talaromyces* and related genera. II. *Studies in Mycology,* 2. Centraalbureau voor Schimmelcultures, Baarn.

STOLK, A. C. and SCOTT, DE B. (1967). Studies on the genus *Eupenicillium* Ludwig. I. Taxonomy and nomenclature of Penicillia in relation to their sclerotioid ascocarpic states. *Persoonia,* **4,** 391–405.

THOM, C. (1910). Cultural studies of species of *Penicillium. US Dept. Agric. Bur. Anim. Ind., Bull.,* **118.**

THOM, C. (1930). *The Penicillia.* Williams & Wilkins, Baltimore.

UDAGAWA, S. (1966). Notes on some Japanese Ascomycetes III. *Trans. mycol. Soc. Japan,* **7,** 91–7.

WEI, R. and LIU, G. (1978). PR Toxin production in different *Penicillium roqueforti* strains. *Appl. and envir. Microbiol.,* **35,** 797–9.

12

Growth Requirements and Ecology

The fungi, in common with other living organisms, possess tools or reagents far more specific, more delicate, and more powerful than those available in the laboratory.

Lilly and Barnett, *Physiology of the Fungi*, 1951

12.1 Introduction

The growth of fungi, like that of all other living things, is profoundly affected by environment. Variations in external conditions may not only affect the rate of growth of a mould, but, in many cases, can bring about differences in type of growth. Ignorance of this fact led many of the older mycologists to describe as different species, or even different genera, what were in reality physiological variants of one and the same species.

One thing which soon becomes obvious is that very few generalizations can be made about these organisms. In their food requirements, response to external stimuli, and tolerance of adverse conditions they show the utmost diversity.

12.2 Food requirements

Water and oxygen are both absolutely necessary for growth of fungi and, in addition, the following macroelements are required – carbon, nitrogen, phosphorus, potassium and magnesium. Iron, zinc, copper, manganese and molybdenum are known as microelements, being required in concentrations of parts per million whilst the macroelements are needed at much higher concentrations. Some additional microelements such as calcium and gallium may be required by some fungi, possibly only when utilizing a particular nutrient (Hutner, 1972).

As all the essential trace elements are required by plants, culture media made from parts of plants are rarely deficient in this respect. Some of the elements are required in such minute amounts that they occur in adequate concentrations in A.R. chemicals or are extracted, during sterilization, from the glass of culture vessels. This being so, it is apparent that very elaborate methods of purification are necessary in studying the effects of trace elements. On the other hand, it has been shown by Smith (1949) that Czapek agar made up with A.R. chemicals and glass-distilled water gives abnormal growth of many species of *Penicillium*, and that amounts of zinc and copper comparable with the amount of iron can be added with advantage to Czapek solution and Czapek agar in order to induce normal growth and abundant production of spores.

Most mould fungi can utilize inorganic sources of all the essential elements except carbon, but a few are unable to use inorganic nitrogen. Others, such as *Rhizopus stolonifer*, grow well on media containing ammonium salts or amino compounds, but cannot utilize nitrogen from nitrates. However, the great variety of natural substrata which are regularly colonized by numerous species of moulds indicates that these organisms can tolerate wide variations in concentrations of essential nutrients, and can utilize both carbon and nitrogen combined in very diverse forms. An interesting observation by Gupta and Nandi (1957) is that the production of perithecia in a strain of *Penicillium vermiculatum* was dependent mainly on the concentration of nitrogen. The favourable concentrations corresponded to 0.05–0.1% of $NaNO_3$.

In addition to essential elements fungi can utilize, or at least take up from the medium, a few other elements whose presence or absence seems to have no effect on growth. Raulin analysed the ash of *Aspergillus niger* and, on the basis of his analyses, devised his well-known culture medium. One of the ingredients, potassium silicate, is certainly not essential, but when it is used, silica is found in the ash of the mould. Chlorine is not necessary for growth, the potassium chloride often added to culture media (as in Czapek-Dox solution) being used as a convenient source of potassium. A number of fungi can, however, utilize the chlorine contained in such media, building up complex organic compounds containing chlorine. Raistrick and Smith (1936) showed that one strain of *Aspergillus terreus* could utilize over 95% of the chlorine contained in Czapek-Dox solution, most of it being found in two compounds, 'geodin', $C_{17}H_{12}O_7Cl_2$, and 'erdin', $C_{16}H_{10}O_7Cl_2$. Clutterbuck *et al.* (1940) have published a survey of the chlorine metabolism of a large number of species of moulds, and shown that the majority can utilize this element to a small extent, whilst a few metabolize over 25% of the total chlorine in the medium, an outstanding species being *Caldariomyces fumago* Woronichin. This species utilizes about 95% of the chlorine in the medium producing a compound named 'caldariomycin', $C_5H_8O_2Cl_2$. Two other interesting chlorine-containing metabolic products of moulds are 'griseofulvin', $C_{17}H_{17}O_6Cl$, from *Penicillium griseofulvum* (Grove, 1964), and 'sclerotiorin', $C_{21}H_{22}O_5Cl$, from *Penicillium sclerotiorum* and *P. multicolor* (Curtin and Reilly, 1940; Birkinshaw, 1952).

It has been found that some fungi require, in addition to the substances listed above, very small amounts of complex organic compounds, some of which are vitamins, better known as essential metabolites in animal nutrition, and all of which resemble these vitamins in producing effects on growth which cannot be attributed to their value as nutrients in the usual sense. Many common moulds grow well on synthetic media made up with glucose and pure salts, and obviously these do not need to be supplied with any of the accessory growth-substances. However, it is not true to say that they do not require such substances. They synthesize sufficient for their own requirements (sometimes more than sufficient) and are therefore independent of outside sources. Other moulds, for example *Penicillium digitatum*, *Sordaria fumicola*, and many other Ascomycetes, grow very poorly or not at all on Czapek agar but spread rapidly on malt agar or other 'natural' medium. Some Ascomycetes grow rapidly on synthetic media and form perithecia, but these are abortive, containing no asci, and only when the medium is supplemented with certain growth substances are fertile perithecia formed.

The first fungus shown to require one of the vitamins already known to students of human and animal nutrition was *Phycomyces blakesleeanus*, which requires thiamin (aneurin, vitamin B_1). Since then it has been found that most fungi which need external supplies of growth-substances require thiamin, with or without other substances. Some require the complete molecules, others, for example *P. blakesleeanus*, can synthesize the vitamin if supplied with the pyrimidine and thiazole components, and some need to be supplied with only one of the two components of the thiamin molecule and hence can synthesize the other.

Biotin (vitamin H) is required by a number of fungi, although deficiencies for this are not nearly as numerous as for thiamin. Pyridoxine (vitamin B_6) is required by even fewer fungi, whilst deficiencies for inositol, riboflavin and p-aminobenzoic acid are rare. All such vitamins are required in concentrations of parts per million. Apart from the above mentioned water soluble B vitamins and vitamin H, other requirements for organic compounds in similarly low concentrations are very rare, although there are reports of stimulatory effects of certain other complex natural substances (Fries, 1965).

The knowledge of fungal requirements of growth substances has been put to good use by employing fungi as test organisms for various vitamins. Such methods are both much less expensive and much quicker than animal feeding experiments. *Phycomyces blakesleeanus* is used for estimating thiamin; a yeast-like fungus, *Nematospora gossypii*, is used to estimate biotin and inositol; and a number of deficient strains of *Neurospora sitophila* and *N. crassa* are available for estimating several vitamins and a number of amino acids. Up to the present, no fungus has been found to be deficient for riboflavin (vitamin B_2), but this can be estimated by means of species of *Lactobacillus*.

12.3 Respiration

All fungi require oxygen for growth. Many yeasts develop characteristically when completely submerged in liquid, where the amount of available oxygen is small, but, while fermentation of sugar may proceed rapidly under these conditions, the amount of new cell substance formed is very small compared with the amount of growth under truly aerobic conditions. Some other fungi produce atypical structures, which can be yeast-like, or slimy, when growing submerged in liquid, but in all such cases growth is slow and, so far as is known, depends on the oxygen dissolving in the fluid. The spores of a great many filamentous fungi will germinate when immersed in a liquid medium, but grow exceedingly slowly until some hyphae reach the surface, after which growth is rapid and typical until the surface is covered.

Normal metabolism results in the breakdown of some of the organic food material to CO_2 and, unless this is continually removed and replaced by a fresh supply of air, growth ceases, whether on account of the diminution of oxygen tension alone or by actual poisoning is not known. Thus it is rather surprising that the means taken to exclude contaminants from cultures do not have more effect on growth and sporulation. Moulds are commonly grown on liquid media in plugged test-tubes 15 cm long, the depth of liquid being approximately 3 cm. It would be reasonable to assume that with the tube in a vertical position CO_2 would accumulate above the surface of the fluid. In a culture flask or in a tube of sloped agar, with a much greater active surface relative to the plugged opening, conditions should be decidedly worse. However, analyses of air drawn from flasks containing actively growing moulds invariably show only very small differences from analyses of ordinary air, proving that there must be a very rapid removal of CO_2 and intake of fresh air.

By passing measured volumes of air through a culture vessel (properly stoppered, of course, and not plugged), or through a number of similar vessels in series, it can readily be shown that lack of an adequate supply of air not only influences rate of growth but, in many cases, alters the type of growth. One common effect of very restricted aeration is to suppress normal colour production in both mycelium and conidia. A striking instance of this is mentioned by Thom in *The Penicillia* (1930, p. 83) – when Roquefort cheese is cut the fresh surface is often colourless, but turns green very rapidly on exposure to air.

Although fungi are unable to use CO_2 for photosynthesis and normally liberate the gas, with some fungi a portion of the CO_2 produced by respiration can be reabsorbed and utilized for building up compounds containing four or more carbon atoms from the 3-carbon primary breakdown products of glucose. Experiments with labelled carbon have shown that some of the common acidic metabolic products of moulds are formed in this way (Krebs, 1943; Smith and Gilbraith, 1971). For more detailed information on fungal respiration the reader is referred to Beevers (1961).

12.4 Reaction of medium

Most of the commonly occurring moulds will tolerate a wide range of hydrogen ion concentration, provided other conditions are favourable, but, as might be expected, different species respond in various ways to changes in reaction. In general, a slightly acid medium is favourable to spore germination and rapid growth of the young colony, just as a slight alkalinity is preferred by the majority of bacteria. Once growth is established the reaction of the medium usually changes, owing to the accumulation of products of metabolism. Many species, particularly of *Aspergillus* and *Penicillium*, form fairly large quantities of organic acids, oxalic, citric, and gluconic acids being very frequently found, and many others more rarely. In the absence of metabolic products of a strongly acid nature, alteration in reaction of the culture medium depends largely on the types of inorganic salts present. For example, the ultilization of nitrogen from sodium or potassium nitrate tends to liberate base and increase the pH, and nitrogen from ammonium sulphate tends to liberate acid and decrease the pH. The presence of appreciable amounts of buffer salts tends to stabilize the reaction, or at least minimize the pH drift. It is not unusual for appreciable quantities of organic acids to be produced during the early stages of growth and then be utilized by the mould as other sources of carbon become exhausted. In such cases the pH of the medium will first decrease and then gradually increase as growth proceeds.

Sometimes the appearance of the mould itself indicates changes in reaction of the medium. In many of the green species of *Aspergillus* and *Penicillium* the bright green of the young colony is very stable, often persisting for several months, on a buffered acid medium such as beer wort. However, on Czapek's medium, which tends to become alkaline unless strong acids are synthesized, the initial green colour gradually changes to brown or grey shades. Many mould pigments, produced either in or on the mycelium or in the medium, act as indicators, showing very distinct colour changes according to alterations in reaction.

12.5   Influence of light

It is impossible to generalize concerning the effect of light on the growth of fungi. Many common species seem to grow equally well and characteristically in light or darkness, others are definitely stimulated or affected by light. For example, many of the Mucoraceae seem to be quite unaffected by moderate illumination while some strains of the common mould *Rhizopus stolonifer* grow appreciably faster in diffused light than in darkness. Some fungi are positively phototropic, i.e. they grow towards the light or orientate their fruit-bodies in the direction of maximum illumination. The sporangiophores of *Phycomyces blakesleeanus* are comparatively short when a culture is fully exposed to the light, but attain a length of 20–30 cm if the mould is grown on a layer of medium placed at the bottom of a tall vessel which allows light to enter only from the top. Figure 12.1 shows a very characteristic growth of *P. blakesleeanus* in a Roux bottle which was covered with opaque material except for a patch on one shoulder – the sporangiophores have all turned towards the point of entry of the light. A number of other species exhibit similar tendencies. The sporangia of *Pilobolus* species, which are violently shot off at maturity, point in the direction of strongest illumination; the stipe of the wood-destroying fungus, *Lentinus lepideus*, grows towards the light and the cap develops only when the intensity of the light exceeds a certain minimum; the necks of the perithecia in many Pyrenomycetes, particularly the species which grow on dung, are regularly orientated towards the point of maximum illumination. In all such cases phototropism represents an adaptation for the more efficient dispersal of spores.

Williams (1959) described the interesting effect of light on spore size. Spores of all species examined were appreciably larger when produced in the dark than when cultures were continuously illuminated. Williams discussed other factors influencing size of spores but found that the response of different species of fungi, to all the factors except light, varied considerably.

The effect of light on many moulds is to stimulate the production of spores or fruit-bodies, short exposures to sunlight or ultraviolet rays often being a useful method of inducing sporulation in cultures which have deteriorated to

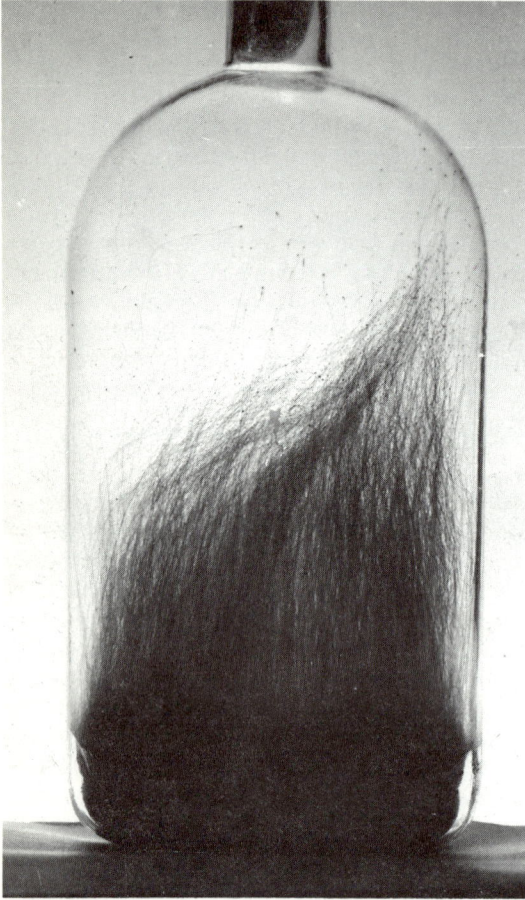

Fig. 12.1 *Phycomyces blakesleeanus* Culture in Roux bottle, illuminated from one shoulder only during period of growth, showing well-marked phototrophism, ×0.4.

the point of sterility (Leach and Trione, 1966). Longer exposures to ultra-violet rays results in death of both mycelium and spores, dark coloured spores being more resistant than pale or colourless ones. Doses of ultraviolet rays just insufficient to kill a culture completely often results in a high rate of mutation in the surviving spores. This method of inducing mutation has been used to obtain strains with altered biochemical characteristics.

Another common effect of moderate illumination is to stimulate produc-tion of mycelial pigments.

12.6 Temperature relationships

Fungi show great differences in their response to temperature changes and in their powers of resistance to heat and cold. Application of heat is, of course, the usual method of sterilizing culture media, vessels, and tools, and of killing unwanted cultures, as there are few moulds which can withstand the action of steam or boiling water for any appreciable length of time. Thermal death points, however, cover a fairly wide range. For example, some strains of *Penicillium brevi-compactum* are killed by prolonged exposure to a temperature of 33°C, while *Byssochlamys fulva*, a fungus responsible for much trouble in the canning industry, can survive the normal sterilization process for canned fruits in which, for a short time, the temperature exceeds 90°C (Olliver and Smith, 1933), and ascospores of *Neurospora sitophila*, the red bread mould, can survive inside a loaf during baking, and actually will not germinate unless previously heated to at least 60°C. On the whole, spores are more resistant than mycelium, and both are less affected by dry than by moist heat. The spores of many moulds are killed by freezing in the presence of water, but when dry their germinative powers are unimpaired by prolonged cooling in liquid air.

The limiting temperatures at which growth will occur, as distinct from mere retention of vitality, are less extreme. Spores of *Cladosporium* sp. and *Sporotrichum* sp. will germinate and grow on meat in cold storage at temperatures below freezing-point (Cochrane, 1958). Ingram (1951) states that osmophilic yeasts grow in concentrated orange juice at least down to −10°C. McCormack (1950) claims that a pink yeast, isolated from frozen oysters, will grow at temperatures down to −18°C, and according to Davis (1951), a temperature of −23°C is required to stop the growth of *Oospora lactis* (= *Geotrichum candidum*). Numerous other workers have confirmed that many moulds can grow at or just below 0°C. At the other end of the scale, some fungi will flourish at temperatures sufficient to inhibit the majority of species and to kill some of them. *Aspergillus fumigatus* grows well at 50°C; La Touche (1950) has described *Chaetomium thermophile* which grows at temperatures up to 62°C, with optimum between 40°C and 50°C; and quite a number of moulds, including species of *Mucor, Absidia, Aspergillus*, and *Penicillium*, flourish at blood heat, 37°C. Cooney and Emerson (1964) have published a monograph on the 'thermophilic fungi'. They restrict the term to fungi which can grow at 50°C or over, and which cannot grow at 20°C or below. Species which can grow below 20°C, such as *Aspergillus fumigatus*, are described as 'thermotolerant'.

Each species is found to grow best at round about some particular temperature, known as the optimum temperature for that species. On the whole, species of *Penicillium* are commonest in temperate countries, and have optima between 20°C and 25°C, while species of the nearly related genus *Aspergillus* grow best at about 30°C, and hence are found most frequently in warm regions. This is a very rough generalization, with a number of exceptions, but is often of value in industrial work, where 90% of the fungi encountered belong to these two genera. One particular section of the Aspergilli, the *A. glaucus* group (see p. 173), deserve special mention, as some of the common species have two temperature optima. Conidia and vegetative growth are produced most freely at somewhat low temperatures and, therefore, optimum temperatures as determined from growth rate are low, for some strains as low as 10°C. Perithecia, on the other hand, are formed most readily and abundantly at higher temperatures, some species being almost entirely perithecial above 30°C, and there is thus a second optimum relating to the perfect stage of the fungus. Although the majority of Penicillia have optima which are comparatively low, there are a number of species, mostly belonging to the section Biverticillata-Symmetrica (see p. 237), which show maximum growth at temperatures near to 30°C and which flourish up to at least 37°C.

A few species of the Mucoraceae have high optimum temperatures, in particular *Absidia corymbifera* and *A. ramosa*, both of which are parasitic to warm-blooded animals and show maximum development in culture at 37°C.

An interesting type of reaction to temperature change is shown by *Thamnidium elegans*. Although this species grows over a wide range of temperature, its optimum for growth is 25–27°C. Hesseltine and Anderson (1956) have, however, shown that in order to obtain zygospores from paired cultures of mating strains it is necessary to incubate at low temperature, 6–7°C.

12.7 Moisture requirements

Most fungi will only grow above 95% relative humidity – fungi are frequently found growing within substrates where there is high humidity. Where fungi are growing on surfaces it should be noted that the humidity at a surface may be considerably higher than a few centimetres away, where a hygrometer might conveniently be placed. In the laboratory where fungi are grown either in liquid culture or on agar these high humidities are easily attained. Some fungi, particularly those associated with stored products, such as members of the *Aspergillus glaucus* group, are able to grow at significantly lower relative humidity. Other fungi, known as xerophiles, can grow at humidities between 65–75% RH (see Chapter 14, section 2.4.).

Increasing the osmotic pressure of solutions is inhibiting to many micro-organisms, including most fungi, but there are some which are tolerant of high osmotic pressures; these are known as osmophiles, or if specifically tolerant of high concentrations of sodium chloride, as halophiles.

At the other end of the scale there are a number of fungi, chiefly plant parasites, which require the presence of liquid water for spore germination. Ingold (1954) has made a special study of a large group of Hyphomycetes which complete their development submerged in the water of ponds and streams. Some wood-deteriorating Ascomycetes often sporulate in submerged marine situations (Jones and Irvine, 1971).

12.8 Distribution

As fungi are heterotrophs they are capable of growing over an extremely wide range of habitats where carbon sources are available for growth. Among the fungi are those able to grow on carbon sources derived from dead animals and plants, fungi growing as parasites, and even in symbiotic association as lichens with algae.

Unlike bacteria, the other major group of mainly heterotrophic micro-organisms, the fungi, by their possession of a hyphal system, can frequently penetrate more effectively into substrates. As fungi, like all other micro-organisms, must utilize nutrients in solution it follows that many fungi must possess a means of solubilizing insoluble nutrients. This is effected by the secretion of extracellular enzymes. It is the combination of the above factors which makes the fungi such effective colonizers of the large masses of polymeric materials produced by plants and animals.

Geographic distribution of fungi, particularly the micro-fungi, is not particularly marked (unlike many other groups of organisms). Instead it is the parameters of specific habitats which appear to be of over-riding importance – tropical temperatures can be found in many parts of the temperate regions, even in winter. As spores are so ubiquitous the mycoflora of a given situation will be determined by the factors mentioned in the earlier parts of this chapter. For many situations the most important determining parameters appear to be carbon source, temperature and pH.

In considering the growth of fungi in a particular habitat, attention should always be given to the source of nutrients. This is particularly important where highly recalcitrant substrates are concerned, as the organisms may be growing on dirt or condensed gaseous organic compounds rather than on the supporting substrate.

The actual quantity of food material required to support mould growth is very small as shown by the following examples. Fungi are frequently to be found growing on polished furniture woods, or even on old iron, where the total amount of available nutrients must be exceedingly small. Textiles made of cellulose (cotton, linen, and some rayons) may be extensively mildewed without the fabrics showing the least sign of weakening or tendering, the only available food material being the small amounts of substances other than cellulose which are contained in all such fabrics. Solutions of many inorganic salts often develop, on standing, slimy growths of mould mycelium and it is common experience that many species can grow and produce spores on plain agar made up with tap-water, a medium which is, since the agar is not utilized, extremely poor in nutrient material.

It is very rare to find only one fungus growing on a substrate, and it may be very difficult to tell whether an organism is actively growing or in a dormant state. Many apparently homogeneous substrates represent to microorganisms a multitude of separate micro-environments, where there may be variations in nutrient contents, pH, aeration and other growth parameters.

The effects of temperature on the growth of fungi in substrates should not be under-estimated. Materials can vary widely in temperature both due to insolation as well as internal and external air temperatures. Thus in many substrates found in nature there are a range of microorganisms, each able to grow at particular temperatures and usually able to respond to those temperatures when they occur.

Although many fungi do have an optimal pH below neutrality it should not be assumed that fungi are not active at alkaline pH. Some cellulose decomposers are very active at a pH as high as 8.5.

12.9   Ecology

Most ecological studies on the fungi have been carried out using soil as a substrate. Individual species of fungi are rarely found growing alone in a substrate, and are frequently found in association with bacteria, and also with algae on lighted surfaces. Although much work has been done on the association of fungi with higher plants, from the industrial mycologist's viewpoint it is the involvement, often passive, of fungi with small insects and other microfauna which will be of importance to him. Unfortunately there is not a great deal known about this; certainly nematodes and mites can carry spores and some specialize in mycophagy, indeed some are associated with a specific fungus.

When a microorganism colonizes a suitable substrate, a pure culture rarely results. If the substrate is exposed to the air a large number of spores from a range of species will settle on the surface and if it is moist they are likely to germinate. If the substrate is in contact with soil the surface will be colonized initially by both mycelium already actively growing in the soil, as well as by the germination of previously dormant spores (Cartwright and Findlay, 1958).

Numerous studies have been carried out to determine the factors which affect the pattern of colonization of substrates. Of course the factors already mentioned in this chapter will influence this pattern, and the intrinsic characteristics of fungal species are also very important. Many substrates are first colonized by what are known as sugar fungi, capable of growing and sporulating rapidly on simple, soluble sugars and nitrogen sources. Later colonizers are able to utilize more highly polymerized compounds such as hemicellulose and cellulose, secreting extracellular enzymes. The latter are frequently associated with secondary fungi, which utilize some of the enzymatically produced sugars. Thus it is often difficult to isolate cellulose and lignin decomposers on standard laboratory media containing simple sugars as they are usually associated with sugar and secondary fungi that grow, sporulate and smother the polymer decomposers before the latter have had a chance to develop any substantial mycelium (Eggins and Pugh, 1962; Eggins, 1965).

From the above it can be assumed that for most substrates there will be a succession of microorganisms, both fungi and bacteria. For many solid substrates the fungi have an advantage over the bacteria in being able to effectively penetrate. The surfaces of substrates in atmospheres below 95% rH (and above 65%) are likely to be colonized by fungi in preference to bacteria as the latter can rarely grow on surfaces below 95% rH.

12.10   Influence of other fungi

If several species of fungi are growing together on the same substratum they may affect each other's growth in various ways. In the simplest case there is competition for the available food material, and the ultimate result is that some species are prevented from spreading, whilst others flourish. It is impossible to predict from knowledge of the growth rates of the individual species which will be dominant in mixed cultures, since rapid growth in pure culture does not necessarily mean ability to acquire food in face of competition. In many cases of mixed infections it is found that some of the moulds grow abnormally, producing freak or dwarfed sporing structures, or remain sterile, and it is therefore very unsafe to attempt to base identifications on examination of such material. Figure 12.2 shows the first four of a set of

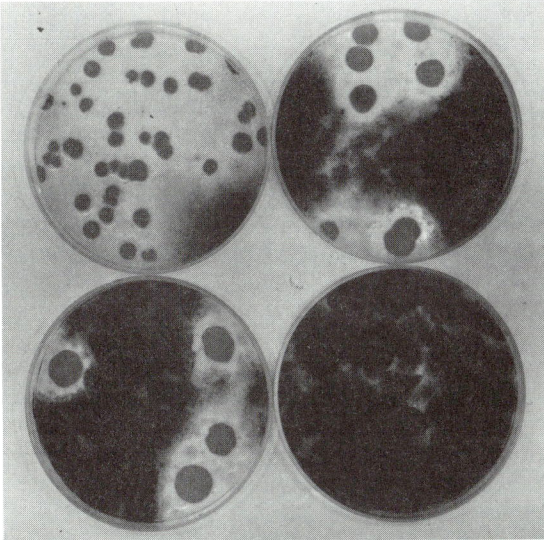

Fig. 12.2   Four of a set of dilution plates containing colonies of *Penicillium* and *Stachybotrys,* ×0.35.

dilution plates made from a mixed culture of *Stachybotrys atra* and a species of *Penicillium*. The fact that the fourth plate still contains scores of colonies of the *Stachybotrys* (black) indicates that the fruiting of this species has been completely suppressed in the vicinity of the *Penicillium* colonies (grey).

In some cases, not only is there competition for food, but active antagonism. Some species, when growing in proximity to other moulds, produce more or less circular colonies surrounded by clear zones in which growth of the competing organisms is inhibited. This indicates, of course, that a toxic substance is being produced and liberated into the culture medium. Such substances, commonly known as antibiotics, are more suitably described in connection with the uses of fungi (Chapter 15).

In contrast to the inter-reactions described above are many cases in which the growth of one fungus assists that of another. Often, when a suitable substrate is exposed to chance infection, there is a well-marked progressive change in the fungal population, correlated with the different enzymatic activities of the various species. One group of species breaks down complex organic material to provide suitable food for another group, while a third group will utilize the metabolic products of the second. Cellulosic materials in nature, particularly lignified cellulose, are usually broken down in stages, and the same applies to dung.

Another case of frequent occurrence is for one species to synthesize, in excess of its own requirements, a growth-substance essential for another species. The first such case to be described was by Heald and Pool (1908) and, in consequence, the phenomenon is often termed 'the Heald-Pool effect'. They found that *Melanospora pampeana* grows very poorly in pure culture and produces no perithecia, but thrives and gives abundant perithecia when grown in mixed culture with *Fusarium moniliforme*. *Melanospora* (now *Sordaria*) *destruens* does not produce perithecia on a medium containing only salts and sugar, but is stimulated by growing in mixed culture with any one of a large number of moulds and bacteria. Hawker (1939) has shown that *S. destruens* requires both thiamin and biotin for production of fruit-bodies and is stimulated by any organism which produces both these growth-substances.

True parasitism amongst the fungi is by no means uncommon. The higher fungi, particularly mushrooms and toadstools, are attacked by a large variety of micro-fungi, and even a few of the species of moulds commonly grown in the laboratory are parasitized, usually by nearly related species. Thus, crude cultures of *Mucor* species, particularly when isolated from dung, are often attacked by species of *Chaetocladium*, and quite a number of the Mucorales are attacked by species of *Piptocephalis*. The *Penicillium* disease of *Aspergillus niger* is another well-known case of parasitism. The parasites are members of the section Biverticillata-Symmetrica (see p. 237), usually *P. rugulosum* or closely related species, and they cover the *Aspergillus* heads with dense clusters of dark green penicilli, eventually killing the host (Fig. 12.3). *Gliocladium roseum* is reported by Barnett and Lilly (1962) to parasitize a number of other fungi, attacking chiefly the conidia.

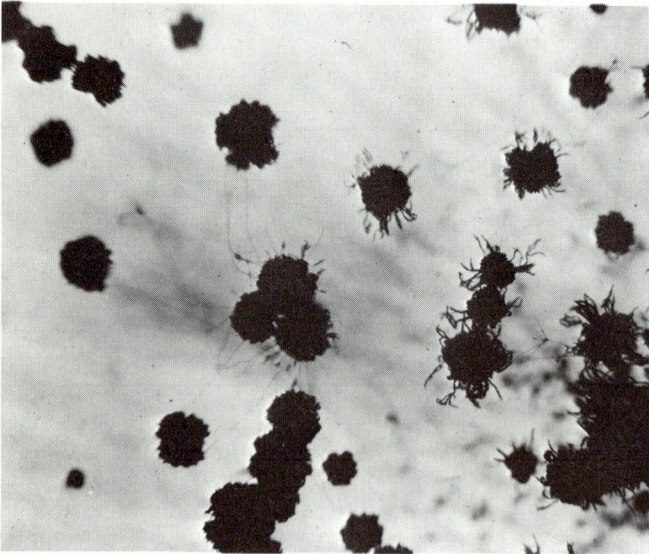

Fig. 12.3   *Penicillium* disease of *Aspergillus niger*   Infected heads as seen in Petri dish, ×25.

12.11 General reading

COOKE, W. B. (1979). *The Ecology of Fungi*. CRC Press Inc., Boca Raton, Florida.
HAWKER, L. E. (1950). *Physiology of the Fungi*. Univ. London Press, London.
PITT, J. I. (1975). Xerophilic fungi and the spoilage of foods of plant origin. In *Water Relations of Foods*. DUCKWORTH R. B. (Ed.). pp. 273–307. Academic Press, London.
ROSS, I.K. (1979). *Biology of the Fungi. Their Development, Regulation and Association*. McGraw-Hill Book Co., N.Y.

12.12 References

BARNETT, H. L. and LILLY, V. G. (1962). A destructive mycoparasite, *Gliocladium roseum*. *Mycologia*, **54**, 72–7.
BEEVERS, H. (1961). *Respiratory Metabolism in Plants*. Row-Peterson, Evanston, Illinois.
BIRKINSHAW, J. H. (1952). Metabolic products of *Penicillium multicolor* G.-M. and P. with special reference to sclerotiorin. *Biochem. J.*, **52**, 283–8.
CARTWRIGHT, K. ST. G. and FINDLAY, W. P. K. (1958). *Decay of Timber and its Prevention*. pp. 250–3. HMSO, London.
CLUTTERBUCK, P. W., MUKHOPADHYAY, S. L., OXFORD, A. E. and RAISTRICK, H. (1940). A. A survey of chlorine metabolism by moulds. B. Caldariomycin, $C_5H_8O_2Cl_2$, a metabolic product of *Caldariomyces fumago* Woronichin. *Biochem. J.*, **34**, 664–77.
COCHRANE, V. W. (1958). *Physiology of Fungi*. Wiley, New York.
COONEY, D. G. and EMERSON, R. (1964). *Thermophilic Fungi*. W. H. Freeman and Co., London.
CURTIN, T. P. and REILLY, J. (1940). Sclerotiorin, $C_{21}H_{22}O_5Cl$, a chlorine-containing metabolic product of *Penicillium sclerotium* Van Beyma. *Biochem, J.*, **34**, 1419–21.
DAVIS, J. G. (1951). The effect of cold on microorganisms in relation to dairying. *Proc. Soc. appl. Bact.*, **14**, 216–42.
EGGINS, H. O. W. (1965). A medium to demonstrate the lignolytic activity of some fungi. *Experimenta*, **21**, 54.
EGGINS, H. O. W. and PUGH, G. J. F. (1962). Isolation of cellulose decomposing fungi from soil. *Nature*, **193** (581), 94–5.
FRIES, N. (1965). *The Chemical Environment*. In *The Fungi, an Advanced Treatise. Vol. 1. The Fungal Cell*. AINSWORTH, G. C. and SUSSMAN, A. S. (Eds). pp. 491–523. Academic Press, New York and London.
GROVE, J. F. (1964). Griseofulvin and some analogues. *Progress in the Chemistry of Organic Natural Products*, **22**, 203–64.
GUPTA, A. D. and NANDI, P. N. (1957). Role of nitrogen concentration on production of perithecia in *Penicillium vermiculatum* Dangeard. *Nature*, **179**, 429–30.
HAWKER, L. E. (1939). The nature of the accessory growth factors influencing growth and fruiting of *Melanospora destruens* Shear, and of some other fungi. *Ann. Bot.*, N.S., **3**, 657–76.

HEALD, F. D. and POOL, V. W. (1908). The influence of chemical stimulation upon the production of perithecia of *Melanspora pampena* Speg. *Rep. Neb. Agric. Exp. Sta.*, **96**, 185–99.

HESSELTINE, C. W. and ANDERSON, A. (1956). The genus *Thamnidium* and a study of the formation of its zygospores. *Amer. J. Bot.*, **43**, 696–703.

HUTNER, S. H. (1972). Inorganic Nutrition. *Annual Review of Microbiology*, **26**, 313–46.

INGOLD, C. T. (1954). Fungi and water. *Trans. Br. mycol. Soc.*, **37**, 97–107.

INGRAM, M. (1951). The effect of cold on microorganisms in relation to food. *Proc. Soc. appl. Bact.*, **14**, 243–60.

JONES, E. B. C. and IRVINE, J. (1971). *The Role of Marine Fungi in the Biodeterioration of Materials*. In *Biodeterioration of Materials, Vol 2*. pp. 422–31, Applied Science, London.

KREBS, H. A. (1943). Carbon dioxide assimilation in heterotrophic organisms. *Ann. Rev. Biochem.*, **12**, 529–50.

LA TOUCHE, C. J. (1950). On a thermophilic species of *Chaetomium*. *Trans. Brit. mycol. Soc.*, **33**, 94–104.

LEACH, C. M. and TRIONE, E. J. (1966). An action spectrum for light induced sporulation of fungi. *Plant Physiol.*, **40**, 808–12.

MCCORMACK, G. (1950). 'Pink yeast' isolated from oysters grown at temperatures below freezing. *Conn. Fish Rev.*, **12**, 28.

OLLIVER, M. and SMITH, G. (1933). *Byssochlamys fulva* sp. nov. *J. Bot. (Lond.)*, **71**, 196–7.

RAISTRICK, H. and SMITH, G. (1936). The metabolic products of *Aspergillus terreus* Thom. Part 11. Two new chlorine-containing mould metabolic products, Geodin and Erdin. *Biochem. J.*, **30**, 1315–22.

SMITH, G. (1949). The effect of adding trace elements to Czapek-Dox medium. *Trans. Br. mycol. Soc.*, **32**, 280–3.

SMITH, J. E. and GILBRAITH, J. C. (1971). Biochemical and physiological aspects of differentiation in the fungi. *Advances in Microbial Physiology*, **5**, 45–134.

THOM, C. (1930). *The Penicillia*. Williams & Wilkins, Baltimore.

WILLIAMS, C. N. (1959). Spore size in relation to culture conditions. *Trans. Br. mycol. Soc.*, **42**, 213–22.

13

Laboratory Culture of Organisms and Industrial Investigations

13.1 Examination and sampling

Before fungi or other organisms can be grown and studied in the laboratory, samples of infected materials must be obtained. Although this pre-requisite is obvious, there are several pitfalls to avoid.

If possible, the material should be examined *in situ* and notes and photographs taken before and after sampling. Data on local environmental conditions should be collected. The numbers of samples taken or the amount of material removed is very much a matter of experience, but suffice it to say that the samples must be adequate for all envisaged investigations and show the typical decayed condition – it is far easier to recognize dry rot on a piece of wood several centimetres square than on a small sliver of wood the size of a matchstick.

The tools used for sampling vary with the material. Most soft materials can be dealt with using a scalpel or sharp knife, forceps and spatula. These can all be flame sterilized on the spot and samples transferred to heat-sterilized glass bottles or plastic bags. Harder materials such as wood, if sampled in depth, may require special augers (which can also be sterilized) or saws. If large samples of materials such as wooden poles need to be taken, excess can be cut first using an ordinary saw, and samples can be cut out later by more aseptic means in the laboratory, outer parts of the original sample being discarded. All samples must be labelled. This is usually best done by labelling the container with a waterproof marker pen.

A hand lens ($\times 10$) and a torch, and particularly a lens with a built-in light (commonly sold for map reading) will prove useful for on-site investigations.

Sealed and labelled samples should then be transferred to the laboratory without delay. If samples are to be sent by post, they should be protected not only from physical damage, but against extremes of temperature and desiccation. Some types of perishable materials cannot be sent by post, and regulations should be checked first.

On arrival in the laboratory, samples should be opened with care to avoid cross-contamination or fouling the laboratory atmosphere. The advent of safety cabinets in laboratories greatly assists with this.

13.2 Culturing – some general hints

Samples often need dividing up before plating out on agar. This may be done simply with scalpel and forceps in a sterile empty petri dish or a glass plate.

Where appropriate, grinding or milling using a blender may also be a useful technique. The small fragments or particles thus obtained can then be introduced into or onto liquid or solid sterile culture media. To assist in later purification and identification of the types of organisms present, it is better to plate out small samples of the material on a large number of plates. If, for example, paper or textile is being investigated, a suitable arrangement would be three or four 2 mm side squares on each petri dish. Some may be taken in a random fashion and others deliberately picked from an obvious infected area from the original sample. The media used for the isolation of fungi from decayed material should be chosen with regard to the original material and its suspected contaminants. If growth is sparse or absent when an obviously decayed material has been plated out, not only the condition of the sample but also the choice of growth medium is suspect.

Temperature of incubation should also relate to original conditions, but in the initial stages of an investigation it may be worthwhile to set up plates at both lower and higher temperatures, especially if original conditions are poorly defined or are known to fluctuate.

As soon as organisms are seen to be growing, sub-cultures should be made as soon as possible by picking off tiny fragments of the colony and plating out again in fresh sterile media. By this means pure cultures may be obtained which can be left to grow to maturity and be identified. Many common storage moulds will show growth visible to the naked eye after 1–2 days, and will cover most of a petri dish, having produced spore-bearing structures after 1–3 weeks, assuming good growth conditions. Plates should therefore be inspected daily, especially in the early stages.

Once a culture appears to be pure, it can then be transferred to an agar slope in a bottle or tube. When the fungus has grown, the cap can be screwed fully down and the culture can then be stored for some time, perhaps for several months under cool conditions, and plates re-inoculated from this culture for subsequent work.

Cultures purchased from a culture collection may arrive as colonies on agar slopes, or in freeze dried form. Slope cultures are transferred to agar plates by means of fragments taken under aseptic conditions using a sterile wire loop or needle; freeze-dried cultures are re-hydrated with sterile water to create a suspension of cells, which is then spread on an agar surface using a sterile pipette and a glass rod.

The basic equipment and techniques required in a mycological laboratory are listed later in this chapter (p. 290) together with details on making slopes and plates.

13.3   Culture examination and slide making – some basic points

Having grown cultures, they will need to be examined and identified. At all stages of the culturing of fungi, descriptions of the colonies on specific media can be listed and the appearance as seen through the bottom of the dish should also be noted. Many fungi produce pigments which are more easily seen from below, and with experience, the appearance of the fungus in the petri dish will give many clues to its final identification.

Several general techniques exist for the preparation of slides for microscopic examination of fungi. These 'temporary preparations' are quick and easy to make, but cannot be stored for as long as the 'permanent' or 'museum' specimens which require more complex preparation. The life of temporary preparations can be greatly extended by sealing the gap between the edge of the coverslip and the slide by the use of ordinary clear nail varnish or quick-drying paint. Nail varnish is usually the best, as the bottles have built-in brushes and the product is readily available.

One method of slide preparation is to use clear adhesive tape (Butler and Mann, 1959). Structure is less disrupted than by using more conventional methods, although clarity may be diminished, especially under high magnification. A small piece of tape, say 1 cm square, is cut with scissors and held in forceps. The sticky side is then dabbed on the colony and then smoothed onto a microscope slide, sticky side downwards. The slide can then be examined. If the material is to be stained, the tape can be placed on the slide, sticky side up, held with another small piece of tape along one edge, and stain and coverslip applied in the usual way.

The usual method is to remove a scraping from the colony with a sterile wire loop. This is placed on a glass microscope slide and teased out if necessary. Water, mounting medium or stain is then added and a coverslip is placed on top. Excess fluid is removed by blotting gently and the slide is then examined. These and other techniques are covered in more detail later in this chapter (p. 311).

13.4   Examples of materials testing

Once a potential material preservative has been chosen by simple agar or other tests (as mentioned in Chapter 14) it may be desirable or necessary to carry out further trials in the laboratory before field trials or actual use. Examples of two such tests are outlined below.

Soil burial tests

In this type of test, beds of soil in tubs or boxes are prepared, and samples of solid materials are buried or partially buried for periods from a few weeks to a few months. After removal, the specimens are examined for evidence of microbial attack. The main attractions of soil burial testing and the reasons for its widespread use are that the test is not only severe and also approaches more closely natural or field conditions than agar techniques, but also that it can subsequently be made quantitative, for as solid test materials are employed, these can then be subjected to tensile and similar quantitative tests after burial. It is, however, widely recognized that this form of testing can be extremely variable (Turner, 1972; Lloyd, 1968). Despite this, such tests are extremely valuable, so long as absolute values from different laboratories are not compared.

Paint film testing

In recent years, comparatively simple equipment and guides to result assessment have been developed specifically to test the resistance of paint films to fungal attack (see Skinner, 1971; Barry *et al.*, 1977; Bravery *et al.*, 1978.) The latter stages of this work have come about due to the co-operative efforts of the International Biodeterioration Research Group (IBRG), which serves to highlight the need for organized co-operation in the development of useful and sensible test methods.

The method centres upon use of a test cabinet into which are inserted glass tubes, painted on their outer surfaces. The painted surfaces are inoculated with fungi by spraying on a spore suspension. A thin film of moisture, conducive to fungal growth, is maintained on the surface of the paint by condensation of the warmer moist air from inside the cabinet when it impinges on the tubes, of which the inner surfaces are exposed to a cooler external atmosphere. After some weeks the tubes are removed and visually assessed for mould growth. By using simple rules, tested over several years of experimentation, good agreement has been obtained between laboratories.

13.5   Freeze-dried cultures

Many culture collections now store and supply fungal cultures in freeze-dried form. The advantages of this method over agar-slope cultures are that all contamination is excluded, storage space is reduced, and most important of all, that as the fungal tissue is dormant, sub-culturing is unnecessary and chances of mutation are minimized, storage life being very much longer than that of agar slope cultures.

Each sealed glass ampoule contains a pellet of fungal tissue, a cotton-wool plug and usually a name or code label. To prepare an active culture the fungus must be transferred to an agar plate in the first instance by the following simple procedure.

The neck of the ampoule is scratched with a triangular file, flamed to sterilize the surface, and then snapped off. The open ampoule is set upright in a holder, which may be a card or plastic box with holes punched in the top. The cotton-wool plug is removed with alcohol-flamed (sterile) forceps and a small quantity of sterile distilled water or isotonic Ringer's solution is introduced by means of a sterile Pasteur pipette. The fungal pellet is allowed to hydrate for a few minutes, and then portions of the resulting fungal suspension are drawn off with a sterile pipette and plated out on prepared agar plates. The drops are distributed across the agar surface with a sterile L-shaped glass rod spreader and plates are then incubated. Slopes can subsequently be made from this culture for preparation of spore suspensions. Although in theory slopes could be made directly from the freeze-dried culture, plates made as an intermediate stage serve as a check on viability, vigour and purity of the culture.

13.6   Microscopy and photography

Earlier editions of this book have paid great attention to details of microscopy and photography. In recent years the technology of optical and associated systems has advanced greatly, and whilst the principles of such systems are still understandable by the non-specialist, the detail can now be firmly placed in the realms of specialist publications, which often have more affinity to optics and electronics than practical biology. The following comments are intended, therefore, to provide a brief guide for those who may be new to microbiology; those wishing to delve further are referred to more specialist books and encouraged to seek the advice of working biologists.

Optical systems

Hand lenses and magnifiers

In addition to a good quality ×10 or ×15 scientific hand lens which should be the 'badge' of most biologists, many of the more general purpose illuminated magnifiers on the market will also prove to be of value. Apart from the obvious value of a light source, the latter often do not require the observer to be as close to his subject as with a normal hand lens.

Low power microscopes

Depending on the lens or lenses fitted, these stereoscopic microscopes give a magnification of around 10 to 100× or more. These developments of what used to be commonly known as 'dissecting microscopes' are very useful for examining decayed materials and fungal growth on plates and slopes. The depth of focus and working distance between objective lens and sample is much greater than a standard high power microscope, and adjustment and illumination is less critical. When fitted with a zoom lens (which keeps in focus at all magnifications) and binocular eyepieces, these microscopes are most useful, especially for routine inspection.

High power microscopes

These are essential for detailed examination and identification of fungi, but do not replace the low power stereoscopic microscopes. Specimens need to be prepared as slides as depth of focus is very small. To obtain high magnification, precision optics and lighting are necessary and general mechanical precision is essential to achieve fine focusing and specimen movement. These requirements make for high cost and a good quality laboratory microscope will today cost anything from £1000 upwards. To obtain full value from such equipment training is required, not only in the operation of the equipment, but in interpretation of the images observed. This is a tool for a trained microbiologist.

Photographic systems

These are usually available within the range of extras available for high power microscopes. They may simply be standard camera bodies which fix onto or replace the viewing head by means of adaptors, or may be built into the microscope. Unless photographs need to be taken very rarely, a camera attachment and light metering system which causes little or no inconvenience to the normal visual system, although usually more costly, is preferred.

A range of camera attachments is available. The simplest types are clip-on 110 format cartridge film cameras of limited versatility, designed for student class work, and relying on fixed-mark focusing.

The simplest of the 'serious' systems consists of a tube containing a photographic eyepiece which is fitted to the microscope after removal of the viewing head. A camera, usually a standard single lens reflex body, is then attached to this tube by means of an adaptor. Light metering is performed through the camera's own system, or via a special sensor which is fitted temporarily instead of the photographic eyepiece before attaching the camera. Focusing is carried out via the camera's viewfinder. Such a system can give good results, but scanning the slide once the camera is fitted is less easy and metering can be a problem.

More sophisticated systems have a camera attached permanently in addition to the monocular or binocular viewing head and metering can be automatic and not interrupt focusing or viewing. The most versatile and accurate systems can be very expensive.

Cameras may use conventional films or the 'instant' types which have a certain appeal in some situations. There is usually little advantage in using colour film for high power work, except where slides for projection are required. Colour, and in many cases tone, under the microscope, is a function of the lighting system and only specifically stained specimens have any meaningful colour.

13.7   Basic equipment, methods and techniques

The first essential in any laboratory experimental work involving micro-fungi is to obtain pure cultures; that is, to grow each species on a sterile substratum, known as the culture medium, free from admixture with any other organism and with suitable precautions to prevent subsequent contamination. Many fungi fail to grow characteristically in the presence of other organisms and identifications are therefore rendered difficult. It is obvious also that studies of nutrition, of metabolic activity, or of the controlling effect of inhibitory substances are without much value unless they are carried out with individual moulds and not with mixtures of species.

Methods of isolation, of culture, and of study, are, in general, similar to those used by bacteriologists, but differ from them in detail. The methods described here are only a selection of those available but should be sufficient for preliminary essays in mycological studies, and the advanced worker and specialist can add to them as experience directs.

13.7.1   General equipment

Cultures for morphological study, and for many other purposes, are grown either in plugged test-tubes or in petri dishes. Cultures for storing in a collection are also often grown in plugged tubes, but nowadays many mycologists prefer screw-top bottles for this purpose.

Tubes   The tubes may, for most purposes, be ordinary test-tubes, but the special bacteriological test-tubes are much to be preferred. They are made of glass which has a high degree of resistance to chemical action and will withstand repeated sterilization. At the same time they are sufficiently strong mechanically to resist a certain amount of rough handling. It is an advantage to have two sizes, the smaller without, the larger with rims. The rimless tubes, which are commonly 150 by 19 mm, pack into baskets and storage boxes much better than the rimmed variety and are used for 'slopes' (see below). The larger tubes with rims (about 150 by 24 mm) are used to hold medium for plating out, when the rim is a distinct help in pouring cleanly, and the greater capacity makes thorough mixing of the contents easier.

Containers used to hold or store molten agar prior to pouring plates is a matter of personal preference. Where many plates are to be prepared there is a distinct advantage in using containers such as 500 ml medical flat bottles. Smaller amounts of media are quicker to melt of course.

Plugs   Tubes of culture medium, and tubes containing pure cultures, are often plugged with cotton-wool so that any air which enters will be filtered from all contaminating organisms. The cotton-wool should be of the non-absorbent variety. It may be obtained in a number of different colours and these are useful for distinguishing the various culture media. A plug should project approximately 2 cm into the tube and should have a tuft outside the tube by which it can be extracted; it should fit accurately and tightly, but not so tightly that it cannot be extracted when gripped between any two fingers of one hand – and it should retain its shape so that, after withdrawal, it can readily be reinserted. There are several methods of making plugs by rolling and shaping before insertion. An easy and quite satisfactory method is to tear a strip of about 6 cm wide from the sheet of cotton-wool. From this a rectangular piece is torn, of a size that can be determined only by trial, and the edges are folded in to give a piece 6 cm long and of a width twice the diameter of the tube. This is laid across the mouth of the tube and its centre gently pushed in by means of a glass rod, forceps or the blunt end of an aluminium needle-holder. If the plug is definitely a tight fit when pushed in, it will be of the correct fit when the rod is withdrawn. The steaming which it receives during sterilization sets the plug more or less permanently to the shape of the tube.

Rectangular baskets of stainless steel or iron wire, galvanized or tinned, are used to hold tubes during sterilization and may be obtained in any convenient size.

Screw-top bottles   The usual size is about 75 by 20 mm and the screw caps are fitted internally with rubber pads to effect an airtight seal. Plastic caps are also available. Such bottles have certain advantages over plugged tubes. During sterilization the caps must, of course, be somewhat loosened, but as soon as the bottles come out of the sterilizer the caps may be screwed down tightly until the medium is required. It is thus feasible to make culture medium in large batches and to store it almost indefinitely without any danger of its drying out. Cultures in these bottles are easy to store, since they stand upright in drawers or filing cabinets, and they can be labelled on the caps so that all particulars are in full view. The bottles are mechanically stronger than tubes and are thus very suitable for sending cultures by post.

On the other hand, bottles are more difficult to manipulate when being inoculated, since it is virtually impossible to hold three bottles between thumb and first finger, as is done when inoculating duplicate cultures in tubes (see p. 000). Tubes are also more convenient than bottles when examining living cultures. A tube is easily laid across the microscope stage, and the growth therein can be examined with a 16 mm objective, or even with an 8 mm objective, and may be photographed if desired. The glass of the screw-top bottle is too thick to allow of any objective of shorter focal length than 16 mm being focused, and the glass is usually too uneven to transmit a useful image. Another disadvantage is that, once the cultures have been sown, the caps must be unscrewed sufficiently to admit air, and it is not easy to do this without running a grave risk of infection getting in. George Smith (the author of previous editions of this book) found that not all moulds grow equally well in tubes and bottles. Many do, but some species definitely benefit from the freer aeration of the plugged tube.

Recent years have seen the advent of disposable clear plastic screw-top bottles, which may be purchased in a sterile condition, either empty, or with pre-poured agar. Light weight and the clarity of the thin plastic can be an advantage in some circumstances.

Petri dishes   These are flat, circular, shallow glass dishes with perpendicular sides, provided with covers of the same shape, but of slightly larger diameter, so that they fit loosely over the dishes. The air which enters, and it must be remembered that fungi require a continual supply of air, is not filtered as in the case of plugged tubes. Owing, however, to the tortuous path it must take to enter the dish, suspended dust and spores are deposited outside and cultures are reasonably safe from contamination, provided that the dishes are handled with care and that the air of the laboratory is not too heavily charged with infection. As an extra precaution Petri dishes should, whenever possible, be stored and incubated upside down. This also prevents disruption of the colony by falling drops of condensation which may occur. Dishes may be obtained in various sizes but, for most purposes, the most convenient size is 10 cm by 1.5 cm. They are sometimes made with plane, polished bottoms, but this is a refinement too costly for ordinary use.

To avoid the problems of breakage and the high cost of purchase, cleaning and sterilizing glass dishes, extensive use is now made of disposable gamma-radiation sterilized plastic dishes. They are virtually unbreakable and, in any case, have very little tendency to slip when piled together. Plastic dishes are now obtainable packed in tens in polythene bags and, owing to their low cost, are mostly used once only and then discarded.

General apparatus   A certain amount of the necessary equipment is such as is to be found in any chemical laboratory. This includes beakers and flasks of various sizes, measuring cylinders, pipettes, funnels, bunsen burners and tripods, balances and weights. It is assumed that the reader is sufficiently familiar with this type of apparatus for it to require no further mention.

A supply of needles and loops will be required for inoculations and the best are made from short lengths of stiff nichrome wire (20-gauge is suitable), permanently fixed into long aluminium handles. Platinum wire cannot be set satisfactorily into aluminium as it becomes loose after a short period, owing to the difference in coefficients of expansion of the two metals. Adjustable needle-holders which grip the needle in a small brass chuck are preferred by some workers, but they have too many hidden surfaces and require much heating to ensure perfect sterilization, particularly when working with liquid media. Ocasionally a very stiff needle is required and a convenient form is triangular in section with fairly sharp edges which can be used for scraping. Fine-pointed scissors and scalpels are useful for cutting up infected material preparatory to making cultures, and the ones made entirely of stainless steel are worth the extra cost, as they stand up to repeated sterilization without corroding. The list of tools should also include three pairs of forceps, one strong and blunt-ended, one with fine points, and the third of the special type for handling cover-glasses.

Some kind of support is necessary for holding a sterile needle whilst both hands are occupied in such tasks as manipulating plugs and labelling culture tubes. The simplest support is a block of wood $2\frac{1}{2}$–5 cm thick, or a large rubber bung with a hole, just large enough to take the handle of the needle comfortably, bored through it. The needle stands upright in this, without serious danger of infection. Such a support is invaluable when sowing large numbers of cultures. Other tools, which are used much less frequently, may be supported, after sterilization, in any way which ensures that the tips of forceps or the cutting edges of scissors and scalpels are not touching anything, for example resting on the lid of a Petri dish or a tripod.

For labelling culture tubes and Petri dishes grease pencils, specially made for writing on glass, are obtainable in several colours. There are also spirit-based marker pens widely available for marking glass, and these give a more permanent label than the grease pencil. Even ordinary ink will mark glass quite readily, provided that the surface is dry.

Incubators   The majority of moulds will grow reasonably well at the temperatures prevailing in most laboratories. However, in order to induce maximum rate of growth, and, in some cases, to promote the formation of certain types of spores, higher or lower temperatures are essential. The containers used for this purpose are known as incubators, these being provided with means of maintaining automatically and continuously a pre-determined temperature. Also, in physiological work it is often desirable to study the effect of temperature on growth, and this, unless the work is to be unduly prolonged, involves the use of several incubators set to different temperatures.

The usual type of incubators obtainable commercially are designed to work at temperatures above that of the laboratory, and are nowadays almost always electrical. Electric incubators are usually not water-jacketed, a feature more commonly found on older gas heated incubators, but consist of a single-walled copper container, insulated by dry packing. They are controlled by the same type of expandable capsule as is used for the gas incubators, this controlling a simple off-and-on switch. Electric incubators are less trouble to maintain than the gas incubators, but tend to show more variation in temperature within the usable space.

Ordinary incubators have one serious fault for mycological work, in that they necessitate incubation in the dark. Light is often more important than warmth for characteristic growth of moulds. Many species spore much better in the light than in the dark. For maintenance of stock cultures it is often best to dispense with incubators, and grow cultures at room temperature in the light. However, the Commonwealth Mycological Institute at Kew has carried out successful trials of specially designed incubators which obviate the difficulty. Illuminated incubators are available commercially but are expensive, and home-made versions may serve equally well. The top, back and sides are of glass, set in a light wooden frame, and the front consists of sliding glass doors. The shelving is also of glass. Along both ends and the bottom plastic-covered resistance wire is laid zig-zag fashion, stretched between supports of plastic-covered curtain wire. The heating wire is connected to the mains through a 'Sunvic' control, which can be adjusted to give a short range of constant temperatures. Above the incubator is a 'daylight' fluorescent tubular lamp, which can be turned on at night if required, or used to provide extra illumination on dull days.

While the maintenance of temperatures above that of the laboratory is easy, the attainment of constant temperatures below 20°C presents some difficulty. Some fungi grow best at comparatively low temperatures, and, in any case, it is often useful to find the effect on growth of a considerable range of temperatures. The domestic type of refrigerator can be set to give temperatures over the range 0°C to about 8°C. In fact there is usually an appreciable variation in temperature, at any setting, in different parts of the machine. The difficulty is to maintain a steady temperature within the range 10–20°C. Some form of cooling is essential, and one method of achieving this is to place a small incubator (or more than one) inside a large refrigerator, when it is not difficult to balance the heat input against the cooling effect of the ambient atmosphere. Cooled incubators are also available commercially but again are very costly.

Sterilizers Culture media are usually sterilized by steam, either at atmospheric or higher pressure, so as to avoid change of concentration by evaporation. While much may be done by judicious adaptation of domestic utensils, the special pieces of apparatus made for the purpose have distinct advantages when cultural work on a large scale is undertaken.

The Koch type of steamer, designed for sterilization at ordinary pressure, is a tall copper vessel, cylindrical in the smaller sizes, the lower portion of which serves as a water-bath and is fitted with a gauge and sometimes with a constant level attachment. A perforated shelf, fixed a little distance above the water level, serves to hold apparatus. The lid is provided with a tubulure to take a thermometer and to act as a steam outlet, and the whole outside of the steamer, except the bottom, is lagged with felt. It is normally supplied with a stand of sheet iron and a ring burner. The internal height of the Koch steamer should be sufficient to accommodate a litre flask and a large funnel of about 15 cm diameter supported above it. Larger sizes of steamers to hold several baskets of tubes or a number of flasks are usually rectangular in shape. In some laboratories very large steamers constructed on the same principle are used for the sterilization of large numbers of culture flasks or bottles.

Some culture media are difficult to sterilize completely at atmospheric pressure and it is necessary to use an autoclave, in which steam is generated under pressure, usually 0.7–1.4 kgf per cm^2 (10–20 pounds per square inch). An autoclave, if used properly, can also be employed, in place of the steamer, for the sterilization of the more usual media, and the time required for sterilization thereby much shortened. Full particulars of autoclaves, as well as of Koch steamers, may be obtained from the catalogues of laboratory furnishers. For small quantities a domestic pressure cooker can be used.

13.7.2   Sterilization

When working with pure cultures it is necessary to sterilize all tools, utensils, containers, and culture media, in order to avoid contamination. The usual method of sterilization is by heating to a sufficiently high temperature, and for a sufficiently long time, to kill all fungus spores and bacteria.

Tools and glassware

Metal tools are sterilized by heating in a bunsen flame, needles until just red hot, cutting tools at a somewhat lower temperature to avoid loss of temper. Cutting tools are best treated by dipping (when cold) in alcohol or methylated spirits and then carefully flaming-off the alcohol. To avoid sputtering from used loops which may contaminate the immediate area or the worker several devices are available, either to fit over a micro-burner or heated by an electric filament which prevents the usual spread of droplets from a naked bunsen flame. Dry glassware, such as tubes, flasks, and Petri dishes, are sterilized by dry heat, and a capacious air-oven should be available capable of being maintained at a temperature of about 160°C. Three hours' heating at this temperature is sufficient to sterilize anything. Petri dishes are preferably packed in boxes before heating, and allowed to cool and remain therein until required for use. Special boxes of sheet iron or copper are made for the purpose, but they are expensive, and the rectangular tins in which biscuits are sold serve the purpose just as well. A domestic gas oven is used for this purpose at the Commonwealth Mycological Institute.

Sterile graduated pipettes are frequently required for sowing liquid culture media and for making adjustments to sterile media. To sterilize pipettes plug the mouthpiece ends, pushing the plugs well into the tubes, wrap separately in brown paper so that they are completely enclosed, and sterilized by dry heat, at a temperature not exceeding 130°C, for five to six hours. At a higher temperature the paper is charred and becomes useless as a protection.

Tubes for use in aeration experiments are first fitted with the necessary rubber bungs, ready for insertion into flasks or tubes, with every open end plugged with cotton-wool. The whole is wrapped in grease-proof paper and sterilized in the autoclave.

Culture media

The sterilization of culture media is apparently a simple matter but is actually beset with a number of pitfalls for the inexperienced. Very few media are completely stable to heat, and it is not unusual for deleterious changes to take place during sterilizing. Most sugars are altered to some extent and may form products which are toxic to fungi. Of the common sugars, glucose is the most and sucrose the least altered. (For a detailed account of the effect of heat on sugars see Davis and Rogers, 1940.) Many media made from vegetables owe their particular value to their content of vitamins and kindred substances, many of which are destroyed to some extent on heating. Also, agar-agar, and more particularly gelatine, which are used for solidifying media, lose their power of setting when overheated and agar, like some of the sugars, may give rise to toxic products (see Robins and McVeigh, 1951). Therefore, while it is essential to heat sufficiently to destroy infections, it is equally essential to avoid over-sterilization. A most instructive experiment is to make up a quantity of any culture medium, divide it into several portions, and sterilize these by autoclaving respectively for 15 minutes, 30 minutes, 1 hour, $1\frac{1}{2}$ hours, 2 hours, etc., then sow all the batches of medium with the same species of mould. The differences in rates of growth, and often in types of growth, are usually most striking.

Steaming at atmospheric pressure needs to be very prolonged in order to kill all possible contaminants, the spore-bearing bacteria being particularly resistant. It is usual, therefore, to use an intermittent process, the medium being steamed for 30–60 minutes on each of three successive days. The theory of the process depends on the fact that vegetative structures are more readily killed by heat than are the spores. The first short 'cooking' destroys most of the vegetative growth but may not kill the spores. These, however, being in a favourable situation for growth, and being already swollen by the moist heat, germinate rapidly. The second heating destroys the new growth and the third day's treatment accounts for any spores whose germination has been delayed.

Bacterial spores which can withstand boiling for a long time are killed comparatively rapidly at somewhat higher temperatures. The autoclave is therefore used for sterilization of most vegetable media, which are more likely to be contaminated with spore-bearers than are synthetic media. It is, of course, easier to over-sterilize and damage constituents of the medium using the autoclave rather than the steamer. To avoid trouble ensure that the pressure-gauge is reading correctly, and time the duration of the process strictly, reckoning from the time the pressure reaches the required value to the time the source of heat is shut off. Very few materials require more than 30 minutes at 1.05 kgf per cm^2 (15 lb per sq inch) and most media require less. The size of the autoclave is also important (see Langeron, 1945). A large autoclave takes too long to heat and to cool down, with the result that some portions of the contents are almost sure to be overcooked. It is much better, in a large laboratory, to have several small autoclaves rather than one big autoclave.

Cold sterilization

There are two methods by which certain culture media can be sterilized without running the risk of damage by heat. Modern membrane filtration techniques have eased the cold sterilization of liquid media and indeed fungal growth on membrane filters is now a common method of estimating fungal populations in fluids. The second method is described by Hansen and Snyder (1947) and is used for the sterilization of seeds, leaves, portions of stems, and thin slices of vegetables. Using this method it has been possible to grow in pure culture a number of fungi which have never been grown on heat-sterilized media. The material to be treated is placed in a screw-top fruit-preserving jar together with a small quantity of propylene oxide, at the rate of approximately 1 ml per litre of space. The lid is screwed down tightly and the jar allowed to stand overnight. The screw is then slackened somewhat to allow the fumigant to escape, and after a few hours the material is ready for use. The most convenient way of handling the sterilized material is to remove portions as required with sterile forceps and transfer to melted plain agar in Petri dishes. The warm agar rapidly drives off the last traces of propylene oxide and, as soon as the agar has set, the material can be inoculated.

Sterilization of rooms

In theory, several methods of room sterilization are possible, but are either impractical or dangerous. With the advent of operator and culture protection safety cabinets and enclosures, the need for regular fumigation of laboratories has greatly diminished. Regular cleaning and good housekeeping, perhaps with the use of ultra-violet lamps when the rooms are unoccupied, are all that is normally required when inoculation cabinets are available, although no amount of modern aids can completely replace good manual handling techniques.

13.7.3 Culture media

Fungi are usually grown in the laboratory on sterile nutrient jelly, made as a water-agar gel with added nutrients. They may also be grown on or in liquid media or on solid substrates, such as paper, soil, twigs, grains, leaves etc. It is possible to buy prepared media as a dried concentrate to which only water has to be added, or even as agar slopes of jelly in Petri dishes. However, on a large scale this can be expensive and the author (AHSO) prefers growth on media prepared in the laboratory. In addition, there is a wider variety of media available and laboratory prepared media can be adapted to one's own needs.

Fungi usually grow best on rich media with high concentrations of carbohydrate, but this may induce over-production of mycelium and loss of sporulation. Indeed, fungi often grow well on simple vegetable extracts with only the addition of agar. Tap water will introduce sufficient trace elements and can be used to advantage in the preparation of many media, though some species (e.g. *Phytophthora* and *Pythium* species) are sometimes sensitive to impurities in tap water and it is necessary to prepare culture media using pure water, glass distilled water or pond water. Peptone is best omitted from most fungus media, as growth often looks a little unusual on, for example, Saboraud's agar. Most fungi grow best at slightly acid pH (about pH 6.5) and some will withstand quite low pH. However, agar does not solidify well in very acid or alkaline media.

Agar is slow to dissolve and requires boiling, though modern granular agar dissolves fairly freely. In order not to over-cook the other constituents (some of them decompose with too much heating, especially carbohydrates and proteins) the agar may be dissolved in half the water, the other ingredients in the other half, and the two mixed. Agar is used as it does not dissolve or melt until about 95°C, while on the other hand it does not solidify until the temperature falls to about 40°C so it can be used for incubation at quite high temperatures. Gelatine gels tend to liquify at about 30°C.

The composition of several useful media are given in Appendix I.

13.7.4   Types of culture and cultural techniques

Slopes

Cultures which have to be kept for some time for study, or which are to be stored as part of a culture collection, are almost invariably made on agar slopes in tubes. For making slopes in tubes of about 19 mm diameter the amount of agar medium should be 5–6 ml. The medium is sterilized in the tubes in the ordinary way and then, while the agar is still hot and molten, the tubes are inclined at such an angle that the medium forms a layer of decreasing thickness from the bottom of the tube to within about 20 mm of the plug. Laying the tubes on the bench with the plugged ends supported on a glass rod about 12 mm in diameter gives approximately the right amount of slope. The object of sloping the agar is to provide a relatively large surface, on which the progress of growth can be watched far better than on a level surface of agar in a comparatively narrow tube. Incidentally, the varying thickness of the layer of medium often reveals interesting cultural characteristics, the type of growth at the shallow end showing marked differences from that at the deep end.

The method of sowing slopes (assuming that the worker is right-handed) is to hold the tubes by their lower ends between the thumb and first finger of the left hand, with the plugs on the palm side of the hand and pointing slightly downwards. The maximum number of tubes which can be conveniently held is three, allowing for two transfers being made in one operation from a tube culture, or three cultures being made from infected material. A needle is sterilized by heating to redness in a bunsen flame. While it is cooling the plugs are removed, one at a time, by the right hand, using a twisting motion, and placed between the other fingers of the left hand so that they are held by their tops only. The bunsen flame is played round and into the mouths of the tubes, an operation known as 'flaming', until the glass is too hot to touch. The tip of the needle is plunged into the agar in one of the unsown tubes in order to make sure that it is cool and to wet it slightly, and is then used to pick up a few spores or a fragment of mycelium from the parent culture. This is deposited, as rapidly as possible and without being allowed to come in contact with the hot glass, on to the fresh agar surface. When both tubes have been sown the mouths of the tubes are again flamed and the plugs re-inserted. Some workers flame the tubes again after insertion of the plugs, but this is not necessary unless the plugs have been out of the tubes for an appreciable time, and it causes the tops of the plugs to become charred and messy to handle.

A slightly different method of handling tubes is claimed by some to be safer, in the sense that it gives more security against contamination. The tubes are held between thumb and first finger but with the palm of the hand pointing directly downwards. When the plugs are placed between the other fingers they are protected by the hand and there is little risk of their picking up infection. Some people find it very difficult to acquire this technique and should use the more usual method described above.

When large numbers of cultures of a single species are required the best method, provided that the parent culture is sporing freely, is to use a spore suspension. With the usual precautions a fairly large mass of spores are picked up on the needle and transferred to about 1 ml of sterile water contained in a very small test-tube. The tube is shaken to distribute the spores and is then supported, as nearly horizontal as possible, in such a way that the mouth of the tube can be flamed periodically during the sowing. The inoculations are made with a wire loop, a tiny drop of suspension being transferred to each fresh tube of medium. Loops of standard size for bacteriological work, made of either platinum or nichrome wire, unmounted or fixed in handles, can be purchased but are readily made from the ordinary wire needles by winding the ends round a metal or glass rod of about 1 mm diameter.

One other precaution should be taken when sowing slopes. The most characteristic growth is obtained when a slope is sown at a single point near the centre. The bacteriologist inoculates in a wavy line from bottom to top of the slope, but this method is quite wrong when handling moulds. With a single spot inoculation not only is the growth more typical but the surface of the slope is actually more quickly covered.

Cultures on liquid media

Such media in tubes are sown in the same way as slopes except, of course, that the tubes cannot be held with mouths pointing downwards. They should be held as nearly horizontal as possible without getting the liquid on to the hot glass near the mouths, in order that air-borne spores cannot fall directly on to the surface of the medium but will fall on the hot glass and be killed.

Plates

For morphological studies microfungi are grown in Petri dishes (commonly known as 'plates'), as well as on slopes. Dishes planted with single colonies are useful for determining rate of growth and colony characteristics such as zonation and sectoring, while plates containing several colonies are more suitable for microscopical examinations. Many species form dense, opaque felts of mycelium, so that the only part of an isolated colony which can be examined by transmitted light is the extreme edge, and this usually shows no ripe fruiting structures. In a dish containing several colonies it is usually found that narrow sterile zones are left along the edges where the colonies approach each other, and, in the more mature portions of the colonies, spore-bearing organs can be clearly viewed as they hang over these gaps (Figs 13.1 and 13.2).

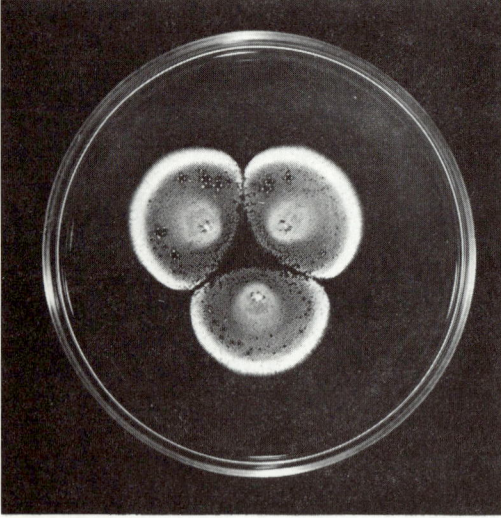

Fig. 13.1 *Penicillium expansum*   Colonies in Petri dish, showing the absence of a sterile edge where the colonies approach one another, ×0.6.

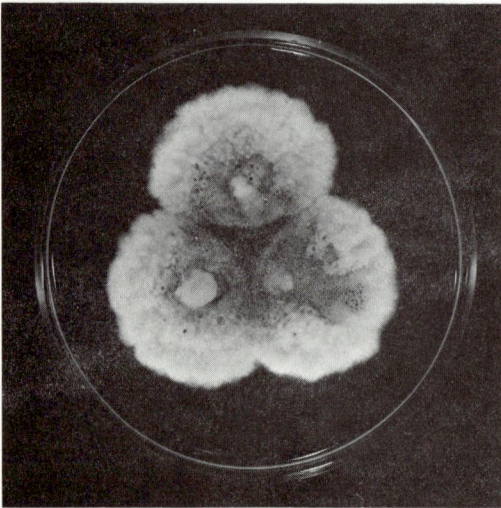

Fig. 13.2 *Penicillium nigricans*   Floccose, tardily sporing but showing the same effect as Fig. 13.1, ×0.6.

The usual 10 cm Petri dish requires approximately 12 ml of medium to give a layer of adequate thickness, and, if comparative cultures of different species, or different strains of the same species, are required, the amount of medium per dish should be fairly accurately standardized. The medium is filled into tubes in correct amounts and sterilized therein. Before pouring the plates the tubes should be allowed to cool down to about 45°C, preferably by leaving them for a few minutes in a water-bath maintained at this temperature, as very hot medium gives off water vapour, which condenses on the cool lid of the dish and then drips back on to the medium. Each tube, as it is lifted from the water-bath, is first quickly wiped free from adhering water. If this precaution is omitted there is a danger of drops of this water, usually far from sterile, getting into the plates. Next the tube is held in a sloping position while the plug is removed and the mouth flamed, and then the medium is poured gently into the dish with one edge of the cover raised as little as is necessary for the purpose. After replacing the lid the dish is carefully tilted to spread the medium and then left to stand on a level surface until the agar has set. Whenever possible it is best to store Petri dishes in the inverted position, both before and after sowing, as this minimizes the chance of infection. When sowing plates or handling them for examination every care must be taken to avoid exposure of the medium more than is absolutely necessary, since protection by flaming, as in the case of tubes, is impossible. If dishes are not inverted it is difficult to sow colonies in predetermined positions, the inoculating needle often leaving a trail of spores right across the surface of the medium. If the dish is inverted, then lifted out of its cover, and the needle approached and receded from directly below, it will be found that very few stray colonies will appear. Another method of sowing colonies exactly where required is to inoculate with loopfuls of a spore suspension.

Slide cultures

For the study of some species which produce very small and fragile sporophores the only satisfactory methods of making preparations for microscopical examination are by means of slide cultures and cultures on cellulose film. There are a number of simple ways of making cultures on microscope slides.

1. If the fungus tends to spread close to the substratum it is grown on a very thin layer of agar spread on the slide. The agar medium should preferably be filtered until clear and transparent. A piece of glass rod, or better, a strip of aluminium sheet not more than 1 cm in width, is bent twice at right angles, in such a way that it will fit into a Petri dish and support a slide clear of the bottom. Shallow slots in the two opposite sides of the bent strip will prevent the slide from slipping about. The dish, with the support and slide in position, is sterilized by dry heat. A drop of melted agar is poured on the centre of the slide while the latter is still warm and spread as evenly as possible by means of a bent glass rod. If the germination of spores is to be studied a little spore material is mixed with the agar before pouring on to the slide, but if spore production is to be observed, it is better to plant the medium at two or three points. About 10 ml of a sterile 20% solution of glycerol is poured into the bottom of the dish. This keeps the thin layer of agar moist, but not wet, as it would become if pure water were used. The slide is sufficiently near to the lid of the dish to allow observation of growth with a low-power objective without removing the lid; if the height of the support has been properly adjusted an 18 mm objective may be used. When the desired stage of growth has been reached the slide is lifted out and placed for a few hours in a similar dish (which need not be sterilized) containing a little formalin, or a solution of osmic acid, in order to kill the fungus and partially fix the structures. The slide may be examined dry, as it is, or, if the growth is too decidedly aerial, with a cover-glass laid on very gently. It is often possible, in this way, to make fairly flat preparations without breaking down fragile structures unduly. With some species which are not quite so fragile permanent mounts may be made in lacto-phenol, afterwards cutting away the agar round the cover-glass and sealing with cement.

2.   Another way of making slide cultures is described by Henrici (Skinner *et al.*, 1947), utilizing a shallow cell, built up on the slide with sealing-wax and a large rectangular cover-glass. With the plane of the slide vertical agar medium is run in to about half the depth of the cell, and the fungus is planted on the narrow surface thus provided. All stages of growth are readily observed owing to the spread of the mould being confined approximately to one plane, and the effect is as if a thin section through a colony were being examined. With moulds of vigorous habit and with large fruiting structures, such as many of the Mucoraceae, a larger cell may be built up on the same lines, using sheets of thin glass, such as old photographic plates stripped free from gelatine, clamped together in pairs with separators made from narrow strips of cardboard.

3.   An elegant and most useful method, particularly for the examination of fragile sporing structures, has been perfected by the late Dr J. T. Duncan and has been used successfully for many years for the examination of der-matophytes in the Department of Medical Mycology, London School of Hygiene and Tropical Medicine. Agar medium is poured into a Petri dish to a depth of about 2 mm. When it has set completely, a small block, about 1 cm square, is cut out with a sterile tool and transferred to the centre of a sterile slide. The block is inoculated on all four *edges*, then covered with a large sterile cover-glass. The slide is incubated in a moist chamber, such as is described above. The fungus spreads out from the agar block and tends to attach itself to the two glass surfaces. The slide may be taken out of the moist chamber from time to time for examination, without much danger of contamination. When growth has reached the desired stage the cover-glass is carefully stripped from the agar block and carefully lowered, fungus side down, on to a drop of mounting fluid (with or without stain) on another slide. Alternatively, the preparation may be stained by any standard procedure before mounting. Next, the agar block is carefully loosened and removed, leaving a second preparation on the original slide. The photograph of *Sporothrix schenckii* (Fig. 9.45) was obtained from such a slide culture. An account of this type of technique is given by Booth (1971), and the basics are given in the *Mycologist's Handbook* by Hawksworth (1974), p. 35.

4. Still another method of making slide cultures has been used at the Commonwealth Mycological Institute. Agar medium is poured into a Petri dish and allowed to set. Using a sterile knife or scalpel, two diametrical slits, at right-angles to one another, are made in the agar. Each section in turn is partially raised, from the central cross, and a flamed cover-glass placed under the agar, which is then allowed to fall back into place. The dish is inverted and the positions of the cover-glasses marked with grease pencil. A small square, about 6 mm side, is also drawn over the centre of each cover-glass. The dish is turned right side up, and pieces of agar, corresponding to the marked squares, are cut out. The medium is now inoculated. In most cases the growth will eventually extend over the bare areas of cover-glass. When such growth is satisfactory the cover-glasses are removed by cutting away the surrounding agar, and mounted in the usual way.

Cultures for biochemical studies

A number of methods have been used for growing moulds on comparatively large volumes of liquid media for studies of biochemical activity and isolation of metabolic products. The widespread interest in such studies at the present time, and the difficulties which have been encountered in some laboratories, are the reasons for describing here the methods which have been found satisfactory in the Department of Biochemistry at the London School of Hygiene and Tropical Medicine, and described by G. Smith in previous editions. There are now many more sophisticated methods and the reader is referred to specialist books on fermentation for these.

Containers   If moulds are grown as surface felts it is necessary to use shallow layers of liquid in order to secure complete utilization of all the nutrients. Trays take up little room but are difficult to handle and to protect from contamination. Conical flasks, although taking up more space in the incubator, have been found to be in every way more satisfactory for total volumes up to 50 or even 100 litres. Cultures can be examined during the period of incubation for uniformity, purity, and vigour; distribution in a number of small containers localizes chance contamination; and the course of metabolism is more satisfactorily followed by examination of the whole contents of a single flask than by drawing off samples from a large container.

Flasks are of good-quality resistance glass and are used solely for this kind of work. For isolation of products 1-litre flasks are used, but smaller sizes, from 100 ml to 750 ml, are used for special purposes and for pilot experiments. The volume of liquid in each flask is such as will give a depth of not more than 4 cm – a normal 1-litre flask should hold approximately 350–400 ml, and a 250 ml flask approximately 100 ml.

Inoculation   Flasks are sown with an aqueous spore suspension (or suspension of mycelium if the fungus is sterile), giving a heavy sowing so as to ensure the rapid establishment of a continuous felt of mycelium. It has been found that the most satisfactory method is to start with a number of cultures on agar slopes, in 24 mm rimmed tubes, allowing one slope for each 4 or 5 flasks if the mould spores well, and one slope for 3 or even 2 flasks if spore production is scanty or the fungus sterile. A similar number of tubes of sterile water,

12–15 ml per tube, are also prepared. The contents of one tube of water are poured, with the usual precautions, into one of the culture tubes, the surface of the agar is lightly scraped with a stiff tool, and the suspension is poured directly into the flasks. The aim should be to distribute the spores evenly, and not necessarily to add the same volume of liquid to each flask. There is usually no need to flame the mouths of the flasks, since the plugs are out for only a few seconds. Finally the flasks are shaken to distribute the spores over the surface of the liquid. The whole process is carried through as rapidly as possible, consistently with efficiency, so as to avoid contamination. Rapid work also ensures that most of the spores are not completely wetted, and therefore float on the surface of the medium in the flasks.

A non-sporing mould is not so easy to handle as a species which spores freely. Flasks must then be sown with a suspension of bits of mycelium, with the disadvantage that most of the inoculum usually sinks, and the formation of a surface felt does not begin until hyphae from the submerged mycelium have extended sufficiently to reach the air, often a matter of several days. It is therefore essential to work in a clean atmosphere and to take every possible aseptic precaution. A method adopted with fungi which will not grow when submerged is to use a long needle to transfer bits of dry mycelium from a slope to the surface of the liquid medium, but to do this successfully requires considerable practice.

In some laboratories adaptations of bacteriological technique are used. Cultures are grown in Roux bottles or flat culture flasks, spore suspensions are then made by scraping off the growth into comparatively large volumes of liquid, and flasks are sown with measured volumes of the suspension delivered from a pipette. This procedure does not usually give such satisfactory results as the method outlined above. Sowing with a pipette certainly means that each flask receives the same volume of suspension but seldom ensures equal distribution of spores. In the rare cases in which the spores of a mould are readily wetted equal distribution is of course achieved, but then most of the spores sink in the culture fluid and the establishment of a surface felt is delayed. If the spores are not easily wetted, as is usually the case, most of them float to the top of the pipette and remain there when the liquid is run out. There is also much greater risk of contamination when many flasks are sown from one suspension than when only a few flasks are sown from one culture.

In certain fermentations which are carried out on a factory scale, such as the manufacture of penicillin, the mould is actually grown submerged in the culture medium, the latter being artificially aerated. In these cases the inoculum should be wetted as completely as possible and, to ensure this, the liquid used for making the spore suspension contains a wetting agent.

Special Media   Most of the culture media used for moulds are slightly acid in reaction, the pH being usually between 4.0 and 5.0. If the mould grows better at a higher initial pH it is necessary to make up and sterilize the medium at a low pH and adjust to the final reaction after sterilization, in order to avoid decomposition of the sugar. The contents of one flask are used for a titration, to determine the requisite amount of alkali to be added. This amount of sterile alkali, usually NaOH, is added to each flask by means of a sterile pipette. It is advisable to have an assistant for this task, to flame the necks of the flasks and hold them in an inclined position while the alkali is being run in.

Media containing both glucose and ammonium salts cannot be sterilized without serious discoloration if the pH is higher than about 4.0. In such cases it is usual to dissolve the ammonium salts separately in a volume which is equivalent to an easily measured volume per flask, say 10 ml, and to dissolve the remaining ingredients in a correspondingly reduced volume. Both solutions are sterilized and, when cold, the solution of ammonium salts is measured into the flasks by means of a sterile pipette. An easier method, and one sufficiently accurate for most purposes, is to pipette the solution of ammonium salts, before sterilization, into a series of plugged tubes. After sterilization, the contents of one tube are poured into each flask.

13.7.5   Methods for isolation and purification of moulds

The method used to isolate a particular mould from a natural substratum and to obtain a pure culture depends somewhat on circumstances. If the fungus is growing more or less luxuriantly and typical aerial fruiting structures can be clearly seen, it is usually easy, working with a fine sterile needle and with the aid of a good hand-lens or dissecting microscope, to pick off a few spores or a single spore-head and transfer to a suitable culture medium. Very often a pure culture results from this first transfer. It is seldom, however, that a mould is found growing in natural conditions entirely free from other organisms, and there is always a danger that direct transfers will carry a contaminant. When the contaminant is a slow-growing species its presence may not be detected for some time and, therefore, cultures made in this way must be watched carefully over a period of several weeks, and purified if at any time there is reason to suspect contamination.

When mould occurs on industrial products such as leather, textiles, and cereals, the presence of the fungus is often betrayed only by a stain or discoloration instead of the more familiar furry growth which one associates with mould. Even when the stain is due entirely to the presence of coloured spores it is difficult, and often impossible, to demonstrate the presence of spore-bearing heads. In such cases direct cultures can seldom be made without introducing gross contamination. A few adventitious spores of a very rapidly growing species, in the presence of a much larger number of those of the causal organism, may result in cultures being completely swamped with a mould which has nothing whatever to do with the damage. With material of this type the best way of isolating the dominant organism is by making a series of dilution cultures, an operation commonly termed 'plating out'.

Dilution cultures

To obtain a dilution culture a number of tubes of agar medium, 10–12 ml in each tube, are heated till the agar is melted and then placed in a water-bath maintained at 45°C until required. While the tubes are cooling a small portion of the mouldy material is reduced to as fine a state of subdivision as possible.

A few fragments are dropped into one of the tubes of medium using sterile forceps, taking the usual precautions in handling the tube. The plug is re-inserted and the tube rotated between the palms of the hands in order to mix the contents. The tube should not be shaken in the usual way as this introduces numerous air bubbles which are very persistent in the viscous fluid. The plug is removed, the mouth of the tube flamed, and the contents poured into a sterile Petri dish. The medium from another tube is then poured into the first, mixed by rotation with the small amount of agar remaining after pouring the plate, and poured into a second Petri dish. The contents of a third tube are poured into the same tube, mixed and poured as before, and this process continued for a number of plates which can be determined only by experience with the particular material; usually five or six plates are sufficient. As soon as the agar has set the plates are inverted and incubated.

The rationale of the process is simple. Thorough mixing of the infected material with the agar in the first tube serves to disseminate the spores of the fungus, or, in some cases, fragments of mycelium, throughout the melted medium. The small amount of agar left in the tube after pouring contains relatively few spores and these are again distributed throughout a considerable volume of medium when the contents of the second tube are added, and so on. Each plate, after the first, will contain only a fraction of the number of spores contained in the previous one and, on incubation, the successive plates will show fewer and fewer colonies. If the spores of a particular mould are very numerous in the infected material they are likely to persist through all the dilutions and give rise to colonies in the later plates, whereas a few purely adventitious spores, which are to be found on almost any material, are eliminated in the first two or three dilutions. The final plate, if the amount of material and the number of plates have been judged correctly, should not show more than nine or ten well-separated colonies. If these are all alike, a pure culture has automatically been obtained and nothing is required but to make transfers to slopes or fresh plates as desired. If more than one mould is present on the final plate, incubation should be continued only sufficiently long for the colonies to be differentiated, and transfers should immediately be made from a colony of any species which it is desired to study or retain. It is advisable to plate out separately each of the moulds isolated, as soon as the transfers are showing spores, in order to check their purity and effect further purification if this is necessary.

It is policy, in most cases, to plate out infected material on two or three different media. An unimportant mould may grow so well on a rich medium, such as wort agar, that it swamps a slow-growing species, or, on the other hand, an important species may grow very poorly or not at all on a synthetic medium, such as Czapek agar, and be completely missed unless a more suitable medium is used as well.

A slightly different method of making dilution cultures is often advocated. Instead of pouring all the plates from one tube, which is refilled for each plate, a little of the thoroughly mixed contents of the first tube is poured into the second tube and the remainder poured into a dish, a little of the second is poured into the third tube, and so on. The method given above is better, unless an exact dilution ratio is necessary (and this involves making the dilutions with sterile graduated pipettes), since it ensures an approximately uniform degree of dilution at each stage, and it is also much easier.

The procedure for the purification of an impure culture is the same as for the isolation of a mould from infected material. The most usual method is by making a series of dilution cultures, as described above. It is necessary to use a very small amount of inoculum to mix with the first tube of agar and, in cases where the spores are small and happen to be easily wetted, it is advisable to make one or two preliminary dilutions in sterile water or saline (a 0.9% aqueous solution of sodium chloride). However, in some cases of infection of cultures the growth of the contaminant is clearly confined to a small area of the slope. It is then frequently possible to obtain a pure culture by picking off a few spores from a portion of the slope remote from the invader.

Hyphal tip cultures

This is a method which is particularly valuable for the purification of a sterile fungus, or one which produces spores only very sparingly. It can be used only when the initial rate of growth of the mould to be purified is greater than that of the contaminant. It happens not infrequently that cultures, particularly when mites have had access to them, become contaminated with *Cephalosporium*. In such cases plating out serves only to isolate the invader, since *Cephalosporium* produces myriads of small, easily wetted spores. Fortunately, species of this genus grow very slowly for the first few days, being easily outpaced by very many species of other genera, and can therefore be eliminated by the hyphal tip method, described below.

A single colony is planted in the centre of a Petri dish containing a medium on which the mould to be purified will grow well. When the colony is about 1 cm in diameter a small piece of agar, containing the tip of one of the radiating hyphae at the edge of the colony, is cut out and transferred to a fresh slope. It is usually possible to find a growing tip which is well separated from other hyphae and which can be cut off cleanly. The most suitable tool for the purpose is the dummy microscope objective described below.

The problem of bacterial contamination

When plating out some materials, particularly soils and industrial effluents, it is not unusual to find all the plates on a series heavily infected with bacteria, often to such an extent that sub-cultures made from colonies of moulds are more likely than not to be contaminated. Another complication is that some common soil bacteria strongly inhibit the growth of moulds and species of fungi present in the infected material do not grow at all.

It is possible to prevent the growth of all but a few species of bacteria, without impeding the development of the fungi, by the use of selective inhibitors in the culture medium. (See Appendix I.) Tests made by G. Smith, using ordinary malt agar and potato agar, showed that there is little advantage to be gained from using antibiotics along with the dyestuffs. Media containing 0.035 g per litre of rose bengal have been found to be very satisfactory for plating out a wide variety of soils, and they involve no additions after sterilization. The amount of rose bengal may be increased, if necessary, to 0.067 g per litre. At this concentration it exerts a distinct anti-spreading effect; above this concentration some species of mould are inhibited.

For the purification of a culture which is infected with bacteria, but is free from contaminating moulds, several methods are useful. Raper (1937) recommends the use of a small glass cylinder (the so-called van Tiegham cell) to one edge of which are fused three small glass beads, $\frac{1}{3}$ to $\frac{1}{2}$ mm in diameter. This is placed in a Petri dish, resting on the beaded edge, and the whole sterilized. Melted agar is poured in to a depth about half-way up the cylinder. When cool, the contaminated fungus is inoculated in the centre of the ring. The fungus grows underneath the free edges of the ring, within the agar, whereas the bacteria cannot spread. Hyphal-tip cultures can be made from hyphae which have spread well outside the ring. In a modification of the method Ark and Dickey (1950) used small pellets of modelling clay instead of glass beads for raising the glass cylinder. Clay has the advantage that it sticks the cylinder to the bottom of the dish, so the latter can be inverted if desired.

Another method is to plate out the fungus on a rose bengal medium. If the infected fungus is not producing spores, a fragment of mycelium can be sown on a medium containing the maximum amount of rose bengal. Usually the bacteria are prevented from spreading, if not suppressed completely, and it is possible to obtain pure hyphal-tip cultures of the mould.

Single-spore cultures

A number of microbiologists regard a culture as pure only when it has been obtained by germination of a single spore. It is true that in some types of studies on fungi single-spore cultures are a *sine qua non*. For example, investigations on the genetics of species of *Neurospora* and yeasts necessitate not merely the making of very numerous single-spore cultures but also the separation in serial order of all the spores in a single ascus. Again, in order to prove the connection between an Ascomycete and its conidial state it has usually been necessary to make cultures derived from both single ascospores and single conidia. However, in industrial work, biochemical studies, and the maintenance of culture collections, the value of single-spore cultures has been much overrated.

Assuming that single-spore cultures are a necessity, there are quite a number of methods available. However, it should be noted that, while any method can be used successfully for isolation of large and/or highly coloured spores, all methods, with the exception of those involving the use of a micro-manipulator, are difficult when applied to moulds which have very small, lightly coloured or colourless spores.

Method 1   The most elegant method of isolating single spores is by means of a good micro-manipulator. A number of types are marketed with which it is possible to pick up any particular spore and transfer it to a fresh substrate, all under a high power of the microscope. Such machines are virtually a necessity in genetical investigations and in any work involving the making of very large numbers of single-spore cultures. Unfortunately they are all very expensive and, for the kind of work done in most mycological laboratories, the outlay is not justified. A book on the subject by El-Badry (1963) gives details of most types of machines, and discusses their application in a number of scientific disciplines.

Method 2   A series of dilution plates is prepared in the usual way, using a clear, filtered agar. The plates are incubated only for sufficient time for the spores to put out the primary germ-tubes and must, therefore, be examined under the microscope at frequent intervals during the first forty-eight hours. A few spores are found which have just germinated and which are individually well separated from all other spores. These are marked with a glass pencil while the dish is on the stage of the microscope, then the agar round them is cut with a sterile scalpel, the flattened end of a needle, or the tool described in Method 3, and the tiny pieces of agar are transferred to tubes of fresh culture medium. The great disadvantage of this method is that the spores do not all lie in one plane and examination of the plates can be extremely tedious. The advantage is that, in a set of dilution plates, there is usually one plate with a particularly favourable distribution of spores, not too few to be difficult to find, and sufficiently spread to make isolation of individual spores fairly easy.

Method 3   Another method is to flood the surface of the agar in a Petri dish with a dilute spore suspension in sterile water, using not more than about 2 ml, allow to stand for a few minutes and then pour off the water. It will be found that quite a number of spores have stuck to the agar and, as these lie approximately in one plane, they are fairly easily found. Otherwise the procedure is the same as that given in Method 1.

A very convenient tool for cutting out minute blocks of agar enclosing single spores was originally described by La Rue (1920). It consists of a dummy microscope objective in which the front lens is replaced by a sharp-edged metal tube, about 5 mm long and 1.5 mm in diameter (this being approximately the diameter of the field using an 18 mm objective and a $\times$ 10 eyepiece). This is screwed on to the nose-piece of the microscope in place of one of the objectives, and, if possible, should be fairly accurately centred with the 18 mm. The plate is examined with the latter lens and a field is found which shows a single spore. The cutter is swung into position and lowered carefully until it just touches the glass at the bottom of the dish. A small circular block is cut out of the agar, but is left behind when the cutter is raised and may be re-examined with the 18 mm to ensure that all is well. The small block is then lifted out with a needle. If the cutter is not perfectly centred with the objective it is only necessary to make a few trial cuts, in order to determine the direction of the error, and thereafter to move the dish the determined amount before making the cut.

A modified form of the cutter, designed particularly for easy sterilization and replacement of the cutter is described by Keyworth (1959).

Method 4   A dilute spore suspension is made by vigorously shaking a small mass of spores in a tube of sterile water or saline. A series of dilutions in water or saline are made from this until a suspension is obtained such that single loopfuls contain usually one spore, otherwise none. This is determined by spreading a series of loopfuls on a slide and examining under the microscope. Single loopfuls are then transferred to agar plates and the presence of not more than one spore in each is confirmed by direct examination. The positions of the drops which do contain a spore are marked on the bottoms of the dishes, so that the spores may be readily located after the drops have evaporated, and so that accidental infection may be recognized by its position on the plate. The plates are incubated, and in those in which the spores are viable pure cultures are obtained.

Method 5   A very ingenious method is described by Hansen (1926). This method works well with species which have large coloured spores, such as species of *Alternaria, Helminthosporium*, etc., but is difficult to apply to most species of *Penicillium* and many of *Aspergillus*. A dilute spore suspension is made in melted agar medium. The suspension is sucked up into a number of fine glass capillary tubes, of bore slightly greater than the diameter of the spores, and the medium is allowed to set therein. Microscopic examination of the capillaries should show short lengths each containing a single spore. These are broken off, sterilized externally with alcohol, and planted in fresh medium. The spores germinate, the germ tubes emerge from the ends of the capillaries and form typical colonies.

An excellent review of methods of single-spore isolation is given by Hildebrand (1938). The paper contains a detailed description of Hildebrand's technique and there is an extensive bibliography. Methods are also described by Booth (1971).

Small items of equipment

In addition to optical and culturing equipment, several small items will be required in microbiological examinations.

At least two needles are required for mounting specimens. The nichrome needles used for sowing cultures are unsuitable, since they are not sufficiently stiff for teasing out material. Fine sewing needles are satisfactory if mounted in wooden or metal handles. If still finer tools are required entomological pins (with the heads cut off) may be used. Special metal holders with very small chucks are obtainable for mounting the latter. Biological equipment suppliers offer a wide range of mounted needles, cutting tools and dissecting instruments, but many more everyday items can be successfully used.

A simple but very useful piece of apparatus is a tile about 150 mm square, half white and half black. The black half is used for manipulating colourless specimens, which are often invisible in the mounting fluid if viewed against a white surface.

Several small dropping-bottles are required for stains and mounting media. The ordinary type of dropping-bottle usually delivers a drop which is far too large. It is preferable to take a short length of glass tubing with one end drawn out to a fine jet, and insert this through the cork of an ordinary bottle.

The requirements include a supply of glass slides, 76 × 26 mm, and cover-glasses. The latter should be of No. 1 thickness and may be circular or square as preferred, convenient sizes being 18 mm diameter or 18 mm square. If larger than this it is difficult to apply a ring of cement all the way round. Both slides and cover-glasses should be cleaned before use by soaking in chromic acid, then thoroughly washing in water and finally in alcohol.

13.7.6 Examination of living cultures

In the study of moulds for purposes of identification a great deal of information can be obtained from observations made on dry living cultures under the compound microscope or a stereomicroscope. Examination in this way should always precede the preparation of slides. Slopes can be placed across the stage and the edges examined by transmitted light with objectives of 18 mm and low power. Petri dishes can be laid flat on the stage, either side up, and examined by incident or transmitted light. As mentioned earlier, when a mould forms dense matted growth the planting of several colonies in one dish will often result in there being narrow sterile zones, where the colonies approach each other, and mature fruiting structures can be observed partially overhanging these clear spaces (Figs 13.1, 13.2, p. 300). In the case of species of *Penicillium* and *Aspergillus* the shape of the spore-heads, the disposition of the chains of spores, and the origin of the conidiophores, whether from submerged or aerial hyphae, have considerable diagnostic value, and the information required for identifications can be obtained only from the study of undisturbed living cultures. In the same way, determination of species of *Mucor* and other members of the same family is possible only when study of slide preparations is combined with examination of living material in Petri dishes. Many other moulds produce conidial structures which fall to pieces at the least touch and, with these, the value of direct observation is apparent.

In the examination of living cultures it is important to remember that a good deal of water is given off by fast-growing species, and some of it appears on the hyphae as small droplets, invisible to the naked eye. It is not uncommon to see the sporangiophores of species of *Mucor* thickly studded with these droplets, and they may readily be mistaken for chlamydospores by the beginner. In tube cultures an appreciable amount of water often collects on the walls of the tube. Sporangia coming in contact with this usually burst and liberate irregular masses of spores. Similarly, false spore-heads held together by mucus, such as those of *Verticillium* and *Cephalosporium*, are broken up. Spores often germinate in such damp situations, even while still attached to phialides or in sporangia, and give rise to structures which are puzzling the first time they are seen under a low power of the microscope.

Preparation of slides

For study of fine detail, and for accurate measurements, slide preparations must be made. With most species of moulds it is difficult, if not impossible, to fix, stain, and mount specimens, as is done for botanical and zoological material, and at the same time preserve structure. In all but a few cases it is general practice to use fluid mounts made with as little manipulation as possible. The method to be adopted often depends on whether the slide is required merely for a temporary examination, or whether it is intended for a permanent collection of typical slides. In the descriptions of methods given here, those which are unsuitable for making permanent slides are indicated.

Success in making slides depends largely on the age of the culture. If the culture is too young there will be few or no fully developed fruiting structures or ripe spores. If the culture is too old, the fruiting structures, particularly in the Hyphomycetales, fall to pieces. For this reason Petri dish cultures are more satisfactory than slopes. So long as the culture is still spreading it is possible to pick off small portions from areas of exactly the right age. If tubes must be used, the right age of culture for the majority of moulds is 5–7 days. Exceptions to this are the species which form balls of spores. In very old cultures which are drying out the mucilage in which the spores are enveloped acts as a cement, and most of the balls remain intact when mounted in fluid medium. If young cultures are examined the spores are dispersed all over the slide, but, on the other hand, the conidiophores are still turgid, whereas in old cultures they usually collapse.

Water has a limited application as a mounting fluid but for the majority of moulds is quite unsuitable. It evaporates rapidly, often causes swelling of hyphae by osmosis, and usually causes the parts of the specimen to adhere together as a tangled mass of hyphae, spores, and air bubbles. Alcohol wets efficiently and makes a fairly satisfactory mountant for a brief and rapid examination, but is too volatile for general use. The most generally useful medium is that known as 'lactophenol'.

A number of formulae for this medium have been published but, for regular use, the original recipe of Amann (1896) is to be preferred.

Lactophenol

Phenol (pure crystals)	*10 g*
Lactic acid (syrup, s. g. 1.21)	*10 g*
Glycerol (pure)	*20 g*
Distilled water	*10 g*

Lactophenol is readily prepared by warming the phenol with the water until dissolved and then adding the lactic acid and glycerol. The refractive index is 1.45.

To prepare a slide place a *small* drop of mounting fluid in the centre of a clean glass slide, with a sterile needle pick off a very small portion of typical material from the culture, place this in the drop of fluid and very gently tease it out with a pair of needles until it is well wetted, then lower a cover-glass on to the preparation in such a way as to avoid air bubbles as far as possible. It is difficult to make slides which contain no air bubbles without teasing out the specimen to such an extent that structure is destroyed and, within reason, their presence does not seriously interfere with observation. Some moulds are wetted by lactophenol only with great difficulty and do not give presentable slides unless some means is taken to accelerate the process. One way which is often effective is to warm the slide over a small flame before putting on the cover-glass. Another way of expelling air, which has been found particularly useful for species of *Penicillium*, is to place the specimen in a drop of alcohol on the slide, wait until most of the liquid has evaporated, then add a fairly large drop of lactophenol, tease out the specimen as required and apply a cover-glass. If even this fails to give a satisfactory slide, as may happen with thick structures such as perithecia, the specimen should be placed in a small watch-glass, completely covered with mounting fluid and the whole placed in any vessel from which the air can be exhausted. A vacuum desiccator, if available, will take several specimens at a time and is admirable for the purpose. To ensure the removal of all air from the specimen, air is admitted to the vessel after 10–15 minutes and suction is again applied for about another 10 minutes. The specimen may now be transferred to a slide and a cover-glass applied.

It is usually impossible to make good slides by using mounting fluid in such quantity that it just fills the space underneath the cover-glass (this is the practice when Canada balsam is used as mounting medium). Generally an appreciable amount oozes out, especially if the cover-glass is gently pressed down with a pair of forceps, and must be absorbed on pieces of filter paper carefully applied to the edge, taking care that no fluid gets on to the upper surface of the cover-glass.

Lactophenol is advantageous as a mounting medium as it rarely causes shrinkage or swelling of the cells of most fungi and it has no tendency to evaporate, so preparations are reasonably permanent, if carefully handled, without any further manipulation. It is also sufficiently viscous to allow the use of an oil-immersion lens without undue movement of parts of the specimen, provided that the immersion oil has not become oxidized. It should be noted that, as lactophenol is a non-swelling medium, measurements of objects in this fluid will not necessarily agree with measurements of the same objects mounted in water, which often causes appreciable swelling of fungal hyphae and spores. Many published diagnoses of fungi include measurements recorded from specimens mounted in water (although there has been for some years an increasing tendency to use lactophenol), and it is necessary to bear this in mind when trying to identify moulds.

In a very few cases it has been found that lactophenol is not a suitable mounting medium. One of its characteristics is that its refractive index is very close to that of fungal hyphae, with the result that colourless structures appear very faint owing to lack of contrast with the background. The majority of colourless and pale coloured moulds can be stained to give adequate contrast, but the yeasts are difficult to stain without causing shrinkage and distortion of the cells. These are best mounted in plain water, in which the cells are clearly visible. Water is also the best medium for mounting a few species of dark coloured moulds whose spores shrink to a serious extent in lactophenol. Slides made with water as the mounting medium may be protected from evaporation by sealing the edges of the cover-glass with shellac cement or nail varnish (see p. 316). Such slides will remain in good condition for periods up to about two days.

Locquin (1952) described a mounting medium of very high refractive index, which he claims to be suitable for mounting fungi directly, that is without any previous manipulation. The medium is made by saturating pure glycerol alternately with potassium iodide and mercuric iodide until no more of either will dissove. It is unfortunate, but not surprising, that this medium causes very serious shrinkage of fungal hyphae and spores, for its refractive index, 1.7–1.8, would otherwise make it valuable for observation of fine structures.

An interesting and completely different method of making slides is described by Skerman (1946). A portion of the mould colony containing sporing structures of suitable age is wetted with a drop of a mixture of three parts ether and one part alcohol. Before this has completely evaporated a drop of 10% collodion in the same mixed solvent is placed on to the moistened area by means of a 4 mm loop. The collodion dries in 1–3 minutes to give a film in which mycelium and conidiophores are embedded. As soon as the film is sufficiently firm it is removed to a slide. The film may be mounted directly in lactophenol containing stain, or may be stained first and then mounted in plain lactophenol. Alternatively, if much mycelium is present, the collodion may be removed with warm solvent and the mycelium then teased out as required. The original paper contains some fine photomicrographs of species of *Penicillium*, showing spore chains *in situ*.

Butler and Mann (1959) have described another useful method of removing portions of colonies without disturbance. A piece of cellulose self-adhesive tape, approximately 8 cm long and 1.3–1.9 cm wide, is lightly pressed on to a mould colony in a Petri dish or on a natural substratum, then placed, sticky side down, in a drop of mounting fluid on a slide. The tape is stretched tight and the ends stuck down to the glass to hold it in position. If it is necessary to wet the colony first with alcohol in order to remove air bubbles, care should be taken to avoid getting any of the spirit on to the ends of the tape. If the ends are wetted with any fluid they will not stick to the glass. The amount of pressure to be applied when the centre of the tape is placed on the colony can be found only by practice. If pressed too hard the tape picks up a dense mass of mycelium and sporing structures. However, one or two trials, with any particular type of colony, should be sufficient to determine the correct procedure.

The latter method is excellent for making temporary mounts, but there does not seem to be any immediate prospect of making permanent slides by this procedure. Commercial varieties of tape all tend to become brittle with age, so that, even if one could seal the edges of the tape, or use a small piece which would lie under a cover-glass, it is improbable that the slides would have any real degree of permanence. Also, lactophenol appears to react with the sticky material on the tape, with the result that the background becomes spotty after the slide has been kept for a few days.

Butler and Mann's paper includes photomicrographs of chains of *Alternaria* spores. The method works very well with such sporing structures where the spores are not too readily deciduous, but with very fragile structures, for example the branched spore chains of *Cladosporium*, the production of a slide showing unbroken structures is a matter of luck.

Staining

In general, moulds tend to take up stains somewhat unevenly, due to the fact that their affinity for colouring matters, which in young structures is considerable, decreases rapidly with age. The simplest, and often the best method of staining is to mix the dye with the mounting fluid, but only comparatively few colouring matters can be used successfully in lactophenol, the chief being cotton blue, picric acid, orange G, picro-nigrosin, acid fuchsin, and trypan blue.

Cotton blue (0.05 g in 100 ml lactophenol) is probably the most widely used of all stains for moulds. It colours very young structures fairly deeply, but, with most specimens, acts very unevenly, a single hypha often showing irregular patches of colour. Another peculiarity is that staining proceeds slowly over a long period, so that a specimen which, when first mounted, shows a satisfactory depth of colour, will often become hopelessly overstained after a few months storage.

The picric acid medium is made by saturating lactophenol with the dye, filtering off any excess before use. It stains comparatively evenly, but the yellow colour appears to many people to be of insufficient intensity. Contrast may be enhanced, especially for photomicrography, by using a blue light filter on the microscope lamp. (The majority of photographs in this book were taken from specimens mounted in this medium.)

Orange G is used as a 1% solution in lactophenol. It usually stains well and reasonably evenly, and gives adequate contrast, which, as in the case of picric acid, may be increased by the use of a blue light filter.

Acid fuchsin is used as a 0.1% solution (see below for comments).

Picro-nigrosin was originally recommended to be used as a 0.4% solution of water-soluble nigrosin (a bacteriological stain) in picro-lactophenol. Nigrosin alone, in aqueous solution or in lactophenol, does not stain fungi, and is actually used in the so-called relief or negative staining, in which the specimen is mounted in a 5–10% aqueous solution of the dye, and appears as a colourless image on a blue-black background. In combination with picric acid it stains readily from lactophenol, the colour being a neutral grey. Picro-nigrosin-lactophenol has, unfortunately, one grave fault. On keeping, often for only a short time, it forms an exceedingly fine precipitate which cannot be completely removed by filtration. The trouble may be avoided by keeping two solutions, picro-lactophenol and 2% aqueous nigrosin. When required, single drops of each of the two solutions are quickly mixed on a slide, and the specimen immersed in the drop immediately. An even better method is first to mordant the specimen with picro-lactophenol, remove the excess as completely as possible, first by drainage and finally by careful application of pieces of filter paper, wash with two or three successive drops of water, stain with 0.4% nigrosin solution, remove excess of this, and mount in plain lactophenol. This method may appear to be complicated, but all the operations are easy and rapid, and, since the stain is a neutral shade and very fast, the results are worthwhile.

Carmichael (1955) recommends a slightly different combined staining and mounting medium, termed lactofuchsin. This is made by dissolving 0.1% acid fuchsin in pure lactic acid. It stains rapidly and evenly, and unlike most other stains, colours the spores of many moulds. The contrast given by the stain is enhanced by the fact that the refractive index of lactic acid is slightly lower than that of lactophenol. This medium, whilst one of the best for making temporary mounts, is, unfortunately, quite unsuitable for the preparation of permanent slides, owing to the instability to light of the dyestuff. Slides exposed to daylight gradually fade, and, if exposed to a powerful light such as is necessary for micro-projection, lose their colour entirely within a few minutes. Another disadvantage is that, whilst the stain gives adequate visual contrast, it is very difficult to obtain sufficient contrast in photomicrography, owing to the peculiar absorption spectrum of the dye.

Another combined stain and mounting medium, which is stated to give a permanent blue-black, is described by Isaac (1958). It is of complicated constitution, containing haematoxylin, bismarck brown, methyl green, ferric and chrome alums, chloral hydrate, glycerol, and gum arabic. At one stage in its preparation the use of an ultra-speed centrifuge is necessary.

All the dyes mentioned so far colour cell contents, but have little or no effect on cell-walls. Boedijn (1956) describes the use of a dye, trypan blue, which he claims will colour the cell-walls of all the species tested. He recommends the use of a 0.1–0.5% solution of the dye in either 45% acetic acid or in lactophenol, accelerating the action by heat if necessary. In the author's (AHSO) experience 0.05% of the dye in lactophenol stains rapidly and gives adequate depth of colour. Staining is very even if the specimen is well teased out in the drop of stain, and the cell-walls are coloured as deeply as the cell contents. The stain appears to be stable to light.

With all these combined staining and mounting fluids most of the stain is absorbed by the specimen, and if the mount is reasonably thin, that is with the specimen well teased out and the cover-glass pressed well down, the 'background' should be almost colourless. With thicker specimens, such as whole perithecia, there is often too much background colour. In such cases it is advisable to allow the stain to act for a few minutes, drain off as much as possible, and mount in a drop of plain lactophenol.

A method of staining which is preferable with some dyes, and which often gives very good results, is to treat the specimen with a solution of the stain in acetic acid or alcohol, and then replace the fluid with plain lactophenol. Boedijn (1956) suggested using trypan blue as a solution in 45% acetic acid, but staining by this method is extremely rapid and it is difficult to avoid overstaining. In any case the dyestuff stains well as a solution in lactophenol. Alcorn and Yeager (1937) recommend Orseillin BB, used as a 0.25% solution in 3% acetic acid. The stain is allowed to act for 3–5 minutes, then the specimen examined under the microscope. If too deeply coloured, the specimen is treated with successive drops of very dilute acetic acid until satisfactory. This gives a pleasant red shade, which gradually turns brownish, particularly if the slide is exposed to strong light. The dye has little affinity for fungal hyphae if applied as a solution in lactophenol. Another stain which works well with young structures, and, incidentally, shows up the nuclei in the cells, is Chlorazol Black E (known as New Black D. E. in some catalogues). This is used as a 1% solution in 90–95% alcohol, and is allowed to act for about one minute, using alcohol to differentiate if staining has been too rapid.

Permanent slides

Any slide which is to be kept as a permanent record should be protected by sealing the edge of the cover-glass with suitable cement. Lactophenol mounts can, it is true, be kept without sealing, but there is always a danger of the cover-glass being accidentally moved. In addition, it is almost impossible to protect slides completely from dust, and dust on an unsealed cover glass is difficult to remove without disturbing the specimen. For the application of fluid sealing cements a turntable is almost a necessity if circular cover-glasses are used. If square covers are preferred, the sealing must of course be done freehand.

There are two cements which are of use for sealing slides of moulds mounted in lactophenol. The first is brown shellac cement, obtainable from any of the dealers in microscopic sundries. The second is nail varnish, and of this the clear unpigmented varieties are best, since the pigments usually incorporated are soluble in lactophenol and often stain the specimens on long standing. Both cements are useful but neither is completely dependable, as sometimes, even when every care is taken in the application, there is some leakage under the cover-glass. The following five points are the main conditions for success.

1. No cement should be applied to a slide until it is certain that the cover-glass is not under strain. If a slide is allowed to stand for a few days, air bubbles often appear at the edge showing that the cover-glass has been bent slightly when making the slide, and has gradually recovered on standing. Such air spaces must be filled in with mounting fluid before applying cement.

2. The cement must be of suitable consistency, i.e. it should run smoothly, but not too readily, from a brush. If shellac cement is too thick it may be brought to the right consistency by the very gradual addition of alcohol. Nail varnish may be thinned by the cautious addition of acetone.

3. Using a turntable, a very thin coat of cement should be put on first, using a fine brush. When this is quite dry, a second thicker coat is added to overlap the first.

4. A necessary precaution with nail varnish is to ensure that the portion of the slide immediately around the cover-glass is clean and dry. This is not so essential when using shellac cement.

5. If the layer of mounting fluid is thick, due to the presence of thick structures which it is unadvisable to squash, any attempt to apply cement with the aid of a turntable will cause movement of the cover-glass. In these cases the first layer of cement must be put on freehand. When this is quite dry the slide may be tidily finished by ringing on the turntable in the usual way.

With these precautions the majority of preparations sealed with either of the cements will keep in good condition for many years. Shellac cement may also be used, as noted above, for sealing temporary mounts in water, for the purpose of retarding evaporation.

A different method of sealing was described originally by Diehl (1929). The object to be studied is placed in a drop of mounting fluid on a 22 mm No. 0 cover-glass and is then covered with a 12 mm No. 2 cover-glass. Any excess fluid is carefully cleaned away from the edge of the small cover, a large drop of Canada balsam is placed in the centre of it and a slide is gently lowered on to it. When the balsam has spread to the edge of the large cover-glass the slide is inverted and allowed to stand until the balsam has set. Linder (1929) claims that this method can be used with lactophenol provided that the mount is thoroughly dehydrated in a desiccator before the final mounting is balsam. However, most workers who have experimented with the method have found that lactophenol and balsam are not compatible, the balsam gradually creeping in between the cover-glasses and eventually ruining the slide. In theory this is an ideal method of making permanent slides and it is to be hoped that someone will discover either a clear cement which does not react with lactophenol, or a mounting medium which is as convenient as lacto- phenol and is, at the same time, compatible with balsam.

Unfortunately, it is true that some slides deteriorate in spite of every care in preparation. As stated above, spoilage may occur because of the leakage of ringing cement under the edge of the cover-glass, but there is another cause of trouble which is entirely unconnected with the method of sealing. The mycelia of many species of fungi contain large amounts of fat and there is a tendency for this to be displaced by mounting fluid, with the consequent appearance of numerous unsightly globules all over the slide. In such cases it is not possible to prepare slides which will keep in good condition unless the mould to be mounted will withstand much more drastic treatment than the simple methods described here.

Whatever the method of preparation, all slides which are to be kept should be labelled as soon as made, giving the name of the species, its number in the culture collection, the stain (if any), the mounting fluid, and the date of preparation.

Artefacts
In the examination of slides beginners are often puzzled by certain artefacts. Perhaps the commonest are very small air bubbles. These are usually circular if the layer of medium is thick, but may be of any shape if the cover-glass has been well pressed down. Circular air bubbles have a very thick dark outline and appear highly refractive, as indeed they are. Thin bubbles have dark edges, but the most noticeable point about them is that portions of hyphae or spores lying inside the bubbles are not in focus when the rest of the slide appears sharp, and also show a greater degree of contrast than similar structures surrounded by fluid.

Another common appearance is a series of circular bodies of very varying size, somewhat refractile and always with smooth edges. These are oily substances exuded from the cells of the fungus, or squeezed out during the mounting process. As mentioned above, these oily drops sometimes increase in number when a slide is stored and may entirely ruin the preparation.

When making slides from cultures of moulds which form thin tough colonies on agar it is not uncommon to carry over bits of culture medium to the mounting fluid. These become flattened out when the cover-glass is pressed down. The first thing noticed about such patches of agar is that they do not take the stain from coloured lactophenol. They usually contain submerged hyphae and often have a few spore-heads pressed into them. The latter always appear somewhat shrivelled and are, like the agar, unstained.

13.7.7   Identification of species

Methods of obtaining pure cultures and general methods of examination having been discussed, it now remains to describe the routine to be followed when an unfamiliar species of fungus has to be identified.

The first necessity is to prepare a full and accurate description of the fungus, as grown on standard culture media. It is assumed that some idea of its general behaviour on one or two media has already been gained during isolation and purification. If the species grows satisfactorily on Czapek agar this should now be used, but if the Czapek medium induces only poor and stunted growth, or if sporing structures are lacking or tardily produced, some more suitable medium must be selected. Several cultures should be made in Petri dishes. One or two of the dishes may be planted with single colonies for measurement of growth rate and for recording the general appearance of colonies, the remainder with three colonies spaced about 2–3 cm apart. It is necessary to make a number of cultures because some of them will be examined, probably with the dishes uncovered before maturity, and are likely to develop infections on continued incubation. Cultures should be examined at frequent intervals and the following details recorded.

1. Rate of growth; described as slow, very slow, moderate, rapid, etc. Some diagnoses give actual diameters of colonies after specified numbers of days.

2. Colony colour and colour changes; whether uniform, in zones, or patchy. Evanescent colours which are often to be observed at the edges of growing colonies should be recorded. Where a standard colour index is available records can be made with a precision which is unattainable by the use of ordinary names of colours.

3. Colour and colour changes of the reverse of the colonies.

4. Colour changes in the medium; whether confined to the area covered by the colony or diffusing.

5. Texture of surface; whether loose or compact; plane, wrinkled or buckled; velvety, matted, floccose, hairy, ropy, gelatinous, leathery, etc.

6. Odour, if any. Odours are usually very difficult to describe. Many species have a smell which can only be described as 'mouldy', but quite a number have characteristic and sometimes fragrant odours, and, of course, many have no odour. Regular sniffing of fungal cultures is not recommended.

7. Character of drops of transpired fluid often found on aerial growth.

8. Character of the submerged hyphae; colour, presence or absence of septa, approximate diameter, characteristics of special structures if any present.

9. The stage at which fruiting structures develop.

10. The character and disposition of the mature fruiting organs; whether sporangia, perithecia, pycnidia, sporodochia, coremia, or detached conidiophores; whether borne in the substratum, on the surface, or on the aerial mycelium. The presence of more than one type of sporing structure should be particularly noted.

11. Colour, size, and shape of mature fruiting organs or fruit-bodies.

12. Details of structure of the fruiting organs, including measurements of essential parts and disposition of the spores thereon.

13. Full details of spores; colour, shape, septation, surface markings, size (including both average and extreme measurements).

Data numbered 1–7 are obtained by examination of cultures with the naked eye or with a hand-lens, 8–11 by observations on living cultures with the aid of a low or moderate power of the microscope. Numbers 12 and 13 necessitate the preparation of slides and the use of the highest powers of the microscope.

The information recorded under numbers 8 and 10 should be sufficient to place the species in its correct Class and Order, and consideration of the rest of the data will lead to the Family and then the genus. The determination of the correct specific epithet is, except in a few genera, a matter of some difficulty. If there is a good monograph of the genus it is usually a question of careful and patient observation, along the lines indicated by the particular authority, then repeated consideration of the data until the unknown fits into its proper place. In the absence of an authoritative treatment of the genus the usual procedure, and often the quickest in the long run, is to consult Saccardo's *Sylloge* and the *Index of Fungi* (Anon, 1940–), look up all the recorded species in the genus, and follow up references to any which seem to be near the one to be identified. Unfortunately, many species have been inadequately described and there are even numerous genera which are ill-defined. In some cases the accepted conception of a genus is more a matter of tradition than of adequate diagnosis, and it is difficult to find any published data sufficiently exact for recognition. However, the genera which are of commonest occurrence are the ones which have been most studied and of which there is an extensive literature. Numerous books and monographs of these have been published in the last 20 years.

Perhaps the greatest difficulty in identifications is to find that, while most of the data from the unknown fit a published description, there is a discrepancy in spore dimensions. It has already been pointed out (p. 312) that the mounting fluid may affect apparent size of spores. It has also been shown (Williams, 1959) that environmental conditions during growth of the fungus can affect spore size. In general, when the fungus is grown in the dark the spores are larger than when cultures are kept in the light. Other factors, such as temperature and nutrition, affect different species in different ways. What this amounts to is that reasonable differences in spore dimensions should not necessarily preclude identity, if other characteristics correspond.

There is one thing the student should guard against. It is doing a great disservice to other mycologists to assume too hastily that an unrecognized fungus is a new species and to publish a description under a new name. The literature is encumbered with a mass of generic names and specified epithets which are nothing more than synonyms of well-known forms, and which ought never to have been bestowed.

13.8   References

ALCORN, G. D. and YEAGER, C. C. (1937). Orseillin BB for staining fungal elements in Sartory's fluid. *Stain Tech.*, **12**, 157–8.
AMANN, J. (1896). Conservirungsflüssigkeiten und Einschlussmedien für Moose, Chloro- und Cyanophyceen. *S. Mikroscopie*, **13**, 18–21.
ANON (1940–). *Index of Fungi*. Commonwealth Mycological Institute, Kew.
ARK, P. A. and DICKEY R. S. (1950). A modification of the Van Tieghem cell for purification of contaminated fungus cultures. *Phytopathol.*, **40**, 389–90.

BARRY, SHIELA, BRAVERY, A. F. and COLEMAN, L. J. (1977). *Int. Biodeterior. Bull.*, **13(2)**, 51–7.

BOEDIJN, K. B. (1956). Trypan blue as a stain for fungi. *Stain technol.*, **31**, 115–16.

BOOTH, C. (1971). Introduction to General Methods. In *Methods in Microbiology*, Vol. 4. BOOTH, C. (Ed.). pp. 1–47. Academic Press, London and New York.

BRAVERY, A. F., BARRY, SHIELA, and COLEMAN, L. J. (1978). Collaborative experiments on testing the mould resistance of paint films. *Int. Biodeterior. Bull.*, **14(1)**, 1–10.

BUTLER, E. E. and MANN, M. P. (1959). Use of cellophane tape for mounting and photographing phytopathogenic fungi. *Phytopathology*, **49**, 231–2.

CARMICHAEL, J. W. (1955). Lacto-fuchsin: a new medium for mounting fungi. *Mycologia*, **47**, 611.

DADE, H. A. (1960). On mounting in fluid media, with special reference to lactophenol. *J. Quekett microscr. Club*, Ser. 4, **5**, 308–17.

DAVIS, J. G. and ROGERS, H. J. (1940). The effect of sterilization upon sugars. *Z. Bakt.*, Abt. II, **101**, 102–10.

DIEHL, W. W. (1929). An improved method for sealing microscopic mounts. *Science*, **69**, 276–7.

EL-BADRY, H. M. (1963). *Monographien aus dem Gebiete der qualitativen Mikroanalyse. Band 3. Micromanipulators und micromanipulation.* Wien, Springer Verlag.

HANSEN, H. N. (1926). A simple method of obtaining single-spore cultures. *Science*, **64**, 384.

HANSEN, H. N. and SNYDER, W. C. (1947). Gaseous sterilization of biological materials used as culture media. *Phytopathol.*, **37**, 369–71.

HAWKSWORTH, D. L. (1974). *Mycologists Handbook*. C.A.B.

HILDEBRAND, E. M. (1938). Techniques for the isolation of single microorganisms. *Bot. Rev.*, **4**, 627–64.

ISAAC, P. K. (1958). A haematoxylin staining mountant for microorganisms. *Stain Technol.*, **33**, 261–4.

KEYWORTH, W. G. (1959). A modified La Rue cutter for selecting single spores and hyphal tips. *Trans. Br. Mycol. Soc.*, **42**, 53–4.

LANGERON, M. (1945). *Précis de mycologie*. Masson et Cie, Paris.

LA RUE, C. D. (1920). Isolating single spores. *Bot. Gaz.*, **70**, 319–20.

LINDER, D. H. (1929). An ideal mounting medium for mycologists. *Science*, **70**, 430.

LLOYD, A. O. (1968). The evaluation of rot resistance of cellulosic textiles. In *Biodeterioration of Materials*. WALTERS, A. H. and ELPHICK, J. J. (Eds.). pp. 170–7. Elsevier, London.

LOCQUIN, M. (1952). Nouveau reactif pour l'etude des structres fines chez les champignons. *Bull. Soc. mycol. Fr.*, **68**, 172–4.

RAPER, J. R. (1937). A method of freeing fungi from bacterial contamination. *Science*, **85**, 342.

ROBBINS, W. J. and MCVEIGH, I. (1951). Observations on the inhibitory action of hydrolyzed agar. *Mycologia*, **43**, 11–15.

SKERMAN, V. B. D. (1946). Simple techniques for the preparation of mould mounts. *Aust. J. exp. Biol. med. Sci.*, **24**, 319–20.

SKINNER, CATHERINE E. (1971). Laboratory test methods for biocidal paints. In *Biodeterioration of Materials*, Vol. 2. HUECK-VAN DER PLAS, E. H. and WALTERS, A. H. (Eds). pp. 346–54. Applied Science, London.

SKINNER, C. E., EMMONS, C. W. and TSUCHIJA, H. M. (1947). *Henrici's Moulds, Yeasts and Actinomycetes*. Chapman and Hall, London.

TURNER, R. L. (1972). Important factors in the soil burial test applied to rot-proof textiles. In *Biodeterioration of Materials, Vol. 2.* HUECK-VAN DER PLAS, E. H. and WALTERS, A. H. (Eds). pp. 218–26.

WILLIAMS, C. N. (1959). Spore size in relation to culture conditions. *Trans. Br. mycol. Soc.*, **42**, 213–22.

14

Materials Deterioration and its Prevention

14.1 Spoilage of materials

Ever since matter has been used by man it has been subject to attack by fungi and other organisms, and as through time the materials of man's economy have become more numerous, more complex and highly processed, the economic impact of such damage has increased.

In the past, the study of materials deterioration has been confined to a fairly small number of major areas which are easy to recognize, and so industries concerned with such topics as the treatment and prevention of timber decay and the mildewing of cotton goods have developed in response to specific needs. Interest in these and similar topics can be traced back to at least biblical times and probably earlier; the problems were certainly there even if we do not have written evidence of them.

In more recent times, where materials and goods are produced in a more complex and composite manner, the techniques for their study and protection have, of necessity, been drawn from more than one area of expertise and the need and convenience of a unified approach to the problems of microbial deterioration of materials has become recognized and accepted.

The **biodeterioration** of materials has been defined by Hueck (1966) as 'Any undesirable change in the properties of a material of economic importance brought about by the vital activities of organisms'. A term often confused with biodeterioration is **biodegradation**. Although in a literal sense this latter term could be considered as encompassing both biodeterioration and all natural breakdown processes in the living world, it has a specific meaning to the scientist. Biodegradation can be considered as the harnessing, by man, of the natural decay abilities of organisms to convert a waste material into a more acceptable and manageable form or to produce a useful end product from waste materials. Examples of such processes are the composting of town waste (garbage) to produce a more compact, drier and hygienic material for sanitary landfill, and the production of fungal protein for animal feed using food-processing wastes as a growth substrate. Further details of biodegradation processes will be found in Chapter 15.

Having defined biodeterioration and separated it from biodegradation, the way is now clear to outline a useful, if artificial, scheme for the classification of types of damage or decay the investigator may encounter.

14.1.1   Types of deterioration

Mechanical deterioration

In this instance, a material is simply disrupted by physical forces exerted by organisms and is not utilized or damaged in any other way. It is rare for this type of damage to be caused by fungi; the lifting of paving by the fruiting bodies of basidiomycetes being a clear but rather trivial example. In other fields, the physical damage caused by rodents, birds and insects can be very extensive and serious.

Chemical assimilatory deterioration

With regard to fungi, this is probably the most significant type of damage caused. Here the organism utilizes the material as a food source, and the widespread saprophytic activity of many fungi in nature is turned towards man's materials. The decay of many forms of cellulose, both relatively unchanged ones such as cotton textiles and building timber, and more processed forms such as cellulose wrapping film and rayon is a clear example of this type of deterioration. The moulding of stored foodstuffs, decay of paint and paint films, adhesives, leather, books and museum specimens are all further examples which serve to show the widespread activity and varied 'diet' of spoilage fungi.

   As the fungus spreads through the material the mycelium liberates enzymes which dissolve susceptible components of the material. Thus rendered into solution the nutrients are able to be absorbed by the fungus, where they are utilized to provide energy and raw materials for cellular growth.

Chemical dissimilatory deterioration

In this type of deterioration, substances liberated from the fungus, not directly concerned with the uptake of nutrients, spoil the material. These substances may include toxins, pigments and corrosive waste products. A classic example of this type of deterioration is the etching of glass lenses in optical equipment, a phenomenon not uncommon in the tropics. The fungus, living on even a very slight film of organic detritus on or near the lens, is able to grow across the surface where its acidic waste products may rapidly damage the fine surface and ruin the lens. The glass itself does not act as a nutrient in any way, but merely acts as a physical support for the fungus. Liberation of toxins from fungi also falls within this category, and the potential significance of this phenomenon warrants further comment.

Mycotoxins
In 1960, following the deaths of poultry and fish, a fascinating series of events was set in motion, which eventually led to the discovery of a range of fungal toxins which have great potential significance in human and animal health. The reader is strongly recommended to refer to the most readable account of these events by Linsell (1977). The article by Moss (1969) gives more chemical details. The source of the first identified toxins was *Aspergillus*

flavus, hence the name aflatoxin. Since then similar toxins have been found to be produced by other fungi and these toxins are grouped together as mycotoxins.

Fungi which have the potential to form mycotoxins are found world-wide in soil and the atmosphere, but the actual ability to synthesize these compounds varies greatly with the strain, growth conditions and type of foodstuff. Humidity and moisture must be very high, and grains and nuts are the most susceptible materials. Chemically, aflatoxins are a group of bis-furano-isocoumarin metabolites synthesized by *A. flavus* and *A. parasiticus*, designated aflatoxin B_1, B_2, G_1 and G_2. Under ultraviolet light B forms are blue and G forms appear green. These toxins are stable under normal food handling conditions, particularly when absorbed on starch or protein surfaces of seeds or cereals. Aflatoxins are very heat stable, far beyond normal food cooking regimes. The only process which has a marked effect is alkali treatment, but this is not common in food preparation.

The toxicity of aflatoxin is high, but varies with animal species. (See Table 14.1.) Apart from being poisonous, these compounds have been shown to be linked with several forms of cancer, especially cancer of the liver and it can be strongly argued that high incidence of this disease in the poorer areas of the world is due to aflatoxins in foods.

Table 14.1 The toxicity of aflatoxin in various animal species. (Data of Linsell (1977).)

Species	Single LD_{50} mg per kg body weight
Rabbit	0.30–0.50
Duckling	0.35–0.56
Dog	1.00
Monkey	2.20
Rainbow trout	0.81
Hamster	10.20
Sheep	2.00

Soiling or fouling

In this case the problem is quite simply the presence of the organism, which causes physical obstruction or impairs the appearance of the material. Fungal mycelium bridging gaps in electrical equipment and causing short-circuits, and the growth of dark-spored fungi living on a film of detritus on plastic or woven glass sheet and textiles, but not using the main materials as nutrients, are examples of this type of spoilage.

The reader will be readily aware that it is unusual for any of the four types of deterioration to occur in isolation. The fungus etching the glass lens with its waste products will also be fouling the surface of the lens by its mere presence. The fungus fouling a paint film will also be utilizing components of the film as a nutrient source and the fungus liberating mycotoxin in grain will at the same time be using that substrate as a nutrient. Nevertheless it is often useful to be able to clarify the processes involved in a case of deterioration, as this will be of help in assessing the significance and importance of the attack and the most effective way of its prevention.

14.1.2   The economics of biodeterioration

Before costs can be attributed to any form of biodeterioration, fungal or otherwise, it is first necessary for biodeterioration to be recognized as such, and the effects delineated from the effects of purely chemical and physical agencies, i.e. corrosion and wear. There is still a great need for education and training in this field, if industrial problems are to be handled in a more controlled and routine manner. Remedial work, even if possible, is usually more costly than preventative work. Some examples may be clear – the replacement of building timbers due to dry-rot attack on the originals can clearly be identified and costed as a case of biodeterioration. The thinning of adhesive solutions in industrial processes, however, is not so easily recognized as biological attack and the cost of such a case is less easy to identify.

Assuming that the problems of the recognition of biodeterioration are able to be solved, criteria for assessment of the cost of attack must still be established. One, or a combination of the following points may be chosen – it should be noted that there is no standardized procedure in practice.

(a)   The cost of prevention of deterioration
This may include chemical rot-proofing, drying, cooling, sealing or sterilizing the material, topics to be covered later in this chapter.

(b)   The cost of replacement
This is most applicable with low-cost materials, which are becoming fewer and fewer as costs increase.

(c)   The cost of remedial treatment
Remedial treatment is almost always difficult and expensive in biodeterioration cases. The best examples of true remedial treatment, as opposed to partial replacement or prevention of further attack are to be found in restored and preserved art objects and museum specimens, i.e. costly and unique items.

The lack of fully comprehensive statistics, even from well known fields such as wood preservation, highlights the problem of encouraging those who are able to recognize biodeterioration to take an interest in the economic aspects. It is even more difficult where the problem is not often even recognized as being biological in origin, for example the disfigurement of paint films by dark-sporing fungi. There are more problems to be considered. One is the natural reluctance of manufacturers to admit to any biological problems, especially in the food and cosmetics trade, and the widespread disbelief that anything so small as microorganisms can cause damage, especially in materials far removed from a human or animal diet. In the tropics, where biodeterioration is often most severe, it is often disregarded or accepted as part of a way of life, as many developing countries have more immediate and pressing problems to solve. Finally, the unpredictability of losses has a nuisance value which is again difficult to cost – how is it possible to attach a cash value to consumer confidence?

14.2   The control of mould growth

The basic principle in control of mould growth is to render the environment in or around the material as hostile as possible to the settlement, germination and spread of the organisms. There are four main ways of achieving this state of affairs.

1.   To ensure continuous sterility by preventing the access of mould spores to an inherently sterile or sterilized product.

2.   To arrange that the material to be protected is kept, or keeps itself, in such physical condition that growth of moulds is severely limited or prevented entirely.

3.   To limit or prevent mould growth by means of toxic substances, known as preservatives or biocides.

4.   To manufacture goods from materials which are highly resistant to attack by moulds.

14.2.1   Sterilization

The most widespread and well-known method employed is that of canning of foodstuffs. In the canning process the main object of sterilization is to destroy bacteria, and since moulds are less resistant to heat than the majority of bacteria, any cooking operation which will destroy the latter will also automatically kill any moulds present. In the fruit canning industry, however, high temperatures and long cooking have to be avoided as they detract from the appearance of the products. Fortunately, the acidity of most fruits is of great assistance in suppressing bacterial growth and processing can therefore be conducted at temperatures below 100°C. As already stated in Chapter 5 one fungus, *Byssochlamys fulva*, has caused considerable trouble in fruit canneries, owing to its ability to withstand a temperature only a little below the maximum attained in processing. This means that there is only a very small temperature difference between incomplete sterility and over-cooking, necessitating accurate temperature control and great care being taken to ensure that all parts of the contents of the cans reach the safe temperature (see Olliver and Rendle, 1934).

Most canning is carried out on an industrial scale, with proper equipment and careful procedures employed but in some parts of the world, notably the U.S.A., home canning is practised and has not completely declined with the advent of the home freezer. Many instances of food poisoning have been traced back to faulty technique and some home-canned produce should be regarded with a certain degree of caution.

A method of partial sterilization that has received considerable attention in recent times utilizes the lethal properties of ultraviolet radiation. It has been found that both mould spores and bacteria are rapidly killed by exposure to a source of light rich in ultraviolet rays, the most resistant spores being, as might be expected, those with dark coloured walls (see Conklin, 1944). It is somewhat doubtful how far the effect is due to the direct action of the rays or due to the action of ionized oxygen produced in the neighbourhood of the source of light. This latter effect has often been assumed when overnight 'sterilization' of laboratory areas has been advocated using ultraviolet lamps. James (1936) claims that direct irradiation is unnecessary and that air circulation during the treatment is beneficial, indicating that it is ionized oxygen which is the inhibitory factor. This may be so when particular wavelengths are used at high intensity. In any case the effect is exerted only on the surface of the irradiated material, since neither ultraviolet rays nor ionized gas can penetrate far into most kinds of matter. This is of little consequence for such materials as bakery products, which are usually sterile as they leave the oven and become infected only on the surface during cooling. Such goods, after irradiation and wrapping in irradiated paper, are virtually sterile throughout, and remain in good condition until they reach the consumer.

With the advent of plastic packaging films, which are not able to be sterilized by heat treatment, interest has again turned to high intensity ultraviolet treatment, and aseptic pouch-filling processes for the food industry are now under development and in use (Maunder, 1977). Many papers on the subject of ultraviolet sterilization have been published, chiefly in the journals of the food industry, but these cannot be adequately summarized here.

One factor with this method is that of psychology, a factor which is rarely considered. The presence of the bright and characteristic coloured lamps on industrial premises serves as a reminder that hygiene is of importance, although if total reliance is placed on lights alone, complacency can set in with regard to other aspects of good housekeeping.

The problems of non-penetration of ultraviolet radiation have been overcome in recent years by the development of gamma irradiation facilities. The gamma rays are usually generated by a cobalt 60 source, and at the time of writing there are at least fifty plants around the world offering this type of sterilization service. Radiation facilities need to be massive if the process is to be economic. However, with the use of gamma rays, products can be sterilized as a last step after final packaging into large cardboard transport containers. A wide range of materials and packages are able to be treated satisfactorily in this way, including disposable plastic goods, such as syringes and laboratory ware, biological media, foodstuffs and animal feeds.

Yet another form of radiation utilized in recent years is that of microwaves. The rapid heating generated by this means makes the method suitable for use in continuous processes. Sealed containers which are permeable to microwaves may be used, provided that they are able to withstand bursting at high temperatures, a processing problem which may be overcome by pressurizing the equipment. The advent of the domestic microwave oven has provided a laboratory tool which may prove of value in the rapid preparation of microbiological media.

14.2.2 Cooling and refrigeration

The preserving effects of lowering temperatures have been recognized for a very long time. In the field of food spoilage the cool cave evolved into the cellar and the ice-house, and ultimately into the domestic refrigerator and deep freezer. The main industrial applications of refrigeration are in the storage of perishable products, usually foods. Even a small degree of cooling can have marked effects on the rate of decomposition. The cooling of freshly-flayed hides simply by blowing air over them to lower the temperature from animal body heat to near atmosphere ambient can very markedly reduce bacterial attack.

A domestic refrigerator can also be a most useful item of equipment in the mycology laboratory. Bulk media and even filled Petri dishes can be stored for longer periods with less chance of contamination, and the growth of laboratory cultures can be slowed to facilitate use or identification when time permits.

The use of refrigeration as a general preservation method is obviously limited; we cannot use such a method for preserving buildings or textiles in use, but there is probably scope for better use of cooling in some situations such as in the use of metal working fluids and the manufacture of some adhesives. Only when microbiological considerations are taken into account at the plant design stage will such methods be able to be employed effectively and economically.

14.2.3   Gas atmospheres

The storage of perishable materials such as grain or fruit in atmospheres modified from atmospheric has applications similar to bulk refrigeration, and in some cases may be more suitable and more economic. The technique differs from fumigation in that the gases are not primarily toxic and leave no residues. The atmosphere may be modified in several ways. An alive and respiring material in a sealed vessel, for example grain, will tend to use up the available oxygen and growth of spoilage organisms will be slowed or stopped. If such a state cannot be achieved before spoilage has started, gases such as CO_2 can be introduced from the start. Inert gases such as nitrogen can also be introduced. For specific uses, modified atmospheres, perhaps coupled with some degree of cooling or chemical protection can be very effective but such techniques are, of course, limited to storage applications.

14.2.4   Control of decay by control of water availability

Many potentially vulnerable materials on which fungi would otherwise grow, are preserved in practice largely because they are dry. In fact, one of the most universally useful methods for preventing mould growth is maintenance of a low moisture content of the material. The longevity of such materials as grains and seeds, dry foods, textiles of natural origin, paper and wood, is very largely due to the fact that these materials are normally stored and transported in a dry condition, although chemical constituents of some (e.g. some kinds of wood) also make a contribution to their immunity.

Water activity (a_w)

Dryness is effective in the control of decay as the chemical potential of water in a dry material is reduced to the level at which microorganisms are unable to obtain sufficient water for growth or normal metabolism. It is well known that some materials which are clearly not dry (e.g. jam) also have a sufficiently low level of chemical potential of water to be unable to support fungal growth (before it is exposed to the atmosphere and is able to take up water). This is often attributed to the osmotic pressure of the solute (sugar in the case of jam) being too great for the microorganism to overcome it. While this is true, it is much more convenient to use a single measure of water availability which can be applied to all materials, regardless of the manner in which the water availability is restricted.

The measure now most commonly adopted is 'water activity', usually symbolized as a_w. This measure can be used for solids, liquids, and gases interchangeably, although it is still common to use relative humidity (which is numerically equal to water activity unless the latter is expressed as a percentage) for gases, especially air. A material of water activity $a_w = 0.7$, for example, is in equilibrium with air of relative humidity 70%. The water activity of a material is the ratio of the vapour pressure of water over the substance to the vapour pressure over pure water at the same temperature. The analogy with relative humidity is obvious.

The adsorption isotherm

For any particular substance there is a relationship between its moisture content and the activity of its water. In relative humidity terms this is often referred to as the 'equilibrium relative humidity' (ERH) of the substance at a particular moisture content. The curve relating the vapour pressure of water over a substance and its moisture content, at a particular temperature, is known as the adsorption isotherm of that substance. The term is also used for the relationship between a_w and moisture content.

As changes in temperature affect the vapour pressure over the substance and over water approximately proportionally, a_w (or relative humidity) over the substance does not change very greatly with change in temperature. The changes which do occur are usually in the direction of increasing a_w with increasing temperature at constant moisture content.

The concept of 'safe moisture content'

As 'dryness' rather than low a_w is commonly understood to be the controlling factor in growth of microorganisms it is usual to speak of a 'safe' moisture content for any commodity. This is reasonable for any single, precisely known commodity at reasonably steady temperature. For example, it is well known in the respective trades that cereal grains at or below 14% moisture content, or softwood in buildings at or below 16% moisture content, are safe, in practice, because they have been found not to be attacked by fungi or other microorganisms. In so far as a_w can be inferred from moisture content the latter is a perfectly valid measure of 'safeness' or resistance to microbiological attack. But there are several reasons why this inference is strictly limited in its application. The more important of these reasons are given below.

1. Moisture content is itself not a precise concept, and it cannot be determined precisely. It is usually determined by loss in weight on drying, but the conditions of drying have a large effect on the amount of water lost. The loss in weight is then expressed as a percentage (usually) of either the original weight (e.g. cereals, foods generally) or of the dry weight (e.g. wood, cotton, building materials). The 'correct' moisture content of any commodity refers only to the method of determination and method of expression which are conventionally accepted in the trade normally concerned with the commodity or which are precisely described in the context.

2. There is considerable hysteresis in the adsorption isotherm. That is, the level of moisture content which is in equilibrium with any particular relative humidity, or in which the a_w has a particular value, depends on whether the substance has approached equilibrium from a higher or a lower level of moisture content. The moisture content will be higher at a given a_w when the substance desorbs (dries) to equilibrium, and will be lower when the substance absorbs water to equilibrium. The amount of hysteresis varies considerably in different materials.

3. Although the adsorption isotherm, expressed in a_w or relative humidity terms, is fairly constant with changing temperature over a narrow range, a wide temperature change will change a_w appreciably while the moisture remains unchanged. The effect is that a substance which will not be attacked at a marginal a_w at room temperature may suffer decay at higher temperatures.

4. The adsorption isotherm of different lots of reputedly the same substance will sometimes be very different because of relatively small differences in composition, method of preparation, degree of subdivision, etc.

a_w limits for fungal growth

The limiting level of a_w for growth of fungi is generally considered to be about 0.65 (65% relative humidity) but only very xerophilic species are able to grow at this level, and then only very slowly. In practice, a_w 0.70 is low enough to ensure freedom from any appreciable growth of fungi on most materials, and, provided that storage is not for very extended periods, an a_w of 0.75 will prevent growth. There have been many laboratory studies designed to determine the limiting level of a_w for growth of various species of fungi on various substrates. Many of these are of doubtful value because the difficulties have not been fully appreciated. Apart from the difficulties which derive from the uncertainty of the a_w/moisture content relationship, there are other important factors which must be considered, some of which are listed below.

(a) The effect of minor temperature variations from place to place in the experimental containers which have a very large effect on the local relative humidity of the atmosphere.

(b) The observation that mixed populations that do occur in practice may be able to tolerate conditions which appear to be limiting to a single species in the laboratory.

(c) In the laboratory, an organism may be found not to develop at a particular relative humidity and this may therefore be judged to be its limit, but this can sometimes merely be the limit for spore germination. In practice, an organism which is limited only in this way is very likely to be able to develop because a very brief exposure to a high relative humidity may enable the spores to germinate, producing a mycelium which is more tolerant of lower levels of a_w than had been supposed from the laboratory experiment.

Control of water activity

In laboratory experiments the most convenient method of control of relative humidity (and hence of the water activity of substances exposed to the controlled atmosphere) is that which depends on confining the atmosphere in contact using a substantial volume of controlling solution. Solutions of sulphuric acid, potassium hydroxide, or sucrose are sometimes used for this purpose. Of these, potassium hydroxide is generally to be preferred for biological purposes because it is able to absorb carbon dioxide produced by respiration without producing a great change in the a_w. This overcomes one of the objections to use of sealed containers for biological experiments with aerobic organisms. However, the objection to use of solutions of fixed strength to control relative humidity is that the changes in solution strength which inevitably result from absorption or loss of water during the controlling process will affect the controlled humidity. In practice, if the volume of solution is large relative to the controlled volume, this is seldom a problem.

The problem of changing strength, and the difficulty of making up solutions very precisely, are easily overcome by the use of saturated solutions in which both solution and undissolved solid are present. This provides a useful buffer against changes in solution strength and obviates the need to make up solutions with great accuracy. A range of solutions can be made from readily available substances which will cover the range of relative humidities fairly completely. This is by far the most popular and convenient method of control. Some solutions maintain a reasonably constant equilibrium relative humidity in spite of temperature change, for others it is necessary to be rather precise in temperature control. In general, the effect of temperature variation on the relative humidity in equilibrium with saturated salt solutions is greatest in the middle range of humidities. The temperature effect is much less towards the extremes of relative humidity.

A selection of salt solutions suitable for general laboratory use is given in Appendix II. (See also Young, 1967.)

In any experimental work it is of vital importance to maintain isothermal conditions throughout any container in which the atmosphere is being controlled by a solution. Unexpected causes for temperature gradients between solution and experimental material include radiation from lights and warm surfaces. Ideally, any such containers should be kept in accurately controlled incubators. In large containers the air should be kept in motion, but any electric motor used for this purpose must be kept well away because of the heating effect.

On a larger scale, relative humidity control becomes an air conditioning problem outside the scope of this book. Control of water activity within substances other than by control of the relative humidity of the atmosphere, i.e. by addition of solutes, is a major part of food technology. Such solutes lower the chemical availability of water exactly as they do in solutions which may be used to control humidity. Salting, many kinds of pickling, drying and the use of strong sucrose solutions, as in jam making, are forms of a_w control for which suitable food technology texts should be consulted.

Establishment of an adsorption isotherm

Provided that a suitable (i.e. repeatable) method is available for determination of moisture content of a material, on a weight/weight basis, it is valuable to establish the adsorption isotherm. Once this has been established, determination of moisture content will enable the water activity to be inferred. Due allowance must be made for the effect of hysteresis – this means that there are two adsorption isotherms which differ according to whether each level of moisture content is attained by drying from a higher level (the desorption isotherm) or by absorption of water from a lower level (the absorption isotherm).

There are two general approaches to the establishment of an adsorption isotherm. In the first, samples of the material are exposed to constant relative humidity and allowed to absorb or desorb water from the air, usually until there is no further change in weight. The moisture content of each sample is then determined. The second approach, which is usually quicker because there is a smaller amount of water vapour to be exchanged, is to allow samples of the material, at a range of differing moisture contents, each to establish a constant humidity in a small volume of air with which it is enclosed. The relative humidity thus established is measured.

For the first type of method it is necessary to have a large volume of controlled atmosphere and/or a generous amount of controlling solution so that the relative humidity is not changed by the amount of water exchanged with the material under investigation. The technique is slow, perhaps taking weeks or months depending on the size and permeability of the material, but it is easy because the only measurement to be made is moisture content. Evidently the method produces points on either the absorption or the desorption isotherm depending on the level of moisture content of the original material.

A variation on this method, which is both quick and convenient for obtaining a single point, or only a few points, on an adsorption isotherm is the graphical interpolation method of Landrock and Proctor. To use this method, a sample of the material of known moisture content is subdivided into a number of portions of similar size and shape. Each of these sub-samples is weighed and then exposed, each to a different relative humidity in a sealed container. At least four different humidities are required, spanning the expected equilibrium level at which the sample should neither gain nor lose weight. After a suitable period, usually about 24 hours, the sub-samples are reweighed, and if the humidities have been suitably chosen some will have gained weight and others will have lost weight. It is not necessary for equilibrium to have been reached. The changes in weight are then plotted against the corresponding relative humidities (Fig. 14.1) and a line, usually approximately straight, drawn through them will cut the zero weight change line at the relative humidity in equilibrium with the original moisture content, thus establishing a single point on the adsorption isotherm. This point will fall between the absorption and desorption lines.

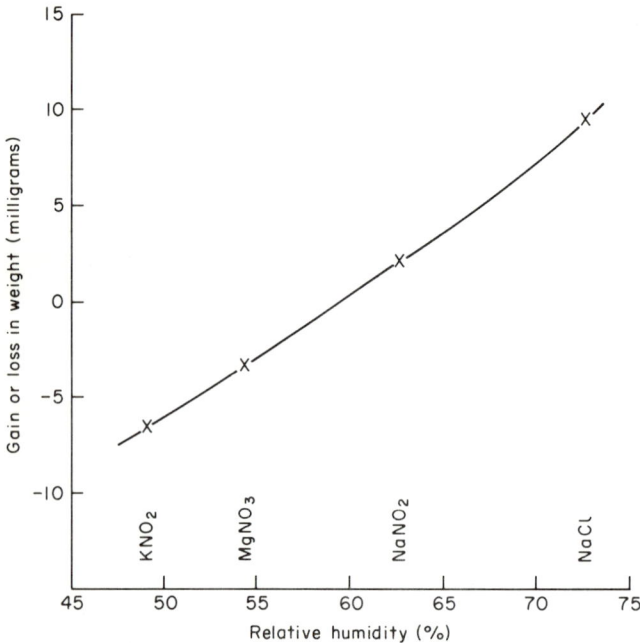

Fig. 14.1 Example of the use of the method of weight change over a range of relative humidities to determine a point on the adsorption isotherm of a cellulosic material. In this example the equilibrium relative humidity is shown to be 59.5%. The relative humidities were controlled by saturated solutions of the solutes indicated on the figure. Temperature 20°C.

The second general method, in which the sample of material is allowed to establish equilibrium relative humidity in the air with which it is enclosed, is attractive, but has hitherto been limited by the difficulty of measuring relative humidity accurately in small volumes of air. Recently, however, much work has been done, first by Ayerst (1965) and subsequently by Pixton and Warburton (1971 *et seq.*), by measuring dew point in the enclosed volume. This method lends itself to rapid and convenient establishment of an adsorption isotherm and both the absorption and the desorption lines may be established by conditioning the samples of material to the desired levels of moisture content by either absorption or desorption of water.

Some of the various commercial electrical humidity measuring devices may be used. Most of these depend on measurements of the electrical properties of a small quantity of material (either a hydrophilic plastic film, a layer of aluminium hydroxide, or a lithium chloride solution held on an inert substrate) exposed to the atmosphere. All these have limitations of range, constancy of calibration, durability of sensors, and differences between different sensors, but, used within their limitations, they give useful results. They are, however, better suited to measurement of humidity in large volumes of air where the effect which their own water exchange (and in some, production of heat) has on the atmosphere being measured is negligible. These may not be negligible in a small volume of air. This is especially true of the type of sensor in which a lithium chloride solution is heated by passage of an electric current to a temperature at which it is in equilibrium with the atmosphere.

The better type of electrical hygrometer for work in small volumes is that which automatically measures dew point using the Peltier effect to cool a mirror surface electrically until dew forms and is detected by a photoelectric cell. Of these, the E.G. & G. instrument which maintains its mirror continuously at or near dew point, and the simpler Protimeter Dewpoint Meter which cycles between ambient and dewpoint, are both suitable. The Protimeter instrument has the advantage of dissipating less heat in a limited environment and is less sensitive to dust contamination of the mirror.

Other methods

Other methods are, in principle, available for determination of equilibrium relative humidity. It is possible to determine the vapour pressure of water above any material by enclosing it in one arm of a manometer, both arms of which are evacuated as in the usual method for determining vapour pressure over solutions, etc. This is very accurate but somewhat impractical for microbiological laboratories. In addition, there is the isopiestic method. This consists in allowing a sample of the material under test to equilibrate at constant temperature with another material of known humidity–moisture content relationship. The changes in weight of each during the equilibration process make it possible to establish a point on the adsorption isotherm, after making due allowance for any hysteresis effect on each material. If the known material is a saturated solution the method is simply the first approach described above, i.e. equilibration with a constant relative humidity.

14.2.5 Fumigation

Considered in its widest sense, fumigation may be performed in several ways. Chemical inhibitors may be dispersed in particulate form as smoke. This technique is mainly confined to insect control. Some other forms of biocide application such as the use of fogs, 'ultra low volume spraying' using concentrated biocides in large volumes of blown air, and the dispersal of blown dusts, also border on fumigation. These methods have been developed mainly for agricultural use and pest (especially insect) control.

For materials, gaseous sterilization using methyl bromide, propylene oxide, ethylene oxide and ozone, and also the use of volatile biocides, is more common.

Gaseous sterilization

It should be stressed from the outset that this form of fumigation needs specialist knowledge and equipment to be carried out safely and effectively.

Ethylene oxide has been the most widely used fumigant in the past twenty years, but there is growing interest in the use of propylene oxide, in part due to the fact that its hydrolysis product, propylene glycol, is more acceptable to such bodies as the Food and Drug Administration in the U.S.A., than the ethylene glycol formed from ethylene oxide. Both compounds may be used on a wide variety of stored materials and foodstuffs. Methyl bromide is used mainly against insect pests, while ozone has been used in applications more concerned with public health than protection of materials. The review by Kereluk (1971) will prove of interest to those readers wishing to explore this topic further.

Volatile biocides

Over the years compounds such as thymol, meta cresol acetate and alkyl mercurials have been used with varying degrees of success in special applications such as the protection of sealed optical instruments. One problem which arises is reaction between the biocides and the components to be protected. Thymol crystals have often been used to protect small objects such as books sealed in containers for storage.

14.2.6   Chemical preservatives

Compounds used to control the growth of microorganisms are known by many general names, many of which are used rather loosely. Thus we may see reference to fungicides, fungistats, antiseptics, biocides, preservatives and material protectants and assume clear differences, when in fact similar types of compounds with similar activity and uses are being described.

In recent years the range of chemical preservatives has increased, much more has been learned of their modes of action, and compounds for specific uses have been developed. However, the number of compounds available is likely to decrease, owing to the present concern over environmental pollution and the stricter regulations which have increased the cost of developing products of proven safety in application and use.

The ideal chemical preservative would have the following characteristics.

(i)   It would be toxic to a wide range of organisms (in fact to all deteriogenic organisms with which it would come into contact). Some preservatives have a narrower spectrum of activity than might be at first thought from their general description as say 'biocides', and may be much more effective against bacteria than fungi. Care should therefore be taken when choosing a preservative, although a restricted spectrum of activity may be an advantage in certain applications.

(ii)   It would not be toxic to humans or animals, by mouth or by absorption through the skin, and it should be safe to use during its application process which may involve spraying and the use of concentrated solutions which involve greater hazards to the applicator than the final user.

(iii) It would not introduce any undesirable colours to the material. This factor may greatly restrict the use of some compounds. The green colour imparted by some copper based compounds tends to restrict their use to wood preservation, where the colour is either not important, can be tolerated or even thought to be decorative in that specific use.

(iv) It would be compatible with any other treatment given to the material being protected. Subsequent heating of a product may inactivate a preservative, or the active ingredients of the preservative may react with components of the material and lose some of their activity. An oily preservative is of little use where the product may subsequently be painted or have other finishes applied which may be upset by the preservative.

(v) It would not affect the physical properties of the material to be protected. The flexibility and feel of textiles, paper and plastic films are important properties and may be altered by the addition of unsuitable preservatives. Some preservatives may weaken the materials by chemical action, or may hasten weakening by other agencies, such as in the case of cotton textiles becoming more sensitive to weakening by the action of sunlight after some phenolic preservatives have been applied.

(vi) It would be resistant to leaching, although if a preservative is totally bound to the material it is generally thought that its effectiveness will be greatly diminished. In practice, the aim is to tailor the amount and solubility of the preservative to the service life and type of use of the product. There is little point in giving ten years microbiological protection to a product which, even with care in use, will wear out or be consumed within two years.

(vii) Finally, after all technical requirements have been met the preservative must be an economic proposition.

The control of mould growth by the use of chemical preservatives has been practised, or at least attempted, for some considerable time. It is, however, only during the last forty years or so that organized research on the subject has put the practice on a scientific basis, explained many of the failures of the past, and made available new compounds of high inhibitory potential and general applicability.

It should be understood that preservatives do not necessarily kill fungi, but often inhibit or retard growth and development to acceptable levels. Fungal spores are often very resistant to toxic substances, and most substances which will kill them are often unavailable for use in industrial products because they are corrosive, give off objectionable odours, or are highly toxic to humans. The action of preservatives is limited by a number of factors, such as the availability of nutrient present along with the preservative, the amount of available moisture, the concentration of the preservative, temperature, time and, in some instances, light.

In certain cases, such as the preservation of wood intended for outdoor or structural use, it is possible to use powerful preservatives and obtain more or less complete protection. However, there are often technical objections to the use of preservatives in really efficacious amounts, as well as questions of cost, and the most that can be achieved is to prevent mould growth under more or less normal conditions of use and storage.

The concentration of preservative required to prevent growth usually increases rapidly as the humidity of the atmosphere increases, and there are some preservatives which, though effective at normal humidities, lose their effectiveness when the relative humidity exceeds 95%. Some preservatives at very low concentrations actually stimulate the growth of fungi. With somewhat higher concentrations germination of spores is delayed, and with still larger amounts, completely inhibited. In a general sense, this may be considered as being analogous with the need of higher plants for trace metals, which when present in large amounts will poison the plant. With some preservatives the degree of inhibition appears to be strictly proportional to concentration, while with others there seems to be a maximum effect, short of complete inhibition, which cannot be exceeded by the use of higher concentrations of the preservative.

An important point which is often overlooked is what may be termed the specificity of preservatives. A compound may give adequate protection so long as the potential infection is limited to certain species of moulds, but may fail completely if spores of another species are introduced. For example, salicylic acid can efficiently inhibit the growth of most of the dark coloured moulds and of many species of *Penicillium*, but can actually be utilized as a source of carbon (a food source) by *Aspergillus niger*. Another example of preservative specificity is zinc chloride, which has been widely used as a preservative for certain kinds of textiles but is unable to inhibit *Aspergillus terreus*. It is advisable, therefore, when considering the question of a preservative for protecting any particular product, to first of all determine the usual species of mould which infect the product and then test various preservatives with these species.

A further kind of specificity, which is often important, is the effect of the substrate on the efficiency of the preservative. A compound which works very well on one material may be much less effective on another. It is, therefore, unsafe for one industry to place too much reliance on the results obtained with the very different materials from another industry. The fact that some preservatives are effective only in hygroscopic materials, i.e. substances which normally contain a certain amount of water which fluctuates with the relative humidity of the air, must also be taken into consideration. These preservatives are quite useless in such materials as oil paints and lacquers.

Another phenomenon, which has increasingly useful applications, is known as 'synergism' – the enhancement of the inhibiting effect of one preservative by the presence of another. Many examples of synergism are known, and many modern preservative formulations contain more than one active ingredient. One example, and probably the classic case, may be quoted as being of special interest. During the second world war, the demand for 'Shirlan' for protecting clothing, canvas, cordage and the like going out to the jungles of Asia, exceeded the potential supply. A search for substitutes showed that a substance used as an accelerator in the vulcanization of rubber, mercaptobenzthiazole, was almost as effective as 'Shirlan'. Further tests showed that if the two compounds were mixed in equal proportions, the total amount could be reduced by about 40% without reducing the degree of protection. A lesser amount of the mixture was therefore as effective as a larger amount of either of the ingredients, and the potential of synergistic mixtures is obvious.

The testing of preservatives to give some indication of performance in use is not a simple matter. A usual method of obtaining a preliminary valuation is to grow a number of species of moulds in the laboratory, on media containing varying amounts of preservative, and measure the growth rates. Ideally, the fungi chosen should be those the material to be protected may encounter or be susceptible to in use. Care must be taken as to how the preservative is incorporated in the medium as dispersion and the effects of sterilizing procedures, especially heat, may drastically alter the preservative.

The basal medium used may be one of the usual agar culture media, or, preferably, a medium containing the particular material to be protected as the sole carbon source. This should be added in as fine a physical form as possible, shredded, powdered or ball-milled, and suspended in a plain agar gel, with or without trace nutrients.

An arbitrary but broadly realistic range of concentrations of preservative should be chosen at first. This range can be made smaller, but with more variations, when the results of the first experiments are seen.

Inoculation is best carried out by cutting plugs of agar from the growing edge of a fungal colony in a Petri dish using a sterile cork-borer a few millimeters in diameter, and placing these plugs, mycelium downwards, in the centre of the test dishes. Rates of growth are then recorded by plotting colony diameters against time, with other comments as to the type of growth, such as density, onset of sporulation and pigment production.

In this way a useful comparison between different preservatives may be achieved, and some idea obtained of the amounts required to suppress growth. This general method, although often the only one available for preliminary tests, is, however, open to the objection that such test media contain a very high percentage of water and the conditions are thus not comparable to those obtaining in practice. It also ignores the fact, mentioned above, that different materials (in this case the culture media and the material to be protected) may require different preservatives.

Thus, before the true value of a preservative can be assessed, it is necessary to make further tests under works conditions. In some industries the manufacturing and packaging processes easily lend themselves to preparing experimental batches, but in other industries such experimentation may be disruptive and costly. It is, therefore, very necessary to obtain as much information as possible from laboratory tests. The final stage is to monitor 'field' performance carefully after any new treatment has been introduced and this requires organizational as well as technical effort.

Types of screening methods which may be regarded as a next step up from agar tests and indeed some which border on field trials, have been described in Chapter 13, with examples being drawn from paint and textile testing.

Despite the widespread use of preservatives, there is no one obvious guide to the types available. This is mainly due to the state of constant change which exists in the field owing to commercial, economic and legislative reasons. Comparative work is also scarce – as it is costly to compile and obviously commercially sensitive it tends to remain in the files of manufacturers. Surveys of preservatives, such as that carried out by Heuck van der Plas (1966), have proved of great interest and indeed are still of use. With the passing of time and new regulations it is hoped that such general surveys will be updated, and indeed this is presently in progress for the above survey.

Preservative manufacturers often have their own lists but these are invariably restricted in availability for commercial reasons and only cover their own fields of interest. The British Wood Preserving Association, 62 Oxford Street, London, W1N 9WD is a useful source of information on wood preservatives.

Although such material dates quickly the reader may find the following lists useful, even though they cannot be claimed to be in any way exhaustive.

Material preservatives

Examples of chemical types which possess antimicrobial activity.

Phenolic compounds
1. *Phenol and its homologues* Phenol, cresols, thymol, diphenyl ortho-phenyl phenol.
2. *Halogenated phenols* Trichlorophenol, pentachlorophenol para-chloro meta cresol.
3. *Halogenated bis-phenols* Dichlorodihydroxy diphenyl methane (di-chlorophen).
4. *Carboxylated phenols and substituted esters* Para-hydroxybenzoic acid, methyl parahydroxybenzoate, pentachlorophenyl laurate.
5. *Nitro and amino substituted phenols* Para-nitrophenol, salicylanilide.

Such compounds have widespread use, especially in the fields of cellulosic materials, including timber and textiles.

Quaternary ammonium halides
1. *Aliphatic* Alkyl trimethylammonium bromide.
2. *Aromatic* Cetyl pyridinium bromide.

Nitrogen containing compounds
1. *Aliphatic and aromatic amines* Dodecylamine, dodecyl amine salicylate.
2. *Imidazolines* 1-decyl-2, 3-dimethyl-imidazolium iodide.
3. *Triazines* 1,3,5-hexa-hydrotriazine derivatives.

Nitrogen and sulphur containing compounds
1. *Carbamates* Dimethyl dithiocarbamate.
2. *Heterocylic compounds* Mercaptobenzthiozole.
3. *Sulphides* Tetramethylthicuram disulphide.

Organo-metallic compounds
1. *Copper* Copper napthenate, copper-8 hydroxyquinolate.
2. *Zinc* Zinc napthenate, zinc dimethyl dithiocarbamate.
3. *Tin* N-tributyltin oxide (TBTO).

4. *Mercury*   Phenyl mercury acetate.
5. *Arsenic*   10,10′-oxybisphenoxarsine.

Such compounds have been widely used, especially for exterior applications and wood preservation. Relatively high mammalian toxicity and pollution considerations have led to usage restrictions on some compounds such as organo-mercury preservatives.

Inorganic compounds
1. *Metal salts*   Copper sulphate, potassium dichromate, copper chrome arsenate (CCA).
2. *Semi-metals*   Boron compounds.
3. *Halides*   Sodium silicofluoride.

The colouration imparted by many metal salts limits their uses, and in some cases high solubility in water can be a disadvantage.

14.2.7   Food preservatives

The number of preservatives allowed in foodstuffs is rightly very limited. It has long been recognized that protection by means of preservatives is no substitute for care and strict hygiene in manufacture, and suitable facilities for low-temperature storage and transport. The official attitude in the United Kingdom has always been that particular preservatives which may be used should be specified, and the allowable amounts defined, rather than that a list of forbidden preservatives should be issued.

For many years, the only preservatives allowed were sulphur dioxide and benzoic acid, and these only in a limited number of foods. A Food Standards Committee was set up in 1947 to advise the appropriate government departments, and in 1951 a Preservatives Sub-committee was appointed. The first report of the latter was published in 1959 and the recommendations therein implemented. Since that time there have been several amendments to the regulations governing preservatives in food and the reader with an interest in this topic is recommended to the paper by Jarvis and Burke (1977) which is a very useful guide. Many foodstuffs are, of course, spoiled by bacterial rather than fungal action, but the number of foods which are attacked by fungi is still considerable. The following list is a brief guide to some of the more widely used food preservatives.

1.   Organic acids and salts

Acetic acid   This is one of the earliest known preservatives, but is little used today because of taste and odour problems, and limited effectiveness in dilute solution. As vinegar, acetic acid used at effective strength is of course an integral part of the flavour of pickled products.

Proprionic acid   This has been used mainly in bakery products and bread. Since the acid itself has a pungent odour, the sodium, calcium or potassium salts are usually used.

Sorbic acid A popular preservative which is also used for materials, including food wrapping and packaging. As sorbic acid has a low solubility in water, it is common practice to use potassium sorbate which is readily soluble and equally effective when kept on the acid side of pH 6.5. Potassium sorbate is also used as a preservative for tobacco products.

Benzoic acid Also used widely in foods, and as with sorbic acid, a salt (sodium benzoate) is usually preferred for solubility reasons. Sodium benzoate has the disadvantages of being relatively more toxic to man and animals than potassium sorbate and being most effective in acid conditions below pH 4.

2. Esters

Parayhydroxybenzoic acid esters Esterification of the carboxyl group of p-hydroxybenzoic acid extends the effective pH range considerably, thus making these compounds (known commercially as Parabens) particularly suitable as preservatives in products with relatively high pH values. The alkyl esters have been shown to inhibit mould growth at pH 7 or higher.

Pyrocarbonic acid diethyl ester When used in aqueous media, this preservative has the important advantage of being hydrolysed to ethanol and carbon dioxide after performing its function, thus leaving no toxic residues. However, since the use of this compound is dependant upon this decomposition, its application is restricted to beverages with pH values less than 4.0.

3. Polyhydric alcohols

Polyhydric alcohols such as glycerol, propylene glycol and sorbitol are added to food and tobacco products for a variety of reasons, which may include preservation. Their preservative action is due mainly to moisture retention effects, and high concentrations are normally required for effectiveness. An important exception is propylene glycol, which has been found to be effective as a mould inhibitor at relatively low concentrations.

4. Inorganic salts

Sulphites The addition of sodium and potassium sulphites to fruit products is a long established practice, the low pH values giving rise to the formation of sulphurous acid which provides the inhibitory activity.

Nitrites These compounds have a history of use in meat curing processes, mainly to enhance the colour of the product. Sodium and potassium nitrites have been shown to exhibit fungistatic action.

5. Essential oils

Certain essential oils, such as cinnamon and cassia oils, are known to exhibit inhibitory properties. The use of such compounds is of course limited to foodstuffs where they are an integral part of the flavour, and can rarely if ever be added in sufficient amounts to ensure full protection. However, it can be important to try to estimate the role such ingredients may play when considering the addition of other preservatives.

6.   Volatile preservatives

These are considered in the section on fumigation in *physico-chemical methods* earlier in this chapter (p. 335).

14.2.8   Use of resistant materials

The ideal method of preventing mould growth is to use only materials which are inherently resistant to attack. In recent years many new synthetic materials have appeared which show a high degree of resistance to attack and the following list includes the more important materials.

Plastics

Thermo-setting plastics, and plastics formulated mainly or entirely from polymers such as polyethylene, are highly resistant to mould growth. Most grades of polyvinyl chloride (PVC) and perspex are also satisfactory, but the degree of resistance shown by these and other plastics depends on the nature of any plasticizers used, since some of these support mould growth rather readily. The widespread use of PVC and similar materials has led to great interest in the microbiological testing of plastics and the reader is referred to the growing literature on the subject (Pantke, 1977; Osmon and Klausmeier, 1977; Pankhurst, Davies and Blake, 1972).

Even if the plastic itself does not support growth, it should be remembered that unless a mobile or volatile preservative is present, mould growth may develop on accumulations of dirt on the surface. This may be of particular importance in the case of plastics where appearance may be one of the reasons for its use.

Care should also be taken when composite materials are employed, especially where plastics are laminated to other more degradable products or where gaps and spaces are filled with wood or fabric. Indeed the use of biodegradable fillers has been considered as a way of alleviating the litter problem caused by plastics.

Textile fibres

Plant fibres are often susceptible to attack by fungi. Some processes improve resistance, such as the scouring of cotton by boiling with soda-ash, but other processes in the production of finished materials such as the addition of starch and other sizes may encourage mould growth, even if only of a superficial nature initially. Most plant fibres require some chemical protection when used in adverse conditions.

Animal fibres tend to be more resistant to mould growth. Pure silk is resistant if completely de-gummed and wool decays only slowly, keratin decomposing fungi being less common and more slow in their action than cellulolytic fungi. These materials, however, are more costly and less versatile or unsuitable for many industrial and commercial uses.

Synthetic fibres are often very resistant to attack by moulds. The completely synthetic ones such as nylon, polyester, polypropylene and pure glass fibre perform very well, but some fibres derived from cellulose, such as some forms of rayon, are easily decayed. Problems of surface growth, mentioned in the section on plastics, and soiling and fouling (see p. 325) also apply to synthetic fibres.

Rubber

Raw rubber and liquid latex are very susceptible to attack by microorganisms, but properly vulcanized rubber is very resistant. Problems occur only in extreme cases, such as rubber used in pipe joint rings in sewers where there is a long period of exposure to organisms, extra nutrients and water. Most synthetic rubbers are also satisfactory.

Adhesives and sealants

The composition of adhesives and sealants covers a very wide range of ingredients. Many adhesives contain preservatives which are primarily intended to protect the product in storage before use, and extra protection may be needed for longer term effectiveness after the adhesive has been applied. Starch, cellulose and animal glues are all susceptible to mould growth, as are many of the more synthetic adhesives which are water based. By the nature of their use adhesives are usually physically covered after application, but problems may occur in wet situations. Sealants are often used specifically to keep out moisture and thus need to be checked for susceptibility to microbial damage before use. This should of course be carried out on materials in the cured state. Many modern and often costly adhesives and sealants such as epoxy resins are very resistant to attack. Hydrocarbon waxes are resistant to attack but beeswax, even when mixed with large amounts of other waxes, is liable to become mouldy.

Leather

Chrome-tanned leather is usually highly resistant to moulds. Vegetable-tanned leather is one of the most easily attacked of all materials, and some very fine supple leathers, such as those used in bookbinding and for very expensive clothing and decorations, may also contain sugars in the finish which increases the risk of attack. Semi-chrome leather is intermediate in its resistance, seldom supporting more than slight growth of mould but never completely resistant.

It is of interest to note that during the conversion of animal skins into leather large losses of both quality and quantity of the material may be experienced as a result of microorganisms. This is mainly due, in the early stages at least, to bacterial action, thus demonstrating the basic susceptibility of such materials. Cooling of freshly-flayed skins and use of biocides have been demonstrated to be of great value in leather production.

14.3   Investigation of mould problems

Space does not permit a long treatise on particular case histories but some basic points are well worth mentioning.

It is essential to ascertain the precise and complete composition of a product under investigation. Often minor components are responsible for observed mould growth and diligence and probing on this topic is usually of great use.

Ideally Koch's Postulates should be followed, and once the organisms thought to be responsible for the observed damage have been isolated in pure culture, they should be re-introduced into sound material in an attempt to reproduce the damage under controlled laboratory conditions. This is often not possible, but every effort should be made to ascertain which of several or many organisms which may be present are in fact the true deteriogens.

Standard tests for mould resistance are few and far between and many in common use are adaptations of tests originally devised for very different materials. Where special tests have to be devised, the uses to which the product may be put and the way in which it may be stored should always be borne in mind. The limitations of laboratory tests, especially those which centre around work with agar in Petri dishes, should be realized, as emphasized earlier in this chapter, and where possible laboratory work should be followed up by field trials.

Sampling is one area which is fraught with difficulty. If at all possible the investigator should insist upon taking samples for study himself, in clean sterile and sealable containers. This may sound obvious but all too often small badly-packed samples of imprecise origin are provided. There is no substitute for a visit by the investigator in person and this should be a cardinal rule for all involved in such industrial work.

14.4   References

AYERST, G. (1965). Determination of the water activity of some hygroscopic food materials by a dew point method. *J. Sci. Fd. Agric.*, **2**, 71–8.

CONKLIN, D. B. (1944). Ultraviolet irradiation of spores of certain moulds collected from bread. *Proc. Iowa Acad. Sci.*, **51**, 185–9.

HUECK, H. J. (1965). The biodeterioration of materials as a part of hylobiology. *Mater. u. Organisms*. **1(1)**, 5–34.

HUECK VAN DER PLAS, E. H. (1966). Survey of commercial products used to protect materials against biological deterioration. *Int. Biodet. Bull.*, **2(2)**, 69–120.

JAMES, R. F. (1936). Moulds and bacteria killed by new lamp. *Food Ind.*, June 1936, 295–7.

JARVIS, B. and BURKE, CAROLE S. (1977). Practical and legislative aspects of the chemical preservation of food. In *S. A. B. Symposium Series – Microbiology in Agriculture, Fisheries and Foods.* SKINNER, F. A. and CARR, J. G. (Eds). pp. 345–67.

KERELUK, K. (1971). Gaseous sterilization: methyl bromide, propylene oxide and ozone. In, *Progress in Industrial Microbiology*, Vol. 10. HOCKENHULL, D. J. D. (Ed.) pp. 105–28. Churchill Livingstone, Edinburgh and London.

LINSELL, C. A. (1977). Aflatoxins. In *Environment and Man. Vol 6. The Chemical Environment*. LENIHAN, J. and FLETCHER, W. W. (Eds). pp. 121–36. Blackie, Glasgow and London.

MAUNDER, D. T. (1977). Possible use of ultraviolet sterilization of containers for aseptic packaging. *Food Technol.*, **31(4)**, 36–7.

MOSS, M. O. (1969). Mycotoxins. *Int. Biodet. Bull.* **5(4)**, 141–7.

OLLIVER, M. and RENDLE, T. (1934). A new problem in fruit preservation. Studies on *Byssochalmys fulva* and its effect on the tissues of processed fruits. *J. Soc. Chem. Ind., Lond.*, **53**, T, 166–72.

OSMON, J. L. and KLAUSMEIER, R. E. (1977). Techniques for assessing biodeterioration of plastics and plasticizers. In *Biodeterioration Investigation Techniques*. WALTERS, A. H. (Ed.). pp. 77–94. Applied Science, London.

PANKHURST, E. S., DAVIES, M. J. and BLAKE, H. M. (1972). The ability of polymers or materials containing polymers to provide a source of carbon for selected microorganisms. In *Biodeterioration of Materials*, Vol. 2. WALTERS, A. H. and HUECK VAN DER PLAS, E. H. (Eds). pp. 76–90. Applied Science, London.

PANTKE, M. (1977). Test methods for evaluation of susceptibility of plasticized PVC and its components to microbial attack. In *Biodeterioration Investigation Techniques*. WALTERS, A. H. (Ed.). pp. 51–76. Applied Science, London.

PIXTON, S. and WARBURTON, S. (1971). Moisture content/relative humidity equilibrium of some cereal grains at different temperatures. *J. Stored prod. Res.*, **6**, 283–93.

YOUNG, J. F. (1967). Humidity control in the laboratory using salt solutions – a review. *J. appl. Cem.*, **17(9)**, 241–4.

15

Industrial Uses of Fungi

As we study and compare the fungi, it becomes more and more evident that these organisms were not created for the innocent amusement and recreation of the taxonomic botanist.

C. L. Shear, *Proc. Int. Congr. Plant Sci.*, 1929

15.1 Introduction

As an offset to the incalculable damage caused by fungi there are a number of industrial processes in which the biochemical activities of moulds are turned to good account. Only very brief descriptions of these can be given here, but there is an extensive literature on the subject and to this the reader is referred for fuller information. A few such processes have already been mentioned in previous chapters, but, for convenience of reference, are described again here.

15.2 Alcoholic fermentation

There are two industries, brewing and baking, which use processes of great antiquity, both dependent on the fact that yeasts convert sugar into alcohol and carbon dioxide. In the brewing of alcoholic beverages alcohol is the important product, but the carbon dioxide, once allowed to escape as useless, is now a valuable by-product, being collected, solidified, and marketed as 'dry ice'. In baking, on the other hand, the production of alcohol is incidental and it is the carbon dioxide which is valuable, causing the dough to rise and giving lightness to the bread. Both industries were, until comparatively recent times, almost entirely empirical, but are now, in most countries, on a strictly scientific basis. During the last half-century both have built up extensive and highly specialized literatures of their own, and hence do not need any further mention here.

However, one effect, rather than a use, of a mould in the wine industry deserves mention. In the manufacture of sweet wines in the Bordeaux district and in the Rhineland the grapes are attacked by *Botrytis cinerea*, a fungus which infects many cultivated plants, and, in wet seasons particularly, can cause serious losses of soft fruits in the United Kingdom. Far from being regarded as a nuisance in the wine-growing areas, the disease is called 'la pourriture noble' in France, and in Germany 'Edelfäule', both terms meaning 'the noble rot'. The fungus causes much of the water of the grapes to evaporate, thus increasing enormously the percentage of sugar, the amount of the latter utilized by the mould being very small. When the wine is made the fermentation stops when the alcohol content reaches 10–12%, and while there is still a considerable amount of residual sugar.

Yeasts secrete the enzyme complex often referred to as 'zymase', which effects the conversion of sugar to alcohol, but lack diastase, the enzyme which breaks down the starch to sugar. There are, however, a number of fungi which secrete a whole range of enzymes and can ferment complex carbohydrates without a preliminary saccharification. In certain processes for the production of alcoholic beverages from starchy materials moulds alone are used. In others, and particularly in processes for producing industrial alcohol, moulds effect the saccharification of the starch, after which a yeast is allowed to act on the sugar produced, since, although the mould can complete the conversion to alcohol, the yield is better when yeast is used for the second stage. Fitz in 1873 was the first to show that alcohol is produced by a mould, *Mucor mucedo* (later identified as *M. racemosus*), and other workers have since shown that similar results are obtained with several other species of this and related genera. In the 'Amylo' process, which is still used in many countries in more or less modified form, *Mucor rouxii* was first used but was later replaced by various species of *Rhizopus*. Good general accounts of earlier practice are given by Lafar (1903), Wehmer (1907) and Galle (1923), and an improved 'Amylo' process is described by Erb and Hildebrandt (1946).

Various Mucoraceae are also active agents in the so-called starters, used for initiating alcoholic fermentations and marketed in various Eastern countries, Chinese rice being the best-known example. Japanese 'Koji' differs from most of the other starters in making use of quite a different group of fungi, strains of the *Aspergillus flavus-oryzae* series. *A. flavus* is also one of the active agents in the production of African native beer. In addition, moulds of the same series are used for the production of soy sauce, made by fermentation of soya beans, the starter being produced by growing the fungus on cooked beans, usually mixed with some other starchy material to aid rapid growth.

15.3   Oriental food fermentations

The use of fungi to change and impart flavours in staple foods is an ancient Oriental art which is still of considerable importance in the modern diet. Solid fermentation is usually used, soya beans being the commonest substrate, although grain and starch roots are substrates in some fermentations. The effect of the fermentations is to increase the digestibility of proteins as well as adding interesting flavours to the substrates (Hesseltine and Wang, 1967; Gray, 1970).

Shoya, or **soy sauce**, is the major fermented foodstuff of Japan, and is also produced in other Oriental countries. Soybeans, sometimes mixed with whole wheat, are inoculated with *Aspergillus oryzae* or *A. soyae*. Eventually the fermented material is mixed with brine, and after some months fermentation, which also involves *Saccharomyces rouxii*, the material is pressed. The resulting liquid is soy sauce.

Miso is also an important commodity in Japan, made by fermentation of soybeans, also involving *A. oryzae* and *S. rouxii*. It is sold as a pale brownish paste and is used to make a potage type of soup which is served as a regular morning dish. Cooked soybeans, 'koji' made from rice, and 10–12% salt are ground together with a small amount of miso from a previous batch. The mixture is packed into closed containers and allowed to ferment at 35–40°C, without aeration, for 3–5 months. It is then allowed to age for a month or more at room temperature before being packaged for sale.

A number of other fermentation products are made in the Far East, some (for example, Tempeh) utilizing certain species of *Mucor*, *Rhizopus* and *Neurospora*, while the food colouring 'Ang-Kak' is produced from *Monascus purpurens* growing on rice (Hesseltine, 1965).

15.4 Mould-ripened cheese

The manufacture of mould-ripened cheese is another industry of unknown origin. Until the present century such cheeses were associated with particular districts, such as the caves of Roquefort and the town of Stilton. The methods used were entirely empirical, the distinct flavours of certain brands being dependent on a combination of slight local peculiarities in the quality of the curd and of a natural local infection of a particular strain of the all-important mould. It was not until 1906 that C. Thom's studies of the cheese moulds led to a proper understanding of the ripening process. Even now, when countless cheeses are being made by pure culture methods, there are certain local blue-veined cheeses whose production is a happy accident rather than a matter to be determined at choice, the nature and source of the marbling being a complete mystery to those who make the cheese.

There are two distinct types of mould-ripened cheese, the soft cheeses of the Camembert and Brie types and the green- or blue-veined cheese, of which Roquefort, Gorgonzola, and Stilton are the best known. In the first type the mould concerned is *Penicillium camembertii* Thom, different isolates giving somewhat different flavours. The isolates were regarded as different species until a recent paper by Samson *et al.* (1977). The curd is made into cakes 3–4 cm thick, salted on the surface and either inoculated with spores of the fungus or placed in an infected room. The initial moisture in the cakes is 55–60%, and the air in the ripening room is maintained at a temperature of 50–60°F and a relative humidity of about 88%. Freedom from infection by undesirable moulds depends on maintaining both temperature and humidity within fairly narrow limits. The mould grows on the surface of the curd and gradually softens the whole mass, the process requiring about four weeks for completion. The most serious source of infection and spoilage is *Scopulariopsis brevicaulis*, which gives the cheese an ammoniacal taste and odour.

For production of the marbled cheese the raw curd is pressed so as to leave irregular cracks. In the pure culture method a sterile curd is inoculated, before pressing, from a culture of the mould (usually on bread), and is later aerated from time to time during the ripening process by piercing with wires. The older process relies on natural infection, and the success of the method depends on the fact that few moulds other than *P. roquefortii* can grow with the small amount of oxygen contained in the narrow air spaces in the curd, and hence the chance of contamination with an undesirable species is small. Although some mycologists consider that distinct species are concerned in the ripening of the various types of cheese, Thom in 1906 and Samson *et al.* (1977) have shown that, for all practical purposes, all the strains may be regarded as one species. However, in view of wide variations in biochemical activity between different strains of single species of other moulds, it is probable that the peculiar characteristics of any of the well-known types of veined cheese is as much due to the strain of *P. roquefortii* used as to the composition and method of preparation of the curd. As far as can be ascertained there has been no adequate study of the effect of using different strains of the mould under otherwise identical conditions (Gray, 1970).

15.5   Edible fungi

Edible fungi, particularly certain higher Ascomycetes have been prized items of human diet for many centuries (the Romans had special gold dishes for serving *Boletus* spp.), but it is only recently that their production has developed from an art to an agriindustrial process (Hayes and Nair, 1975). *Agaricus bisporus* is the cultivated mushroom, the only edible fungus industrially exploited.

A. bisporus must be in the heterothallic stage for production of the fruit bodies. Growers now use carefully selected strains derived from pure cultures. These are used to inoculate either trays or wide rows of compost previously pasteurized by steam or fumigated with methyl bromide. Various mixtures of compost are used today, based on sawdust, spent grains, and cellulose wastes, supplemented with nitrogen and phosphorus although the traditional horse manure or part horse manure is still frequently used. The pH is usually adjusted to between 7 and 8 and the substrate is kept for a week at 21°C, during which the inoculum grows to produce extensive mycelium. Casing is an important preliminary to sporophore production, the composition of the casing soil being of considerable significance. Sporophores are produced three weeks later at a temperature of 15–16°C.

Volvariella volvacea (the paddy straw mushroom), *Lentinus edodes* (Shiitake), *Tuber melonosporum* (the truffle) and *Pleurotus ostreatus* (the oyster fungus) are other edible fungi which are cultivated. However, *P. ostreatus* is the only one with the likely potential for industrial production. Submerged growth of *Agricus bisporus* and *Morchella esculenta* and *M. hortensis* has been investigated by Szeucs (1956), Szeucs (1958) and Litchfield (1968).

15.6   Production of organic acids

When fungi are grown in culture organic acids may, under certain conditions, accumulate in the medium.

15.6.1   Oxalic acid

There are a few modern processes which have been developed as the direct result of purely academic research into the biochemical activities of moulds. The first worker to make substantial progress in this field was C. Wehmer. Wehmer (1891) proved that oxalic acid is a fermentation product of *Aspergillus niger* and made an extended study of its formation from various sugars. Oxalic acid, however, has never been made commercially by this method, as the more usual chemical methods of preparation are cheaper.

15.6.2   Citric acid

In 1893 Wehmer described the production of citric acid by two species of moulds which were made the types of a new genus, *Citromyces* (now included in *Penicillium*). Other investigators have found that citric acid is a fermentation product of many species of *Penicillium*, but in no case is the yield sufficiently high to enable this method of production to compete with the old-established process of extracting the acid from citrus fruits. In 1917, however, Currie showed that in the fermentation of sugar by species of the *Aspergillus niger* group there is a distinct lag between the acidity of the medium due to oxalic acid and total acidity, the difference representing citric acid.

Methods were worked out for suppressing the formation of oxalic acid and increasing the yield of citric acid, the chief essentials being a high initial concentration of sugar (about 15%), low concentrations of ammonium nitrate as a source of nitrogen, and an acid reaction of the culture medium (pH about 3.5 or less). In addition, of course, it is necessary to use a strain of *A. niger* specially selected for the purpose. However, a number of difficulties have had to be overcome before such a process could be worked commercially. Sterilization of culture media on a large scale is an expensive operation; an abundant supply of air is required and must be supplied in a sterile condition, or, alternatively, the fermentation must be carried out in shallow layers of liquid with free aeration and suitable protection from infection, requiring expensive plant and expert supervision; and, perhaps most important of all, any organism is liable to be erratic in its behaviour, making the question of yields uncertain. A large number of patents have been taken out in this field, but the actual methods presently in use have not been made public in their entirety. Citric acid is certainly being produced successfully by mould fermentation, in this and several other countries, sufficiently cheaply and in sufficient quantities to make them independent of imported acid made from citrus fruits.

15.6.3   Gluconic acid

Gluconic acid is formed from sugars by the action of a large number of species of moulds, chiefly species of *Aspergillus* and *Penicillium*, and a considerable amount of work has been carried out in the United States in an attempt to develop its large-scale production by mould fermentation. The first fungus to be used was *Penicillium purpurogenum* var. *rubrisclerotium*, and the fermentation was carried out in shallow pans of pure aluminium (May *et al.*, 1929). Later it was found that certain strains of *P. chrysogenum* gave better results, and still later selected strains of *Aspergillus niger* were used (Wells *et al.*, 1937; Gastrock *et al.*, 1938; Porges *et al.*, 1941; Ward, 1967). In the meantime, in connection with this investigation, Herrick, Hellbach, and May (1935) described a semi-large-scale plant for growing moulds in submerged culture with forced aeration. It was largely the experience with this plant for gluconic acid production which later led to the rapid development of the production of penicillin by the submerged culture method.

Gluconic acid is used in moderate quantities, chiefly as the calcium salt, in place of calcium lactate in medicinal and food preparations.

15.6.4   Itaconic acid

Itaconic acid was first obtained as a mould metabolic product by Kinoshita (1929), using a species which he called *Aspergillus itaconicus*, but the amounts obtained were too small for successful commercial development. Of more interest in this connection was a paper by Calam, Oxford, and Raistrick (1939) describing the production of itaconic acid by a strain of *Aspergillus terreus*. In America selection of strains of *A. terreus* and careful study of culture media and environmental factors have resulted in itaconic acid being obtained in reasonable yield. In addition, selected strains of the mould have been irradiated with ultraviolet rays, and mutants obtained which give enhanced yields. As in other mould fermentation processes, attention has been directed to the possibility of obtaining the acid by submerged culture methods. For example, Kobayashi (1967) reported yields of over 50% of sugar consumed in three days on a medium containing 6% glucose.

Itaconic acid is used in industry as a copolymer with acrylic resins to bond printer's inks.

15.6.5   Other acids

Methods for the production of other acids such as fumaric, lactic and gallic have been determined but either the commercial demand, or the competition from non-biological methods, have precluded present day production by fermentation (Takahasi and Sakaguchi, 1927; Ikeda *et al.*, 1972). Quite extensive discussion on the production of organic acids by Aspergilli is to be found in Lockwood (1975).

15.7   Vitamins

One of the best sources of the vitamin B complex is yeast, which is, in fact, one of the very few readily accessible foods containing the majority of the known substances comprising the vitamin B group. The increasing recognition of the importance of adequate supplies of these vitamins in the diet has led food manufacturers to market a number of preparations of high potency, made from dried yeast, yeast extracts, or autolysed yeast.

Riboflavin, one of the B group of vitamins, is now made in pure form by fermentation. Two closely related yeast-like organisms are used, *Nematospora gossypii* Ashby and Nowell and *Eremothecium ashbyi* (Routien) Batra. An interesting account of the occurrence, systematic position, and uses of *N. gossypii* is given by Pridham and Raper (1950), and details of the manufacturing process are given by Pfeifer *et al*. (1950). Yaw (1952) describes the production of riboflavin by *Eremothecium ashbyi* when grown on a synthetic medium, from which the vitamin is readily isolated. A more recent review of methods of production and biosynthesis is given by Goodwin (1959).

Ergosterol, the precursor of vitamin D, is synthesized by a number of moulds as well as by yeasts (Pruess *et al*., 1931, 1932; Birkinshaw *et al*., 1931) and there are a number of manufactured preparations of irradiated ergosterol, mostly made from yeast.

Vitamin C (ascorbic acid) has been reported by Geiger-huber and Galli (1945) as a metabolic product of a strain of *Aspergillus niger*. This is more a curiosity than a possible means of manufacturing the vitamin by fermentation, since there are other and easier methods of preparation.

β-**carotene** is produced commercially by fermentation using a member of the Choanephoracea, *Blakeslea trispora*. Although the production of *β*-carotene was first investigated in *Phycomyces blakesleanus* Barnett *et al*. (1956) showed that it combined + and − strains of the members of the allied Choanephoracea which produce significant quantities, particularly *Blakeslea trispora*.

Although *β*-carotene is produced commercially by fermentation with *B. trispora* it has not been possible to convert it economically into the more valuable A vitamin.

15.8   Plant hormones

The **gibberellins** are plant hormones, but were first discovered as a product of *Gibberella fujikuroi*, a plant pathogen, whose imperfect state is *Fusarium moniliforme*. A number of gibberellins have been identified but gibberellic acid is the one mainly produced commercially for use in horticulture.

15.9   Antibiotics

Most plants produce secondary metabolites, that is, compounds which are of no obvious use to the cells that produce them. On the whole, secondary metabolites are only of theoretical interest, but one group of these substances in particular is of great value due to its ability to inhibit the growth of other organisms, especially pathogenic bacteria and viruses, and so has enormous potential in medical science. These substances are the **antibiotics**.

All antibiotics currently in use originate from microorganisms, and some groups of microorganisms in particular, such as the actinomycetes (bacteria) and fungi are very active producers of antibiotics. Korzybski *et al.* (1967) have listed over 600 antibiotics derived from bacteria and over 150 from fungi; and this list has been greatly added to since that date.

15.9.1   Penicillin

The first antibiotic to come into widespread use was penicillin, and the literature covering its discovery and subsequent medical application is vast. (Demain, 1966; Korzybski *et al.*, 1967; and Abraham and Newton, 1967). The early work on penicillin was carried out on Fleming's strain of *Penicillium notatum* (= *P. chrysogenum*), but subsequently penicillin has been isolated from a number of species of *Penicillium* and *Aspergillus* and from a species of *Cephalosporium*. The organisms used in the commercial production of penicillin are mutants of *P. chrysogenum*. The penicillin production of *P. chrysogenum* has been increased dramatically over recent years by the careful culturing of selected strains and the use of chemically induced mutants. In 1945 the highest known yield of penicillin was 250 units per cm^3; in 1972 McCann and Calam reported yields of $10–13 \times 10^3$ units per cm^3.

Penicillin is not one chemical compound but in fact a mixture of several, each with varying therapeutic properties. The proportions in which the various penicillins are produced depends mainly upon the availability of the necessary precursors in the fermentation media, and only to a lesser extent on the species and strains used. By adding a variety of chemicals which are not natural precursors of penicillin to the media a wide range of new semi-synthetic penicillins can be produced; these include methicillin, which has a higher penicillinase resistance than natural penicillin; and ampicillin, which is active against gram negative bacteria as well as gram positive. A review of these semi-synthetic penicillins is given by Naylor (1971).

Before semi-synthetic penicillins were discovered the limitations of natural penicillin initiated research into many other substances with suspected antibiotic properties. The genus *Streptomyces* has yielded a wide range of useful antibiotics, such as streptomycin, aureomycin, chloromycetin and terramycin.

15.9.2   Cephalosporins

The cephalosporins are a group of antibiotics derived from *Cephalosporium* spp., and are closely related in structure to the penicillins. As with the penicillins, a wide range of semi-synthetic cephalosporins have been produced, for example cephaloridine and cephalexin. Van Heyningen (1967) and Morin and Jackson (1970) have produced reviews of the cephalosporins.

15.9.3   Griseofulvin

Griseofulvin is an antibiotic first isolated from *P. griseofulvum*, and later from other species such as *P. patulum* (now regarded as a synonym of

P. griseofulvum), strains of which are used in the commercial production of griseofulvin. It has also been isolated from *Khuskia oryzae*, a fungus quite distinct from the penicillia. Griseofulvin is the only effective antibiotic available for the systemic treatment of fungal infections of skin, hair, and nails.

15.9.4   Fusidic acid

Fusidic acid is an antibiotic used mainly against penicillin-resistant staphylo-coccal infections. It has been isolated from *Fusidium coccineum, Mucor ramannianus*, and a *Cephalosporium* species.

The search for new antibiotics still goes on, and new ones are frequently announced. However, we are still waiting for a real cure for chronic tuberculosis, for a substance which will cure, or at least ameliorate, influenza and for a reliable treatment for the common cold.

15.10   Enzyme preparations

As a result of an intensive study of the enzymes of the *Aspergillus flavus-oryzae* series, Takamine introduced into commerce a number of products of high enzymic activity, particularly suitable for the dextrinization of starch and desizing of textiles. The products have been sold under the names of 'Takadiastase', 'Polyzime', 'Digestin', and 'Oryzyme'. Takamine (1914) summarized his work in a short paper.

Today enzyme preparations are widely used commercially, a number of which are derived from fungi, either grown as surface, or, more commonly, in submerged fermentation. A list of enzyme preparations and their manufac-ture has been composed by de Beeze (1970).

Aspergillus oryzae and *A. niger* are commercial sources of α-**amylase** and **amyloglucosidase**. α-amylase is used in bread making, and amyloglucosidase is used as a substitute for malt in the production of beer and spirits. Recently, the production of glucose syrup by the action of such saccharifying enzymes on temperate produced starches is becoming a challenge to cane or beet sugar. **Invertase** from yeast is also used commercially, and **pectolytic enzymes** are produced from a number of fungi for use in fruit processing (Rombouts and Pilnik, 1972). *Trichoderma viride* and *Aspergillus niger* are commercial sources of **cellulases**, used mainly for treating various feedstuffs and **proteases** are used in cheese making and other food processes.

15.11   Transformations

An alternative to the conversion of substrates by the metabolic process of growing mycelium in fermenters has been developed using dormant fungal spores (Gehrig and Knight, 1958; Vezina *et al.* 1968; Vezina and Singh, 1975). It has been found that transformations of compounds in solution or suspension can be made by suspensions of appropriate spores. Thus 11 α-hydroxylation of steroids can be mediated by spore suspensions of *Aspergillus ochraceus*. An advantage of this technique is that the transformation can occur without the introduction of the range of nutrients normally required in fermentations, and which complicate extraction procedures. Another advantage is that the process can be carried out in non-sterile conditions.

Although steroid conversions are commercially important a number of other transformations involving carbohydrates, antibiotics and flavonoids are known. Commercially this process is valuable when the products are expensive, enabling transformations to be made using the spores as a chemical, while the production of the spores may be made at another site. Clearly the process can only be successful where a species produces spores abundantly and care must be taken in handling spores.

15.12 Composting

15.12.1 Organic wastes

The return of organic wastes to the soil is a natural process which was gradually harnessed by agriculturalists and horticulturalists as their activities became more concentrated. Plant wastes, with their high cellulose content, are of importance in both agriculture and horticulture, and cellulolytic and thermophilic fungi are active in breaking down these wastes. Horticulture has for long placed much emphasis on the correct construction of compost heaps, where a combination of good aeration, adequate moisture and insulation encourages the rapid growth of plant degrading fungi, including thermophiles. Before modern methods of steam and electrical heating were used in frames, the development of thermophilic processes in composting hot beds of plant material and manure was an important procedure in temperate horticulture (Golueke, 1972; Senn, 1974). Such a process is still the basis of the production of mushroom composts where thermophilic processes cause self pasteurization of the compost, giving a selective advantage to the implanted mushroom mycelium.

15.12.2 Town wastes

Controlled systems of composting for the speedy degradation of town waste have recently received some attention. This work has been encouraged for three reasons – it has been assumed that modern agriculture would require compost additional to that normally generated on the farm; the bulk of refuse is increasing and other means of disposal are becoming less acceptable; and there is a marked decrease in mineral matter with a significant proportion of degradable organic material (often over 50%). As over one third of town waste is frequently cellulose based such processes have relied principally on the thermophilic cellulolytic fungi such as *Chaetomium thermophile* and *Humicola lanuginosa* to effect the major degradation of the waste (Schulze, 1962; McFarland, 1972; Fulbrook *et al.*, 1973).

At present, due to the high cost of equipment and the low and uncertain value of compost, few town waste systems are operating effectively.

15.12.3 Agricultural wastes

In agriculture, intensive husbandry has led to the accumulation of animal wastes which cannot be used economically as sources of crop manure, and crop monoculture has given rise to accumulations of cellulosic wastes which cannot be used as animal feed. The fungi have been considered as organisms able to transform these materials, principally to animal feeds, but so far controlled transformations have not been used commercially. For reviews of this process see Eggins and Seal (1975).

Particular emphasis has been placed on the controlled conversion of straw and wood wastes by ligninolytic fungi (Eggins and Seal, 1978; Burrows and Seal, 1979), where the protective action of lignin precludes the efficient utilization of the cellulose by ruminant microorganisms. Unfortunately, ligninolytic fungi are also cellulolytic and a process is not commercially available where the unwanted lignin can be removed by fungi.

15.12.4 Other fungal waste treatments

Fungi have been considered as liquid waste treatment organisms. For this process tower fermenters have been suggested for holding mycelial fungi used to remove low concentrations of wastes such as milk and starch solids in food processing waste matter. (See Imrie and Righelato, 1976 for reviews.)

15.13 Liquid culture of food fungal biomass

Much consideration has been given to the production of fungal mycelium in submerged sterile culture for consumption. Obviously fungi used comestibly, such as *Agaricus bisporus* and *Morchella esculenta*, have been examined. Unfortunately the flavour of vegetative mycelium is poor compared to that of the fructifications, and until better flavours can be induced there is little market for such products. (For reviews see Litchfield, 1968.)

Fungi have been widely investigated for the production of single cell protein, although only the yeasts properly qualify for this term. The basic aim of single cell protein production is to speed the production of protein from inorganic nitrogen, using either a hydrocarbon such as mineral oil or a carbohydrate as sources of both carbon and energy. Although much work has been done on the bacteria, the fungi offer the advantage of having a lower proportion of nucleic acids, which must be restricted in mammalian diets. Research has been carried out to ascertain whether mycotoxins or other objectionable metabolites are produced with these fermentations for fungal protein (Moss, 1968).

There is considerable literature on the use of fungi in the upgrading of feedstuffs and cellulosic waste. A review of some of these is given by Han (1978).

15.14 General reading

KORZYBSKI, T., KOWZYK, Z. and KURYLOWICZ, W. (1979). *Antibiotics, Origin, Nature and Properties*. Vols 1–3. American Society for Microbiology.

SMITH, J. E. and BERRY, D. R. (1975). *Filamentous Fungi. Vol. 1. Industrial Mycology*. Edward Arnold, London.

SMITH, J. E. and PATEMAN, J. A. (1977). *Genetics and Physiology of Aspergillus*. British Mycological Society Symposium Series, No. 1. Academic Press, London, New York and San Fransisco.

15.15 References

ABRAHAM, E. P. and NEWTON, G. G. F. (1967). Penicillins and cephalosporins. In *Antibiotics*, Vol. 11. 1–16, GOTTLIEB, D. and SHAW, P. D. (Eds). pp. 1–16. Springer-Verlag, New York.

BARNETT, H. L., LILLY, V. G. and KRAUSE, R. F. (1956). Increased production of carotene by mixed positive and negative cultures of *Choanephora cucurbitarum*. *Science*, **123**, 141.

BIRKINSHAW, J. H., CALLOW, R. K. and FISCHMANN, C. F. (1931). The isolation and characterization of ergosterol from *Penicillium puberulum* Bainer grown on a synthetic medium with glucose as the sole source of carbon. *Biochem. J.*, **25**, 1977–80.

BURROWS, I., SEAL, K. J. and EGGINS, H. O. W. (1979). The biodegradation of barley straw by *Coprinus cinereus* for the production of ruminant feed. In *Straw decay and its effect on disposal and utilization*. Grossband, E. (Ed.). pp. 147–54. Wiley, Chichester.

CALAM, C. T., OXFORD, A. E. and RAISTRICK, H. (1939). Itaconic acid, a metabolic product of a strain of *Aspergillus terreus* Thom. *Biochem. J.*, **33**, 1488–95.

CURRIE, J. N. (1917). The citric acid fermentation of *Aspergillus niger*. *J. biol. Chem.*, **31**, 15–37.

DE BECZE, G. I. (1970). Food enzymes. *Critical Reviews in Food Technology*, **1(4)**, 479–518.

DEMAIN, A. L. (1966). Biosynthesis of Penicillins and Cephalosporins. In *Biosynthesis of Antibiotics*, Vol. 1. SNELL, J. F. (Ed.). pp. 29–94. Academic Press, London and New York.

EGGINS, H. O. W. and SEAL, K. J. (1975). The biodegradation of waste agricultural products using recycled nutrients. *Proc. 3rd Int. Biodeg. Symp.*, pp. 679–86.

ERB, N. M. and HILDEBRANDT, F. M. (1946). Mould as an adjunct to malt in grain fermentation. *Indust. engng Chem.*, **38**, 792–4.

FULBROOK, F. A., BARNES, T. G., BENNETT, A. J., EGGINS, H. O. W. and SEAL, K. J. (1973). Upgrading cellulolytic wastes for recycling. *Surveyor*, 12 Jan., pp. 24–7.

GALLE, E. (1923). Das Amyloverfahren und seine Anwendungsmöglichkeiten. *Zeit, angew, Chem.*, **36**, 17–19.

GASTROCK, E. A., PORGES, N., WELLS, P. A. and MOYER, A. J. (1938). Gluconic acid production on pilot plant scale: effect of variables on production by submerged mold growth. *Industr. engng Chem.*, **30**, 782–9.

GEHRIG, R. F., and KNIGHT, S. G. (1958). Formation of ketones from fatty acids by spores of *Penicillium roquefortii*, *Nature*, **182**, 1237.

GEIGER-HUBER, M. and GALLI, H. (1945). Uber den Nachweis der l'Ascorbinsaure als Stoffwechselprodukt von *Aspergillus niger*. *Helv. chim. Acta*, **28**, 248–50.

GOLUEKE, C. G. (1972). *Compositing: a Study of the Process and its Principles*. Rodale Press, U.S.A.

GOODWIN, T. W. (1959). Production and biosynthesis of riboflavin in microorganisms. *Progr. industr. Microbiol.*, **1**, 139–77.

GRAY, W. D. (1970). The use of fungi in food and in food processing. *Chemical Rubber Co. Critical Reviews in Food Technology*, **1(2)**, 225–329.

HAN, Y. W. (1978). Microbial utilization of straw (a review). *Advances in Applied Microbiology*, **23**, 119–53.

HAYES, W. A. and NAIR, N. G. (1975). The cultivation of *Agaricus bisporus* and other edible mushrooms. In *The Filamentous Fungi, Vol. 1, Industrial Mycology*. SMITH, J. E. and BERRY, D. R. (Eds). pp. 212–48. Edward Arnold, London.

HERRICK, H. T. and MAY, O. E. (1928). The production of gluconic acid by the *Penicillium luteum-purpurogenum* group. II. Some optimal conditions for acid formation. *J. biol. Chem.*, **77**, 185–95.

HERRICK, H. T., HELLBACH, R. and MAY, O. E. (1935). Apparatus for the application of submerged mould fermentations under pressure. *Industr. engng Chem.*, **27**, 681–3.

HESSELTINE, C. W. (1965). A millennium of fungi, food and fermentation. *Mycologia*, **57**, 149–97.

HESSELTINE, C. W. and WANG, H. L. (1967). Traditional fermented foods. *Biotechnology and Bioengineering*, **9**, 275–88.

IKEDA, Y., TAKAHASHI, E., YOKOGAWA, K. and YOSHIMURE, Y. (1972). Screening for microorganisms producing gallic acid from Chinese and Tara tannins. *J. Fermentation Technol*, **50**, 361–70.

IMRIE, F. K. E. and RIGHELATO, R. C. (1976). Production of microbial protein from carbohydrate wastes in developing countries. In *Food From Waste*. BIRCH, G. G., PARKER, K. J. and WORGAN, J. T. (Eds). Applied Science Publishers, Barking, London.

KINOSHITA, K. (1929). Formation of itaconic acid and mannitol by a new filamentous fungus. *J. chem. Soc. Japan*, **50**, 583–93.

KOBAYASHI, T. (1967). Itaconic acid fermentation. *Process Biochemistry*, **2**, 61–5.

KORZYBSKI, T., KOWZYK-GRINDIFER, Z. and KURYLOWICZ, W. (1967). *Antibiotics*. Vol. 2. pp. 1146–93. Pergamon Press, Oxford.

LAFAR, F. (1903). *Technical mycology*. Vol. 2, Part 1. English translation by C. T. C. Salter. Griffin, London.

LITCHFIELD, J. H. (1968). The production of fungi. In *Single Cell Protein*. MATELES, R. I. and TANNENBAUM, S. R. (Eds). pp. 309–29. MIT Press, London and Cambridge.

LOCKWOOD, L. B. (1975). Organic acid production. In *Filamentous Fungi, Vol. I, Industrial Mycology*. SMITH, J. E. and BERRY, D. R. (Eds). pp. 140–57. Edward Arnold, London.

MAY, O. E., HERRICK, H. T., THOM, C., and CHURCH, M. B. (1927). The production of gluconic acid by the *Penicillium luteum-purpurogenum* group. I. *J. biol. Chem.*, **75**, 417–22.

MAY, O. E., HERRICK, H. T., MOYER, A. J. and HELLBACH, R. (1929). Semi-plant scale production of gluconic acid by mould fermentation. *Industr. engng Chem.*, **21**, 1198–1203.

MCFARLAND, J. (1972). The economics of composting municipal wastes, *Compost Sci*, **13(4)**, 10–12.

MCCANN, E. P. and CALAM, C. T. (1972). The metabolism of *Penicillium chrysogenum* and the production of Penicillin using a high yielding strain at different temperatures. *J. Appl. Chem. Biotechnol*, **22**, 1201–8.

MORIN, R. B. and JACKSON, B. G. (1970). Chemistry of the cephalosporium antibiotics. *Progress in the Chemistry of Organic Natural Products*, **28**, 344–403.

MOSS, M. O. (1969). Mycotoxins. *Int. Biodetn. Bull.*, **5(4)**, 141–7.

NAYLOR, J. H. C. (1971). Structure-activity relationships in semi-synthetic penicillins. *Proc. Roy. Soc., Series B*, **179**, 357–67.

PFEIFER, V. F., TANNER, F. W., VOJNOVICH, C. and TRAUFLER, D. H. (1950). Riboflavin by fermentation with *Ashbya gossypii. Industr. engng Chem.*, **42**, 1776–81.

PORGES, N., CLARK, T. F. and ARONOVSKY, S. I. (1941). Gluconic acid production: repeated recovery and re-use of submerged *Aspergillus niger* by filtration. *Industr. engng Chem.*, **33**, 1065–7.

PRIDHAM, T. G. and RAPER, K. B. (1950). *Ashyba gossypii* – its significance in Nature and in the laboratory. *Mycologia*, **42**, 603–23.

PREUSS, L. M., PETERSON, W. H., STEENBOCK, H. and FRED, E. B. (1931). Sterol content and antirachitic activatibility of mould mycelia. *J. biol. Chem.*, **90**, 369–84.

PREUSS, L. M., PETERSON, W. H. and FRED, E. B. (1932). Isolation and identification of ergosterol and mannitol from *Aspergillus fischeri. J. biol. Chem.*, **97**, 483–9.

ROMBOUTS, F. M. and PILNIK, W. (1972). Research on pectin depolymerases in the sixties – a literature review. *Critical Reviews in Food Technol*, **3(1)**, 1–26.

ORTH, R. (1977). The taxonomy of *Penicillium* species from cheeses. *Antonie van Leeuwenhoek*, **43**, 34–350.

SCHULZE, K. L. (1962). Continuous thermophilic composting. *Appl. Microbiol.*, **10**, 108–22.

SENN, C. L. (1974). The role of composting in waste utilization. *Compost Sci.*, **15(4)**, 24–8.

SZUECS, J. (1956). Mushroom Culture. *U.S. Pat. 2 761 246*.

SZUECS, J. (1958). Methods of growing mushroom mycelium and the resulting products. *U.S. Pat. 2 850 811*.

TAKAHASI, T. and SAKAGUCHI, K. (1927). Acids produced by *Rhizopus sp. Bull. of the Agricultural Chemical Soc. of Japan*, **3**, 59–62.

TAKAMINE, J. (1914). Enzymes of *Aspergillus oryzae* and the application of its amyloclastic enzyme to the fermentation industry. *Chem. News*, **110**, 215–18.

VAN HEYNINGEN, E. (1967). Cephalosporins. In *Advances in Drug Research*.

Vol. 4. HARPER, N. J. and SIMONDS, A. B. (Eds). pp. 1–70. Academic Press, New York and London.

VEZINA, C., SEHGAL, S. N. and SINGH, K. (1968). Transformation of organic compounds by fungal spores. In *Advances in Applied Microbiology*, 10, 221–68.

VEZINA, C. and SINGH, K. (1975). Transformation of organic compounds by fungal spores. In *Filamentous Fungi, Vol. I, Industrial Mycology*. SMITH, J. E. and BERRY, D. R. (Eds). pp. 158–92. Edward Arnold, London.

WARD, G. E. (1967). Production of gluconic acid, glucose oxidase, fructose and sorbose. In PEPPLER, H. J. (Ed.) *Microbial Technology*, 200–221. Reinhold Publishing Company, London, New York.

WEHMER, C. (1891). Enstehung und physiologische Bedeutung der oxalsüre im Stoffwechsel einiger Pilze. Bot. Ztg. **49**, 233–638.

WEHMER, C. (1893). *Beitrage zur Kenntnis einheimischer Pilze*. Hansche Buchhandlung, Hannover and Jena.

WEHMER, C. (1907). Chemische Wirkungen der Mucoreen. Alcoholische Garung. Lafar's *Handbuch der technischen Mykologie*, Vol. IV, 506–28.

WELLS, P. A., MOYER, A. J., STUBBS, J. J., HERRICK, H. T. and MAY, O. E. (1937). Gluconic acid production. Effect of pressure, air flow and agitation on gluconic acid production by submerged mould growths. *Industr. engng Chem.*, **29**, 777–8.

YAW, K. E. (1952). Production of riboflavin by *Eremothecium ashbyi* grown in a synthetic medium. *Mycologia*, **44**, 307–17.

16

Maintenance of a Culture Collection

'The cultivation of micro-fungi is still almost wholly an empirical art'
H. A. Dade, Herb. I. M. I., Handbook, 1960.

16.1   Introduction

In most laboratories where serious work on micro-fungi is carried out it is necessary to maintain a collection of moulds. If the work in hand aims at controlling the harmful activities of fungi on an industrial product most of the preliminary work will have to be done on pure strains of moulds isolated from actual cases of damage, and very important species will have to be kept in pure culture so as to be available for numerous experiments extending perhaps over a period of years. Those who study the biochemical activities of fungi, with a view to isolating or manufacturing special products of metabolism, usually find it necessary to test many strains of the same or related species, and to grow them all on many different media and under varying conditions. Again a type collection is essential.

Requirements of different laboratories, as regards size and type of collection, will vary greatly, but there are a number of general methods and precautions whose consideration is applicable to all cases. The main essentials for successfully maintaining a culture collection are briefly summarized below.

1.   All cultures must be kept alive as long as required.
2.   Every strain must be maintained in a state of purity.
3.   As far as possible, each species or strain must be kept true to type, that is, with all the characteristics it had when first added to the collection.

16.2   Growth of cultures

The object of any culture collection is to maintain good healthy cultures. In order to do this it is essential to start with good cultures, obtained either as fresh isolates and/or by using the correct growth methods. For practical purposes it is usual to grow cultures on agar slants and to try to find optimum growth conditions which depend on temperature, light, water activity, nutrients (media), pH and available oxygen. These have been dealt with elsewhere, but are summarized briefly here.

Temperature

Optimum growth temperatures vary with different cultures. Most will grow well or satisfactorily at 20–25°C, or at room temperature, but some spore better at temperatures below 20°C, for example, certain species of the *Aspergillus glaucus* group, and others are thermophilic with optimum growth at higher temperatures. Most cultures have increased longevity when stored at lower temperatures, for example, in a refrigerator at 4–8°C, though a few thermophilic species may be killed by cold.

Light

Many species, especially those with an exposed natural habitat, will spore better in light and it is usually possible to grow the majority of species in the laboratory in day light. Many moulds and soil organisms are not particularly sensitive to light. However, light sensitive organisms, such as most Dematiaceous Hyphomycetes, Ascomycetes and Sphaeropsidales, often spore better in the presence of the additional stimulus of 'Black Light'.

Water activity

In ordinary culture work it should be remembered that some species will not grow well at normal osmotic pressures, and come from habitats with very dry conditions or little available water, for example *Aspergillus penicillioides*. Media with low water activity, such as GYA (Glycerol Yeast Agar) or malt plus 20% sucrose, should always be available.

Media

Media for stock cultures should encourage growth which is characteristic, but not too luxurious and which induces sporulation. For most Penicillia and Aspergilli Czapek agar is ideal but for other species Malt is more satisfactory. This may be too rich and media made from extracts of vegetables such as potato, carrot, mixed carrot and potato, and V8 are suitable, since they provide the necessary growth factors without encouraging rampant growth. For most dark coloured Hyphomycetes corn-meal or oat-meal are two of the best media, usually inducing adequate spore production without excessive aerial mycelium. Tap-water agar with pieces of wheat straw or other vegetable matter is often useful.

pH

Most moulds will grow well at pH 4–7 but occasional isolates are tolerant of pH 2, for example *Aspergillus niger*.

Oxygen

Fungi are aerobic and require oxygen to grow. Closures such as cotton wool plugs or loose caps allow sufficient exchange of gases for most species. However, a few water fungi require bubbling of air in order to spore. These fungi are not of industrial interest, but forced aeration is often used in industrial fermentations.

16.3   Methods of culture maintenance

Microgardening

The simplest method of maintaining fungus cultures is to grow them on agar slants and to transfer them to fresh media before they exhaust all the nutrients or dry out. For some isolates no other methods have been found satisfactory, but for the majority of species other methods are available.

Problems of repeated culture growth

Under normal conditions cultures have to be regrown at fairly frequent intervals for example, every four, six or eight months. With a large collection this requires much labour, and every time a culture is handled it is at risk owing to the following:

1.   Selection favourable to the conditions of culture can occur on frequent transfer.

2.   Some isolates become progressively fluffy or weaker with each transfer and may eventually die.

3.   The operator selects the inoculum and thus the isolate is exposed to manual selection. Specialist operators are therefore desirable, and careful checking of newly processed cultures is essential.

4.   Each time a culture is opened it is exposed to contamination, infection with other organisms or mites, etc.

To cut down the handling of the cultures it is therefore desirable to prolong the intervals between subculturing, and various means are employed to do this.

Cold storage

Good healthy cultures in bottles or tubes will survive for longer if stored at lower temperatures, either in a refrigerator at about 4–8°C or in a deep freeze.

Oil storage

Healthy cultures may be covered with mineral oil (British Pharmocopoea quality Specific Gravity 0.85–0.89) to a depth of about 1 cm above the top of the culture to prevent drying out. The oil allows slow diffusion of gases so growth continues at a reduced rate. This may induce change due to adaption to growth in oil. Some isolates appear stable and survivals of over 20 years have been obtained at the Commonwealth Mycological Institute. Others change rapidly, producing atypical cultures in a few months, for exampie, *Fusarium* species. The cultures can be stored at room temperature but better viability is obtained when stored at lower temperatures, for example 15°C. Oil storage can be regarded as lengthening the life of the cultures, thus long lived ones will live longer while the short lived aquatic fungi such as *Achlya* and the Chytridales may last 12–18 months instead of the usual 4–6 weeks. In this latter case regular subculturing every six months is preferable to every four weeks.

Suspended metabolism

It is possible by applying modern techniques to prepare and store fungal spores and cultures in such a way that metabolism is temporarily stopped. Revival of the isolate at a later time is done with the minimum amount of change.

Freeze drying
Freeze drying (lyophil) is perhaps the most popular form of suspended metabolism. A spore suspension is prepared in a suspending medium (at the Commonwealth Mycological Institute 10% skimmed milk and 5% inositol in distilled water is found suitable). This is distributed in glass ampoules, frozen and dried under vacuum from the frozen state. The ampoules are sealed retaining the vacuum. Some workers prefer to fill the ampoules with sterile nitrogen before sealing. Relatively simple apparatus can be constructed for processing a few ampoules, though quite complex machines are available for larger scale work. The method is very satisfactory as the prepared ampoules are easily stored, not readily broken and most species remain viable for many years. About 60–70% of species normally processed at the Commonwealth Mycological Institute survive the treatment, including most species that produce simple spores. Mycelial cultures, the genera *Pythium* and *Phytophthora*, and species with very large fragile spores are killed by the process. For further details see Onions (1971), Wickerham and Andreasen (1942), Thom and Raper (1945) (a simple method) and Mellor (1978) (a modern monograph on freeze-drying).

Cryogenic storage
Storage of microorganisms at ultra low temperatures is becoming increasingly common and has been shown to be satisfactory for fungi (Hwang, 1960, 1966, 1968a, b). Suspensions of spores or mycelium in a suitable suspending medium (10% glycerol is used at the Commonwealth Mycological Institute) are frozen and immersed in or suspended over liquid nitrogen in special vacuum containers. The cultures are revived by removing from the container, rapidly thawing and culturing in the usual way. Life is regarded as 'at a stand still' at temperatures below $-130°C$, so at the temperature of liquid nitrogen ($-196°C$), provided the cultures survive the treatment, the period of preservation should be indefinite. Thus long-term storage without change is now available. The process is expensive, only as safe as the supply of nitrogen, and is still not fully understood. A recent short review of current research has been made by Calcott (1978). For a simple method of procedure see Onions (1971).

Other methods of storage
Not every laboratory can undertake freeze drying or liquid nitrogen storage. Many workers lack the time for the labour of frequent transfer and find oil storage messy. Indeed, in these times of safety considerations, the spatter made when sterilizing oily needles is undesirable, so various other means of preservation have been introduced.

Soil cultures
A spore suspension is added to sterilized loam and allowed to grow for about 10 days. The cultures are then stored in a refrigerator at about 5–8°C. This

method is particularly suitable for *Fusarium* species, which normally deteriorate rapidly when kept in culture, but by this means remain alive and typical
for long periods. (Gordon, 1952.)

Silica gel cultures
Refer to Ogata (1962, 1968), Perkins (1962). A spore suspension in 5%
skimmed milk is added to anhydrous silica gel in screw capped bottles. This is
allowed to dry for about 14 days, the caps screwed down and the bottles
stored in a refrigerator, though storage at room temperature can be satisfactory. As heat is liberated when water is added to silica gel, everything must be
well cooled before starting the process. Very good revivals are obtained over
several years. This is an adaption of the method of Roberts following Ogata
(1962) who used it for the preservation of genetic stocks of *Neurospora* (see
Onions, 1971, 1977).

Water storage
Surprising success has been reported (Boeswinkel, 1976; Marx and Daniel,
1976) for storage of fungal cultures in water. The fungus is grown on weak
medium in a Petri dish and then cut into small pieces. The pieces are
transferred, using sterile techniques, to sterile distilled water in small screw
capped bottles. The cap is screwed down and the cultures stored, preferably
at a low temperature. Cultures of *Pythium* and *Phytophthora* stored in this
way at 15°C at the Commonwealth Mycological Institute have been found to
be in good condition after two years storage. As these fungi do not survive
storage by freeze-drying or in silica gel this is a useful means of storage.

Other methods of preservation include storage of spores in sand, drying
cultures on filter paper and sealing tubes with paraffin wax.

Reviews of methods of culture preservation have been made by Fennell
(1960), Simmons (1963), Clark and Loegering (1967) and Onions (1971).
Onions (1977) compared various methods of storage including oil storage,
freeze-drying, silica gel storage and liquid nitrogen storage.

Reports of the periodic International Conferences on Culture Collections
are given in Martin (1963), Iizuka and Hasegawa (1970), Pestana de Castro
et al. (1977) and Fernandes and Pereira (1977).

When a collection is constantly in use and when any culture may be
required at short notice, there is little doubt that agar slopes provide the most
convenient method of storage, particularly when supplemented by the use of
silica gel and sand or soil cultures, for preserving critical cultures. The
freeze-drying (lyophil) method is best adapted to larger collections and those
from which cultures are regularly sent out to other laboratories.

16.4 Mites

Mites are a most annoying pest to have in the laboratory, and are one of the
most frequent causes of contamination. A collection may remain free from
their depradations for years and then, suddenly, scores or hundreds of
cultures are found to be infested. Unless the mycologist is constantly on the
look-out, and unless he knows what to look for, a serious attack may go

unnoticed for a long time and may result in wholesale contamination and loss of many species. Figure 16.1 is a photomicrograph of a typical mite. The

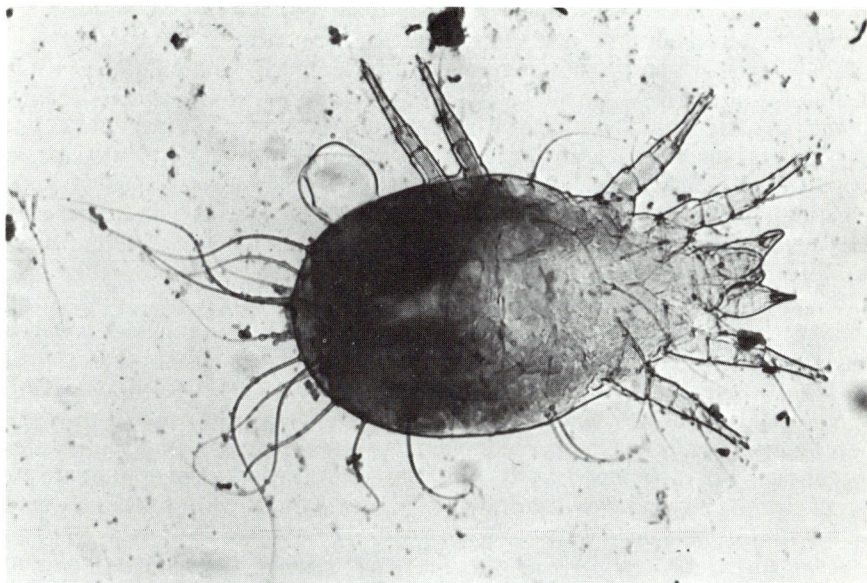

Fig. 16.1   Culture mite, ×100.

different species vary somewhat in shape and size, but, when once seen, can never be mistaken for anything else. An adult mite is usually about 0.02 mm in length, and is thus almost at the limit of unaided vision. When alive and in motion they are readily seen as small whitish specks by anyone with normal sight. However if any cultures are known to be infested, or there is any reason to suspect the presence of mites, all cultures should be carefully examined under the stereo-microscope.

The presence of eggs should be particularly noted, as these are more difficult to kill than the adult mites. Like the adults and larvae, they are pale coloured, but appear to be dark brown when infested cultures are examined by transmitted light. They are surprisingly large, in comparison with the adults, and are readily seen, even when lying amongst masses of hyphae and spores. The evil mites do is two fold. In the first place they eat the cultures and if unchecked, may even destroy them entirely. In addition they crawl from one culture to another with spores adhering to their hairy bodies, spreading infection wherever they go. Petri dishes are readily entered and become contaminated with amazing rapidity when mites appear in the vicinity. Cultures in tubes are just as readily invaded, since the mites find no difficulty in crawling through the tightest cotton-wool plugs and according to Thom (1930), have even been known to find their way through paraffined plugs. If plugs are a good fit the mites shed most of the adhering spores en route to the cultures, the major portion of the damage being then due to their gastronomic activities, but mixing of cultures does occur and is presumably due to spores which pass unchanged through the digestive tract. The commonest infections introduced by these creatures are species of *Acremonium*, and these, with their production of myriads of tiny spores, are exceedingly difficult, and often impossible to eliminate.

The conrol of mites is not easy, they are brought in with all kinds of raw materials, they may be introduced with new laboratory fittings and apparatus, or they may be present with cultures sent from other laboratories, and hence hygiene measures sufficiently stringent to prevent their access to the culture room are difficult to realize in practice. Constant vigilance is necessary, especially during hot humid weather, with prompt application of suppressive measures at the first indication of an attack. The usual method of killing mites is to expose the cultures to the vapour of some substance which is highly toxic to the mites and relatively non-toxic to moulds. Previously *p*-dichlorobenzene and crude kerosene were regarded as the most effective, but the former is regarded as a carcinogen and the latter is now so highly refined that it is no longer effective. At present various acaricides are used at the Commonwealth Mycological Institute including 'Kelthane' (4:4′-dichloro-trichloromethyl-benzhydrol), Tedion, Crypo recommended by Smith (1967) and actellic, a recent acaricide developed by ICI for agricultural use. All these should be handled with care and gloves worn when applying them. Most fungi appear to be tolerant to actellic.

Apart from hygiene various barrier methods have been devised. Some workers poison culture plugs with coloured mercuric chloride solution, but this is dangerous to handle. At the CMI, cigarette paper (see Snyder and Hansen, 1946) is used to seal the tubes above the cotton plugs which are pushed down or under the screw caps of bottles. Small squares of the cigarette paper are glued to the rims of the tubes or bottles by means of a copper sulphate gelatine glue (20% copper sulphate, 20% gelatine, 60% water). If bottles are used and there is no plug to protect the culture from infection the papers must be sterilized. To do this the papers are placed in a honey jar and one or two drops of propylene oxide added, the lid screwed down and left over night. Air passes freely through the paper but the pores in the paper are too small to allow the passage of mites. Other barriers used include standing in water or inside a vaseline barrier. Petri dishes may be sealed using sticky tape. This also prevents drying out. However, although it can act as a deterrent it is seldom a complete seal and a heavy infestation usually penetrates the dishes.

Another excellent way of preserving cultures from infestation is to store them in a refrigerator at a temperature not exceeding about 8°C. Mites are not attracted to cold areas and, even if by some chance an infected culture is put into store, the mites, at this low temperature, move very slowly and do not breed, so that there is little danger of infestation spreading, but they become active again when the cultures are brought into normal temperatures. Mites do not attack cultures stored under oil.

Insect pests. In the previous edition Smith included a section on insect pests in cultures, but in my experience (AHSO) I have only once seen a plate seriously infested with insects, so I regard this as rare. Precautions similar to those used for control of mites should be effective.

16.5   Records

It is most important that adequate records be kept in a culture collection. In the first place, the history of the culture, including its name, source, place of isolation, date and any other known information, while in the second place full notes on the culture should be kept including a description of it when it was received and any changes that may occur during its life in the collection. This becomes increasingly important as the collection grows and for reference when the collection is passed to a new curator. Conditions for growth should be recorded. The means of keeping the records is a matter of preference and can be a note book, loose leaf file, cards or a computor, and will depend on the facilities available.

16.6   Culture collections

There exist a number of large collections of cultures of microfungi, maintained in what are primarily centres of research, but which are organized for the regular supply of cultures to other institutions, firms, and individual workers. The largest and best known of these is the Centraalbureau voor Schimmelcultures, at Baarn, Netherlands.

In the UK fungi are supplied by the Commonwealth Mycological Institute, Kew which issues a catalogue from time to time. Other microorganisms are housed in collections held in appropriate specialist institutions, and many are listed in the Directory of Collections of Microorganisms maintained in the United Kingdom (C.A.B., 1978). Other countries have their own national, industrial and research collections, many of which are listed in the World Directory (Martin and Skerman, 1972).

16.7   References

BOESWINKEL, H. J. (1976). Storage of fungal cultures in water. *Trans. Brit. mycol. Soc.*, **66**, 183–5.

COMMONWEALTH AGRICULTURAL BUREAUX (1978). *Director of Collections of Microorganisms maintained in the United Kingdom*. ONIONS, A. H. S. (Ed.).

CALCOTT, P. H. (1978). *Freezing and thawing microbes*. Meadowsfield Press, Shildon, UK.

CLARK, W. A. and LOEGERING, W. Q. (1967). Functions and maintenance of a type culture collection. *Annual Review of Phytopathology*, **5**, 319–42.

FERNANDES, F. and PEREIRA, R. C. (Eds) (1977). *Proceedings of the Third International Conference on Culture Collections*. University of Bombay, India.

FENNELL, D. I. (1960). Conservation of fungus cultures. *Botany Review*, **26**, 79–141.

GORDON, W. L. (1952). The occurrence of *Fusarium* species in Canada. *Canad. J. Bot.*, **30**, 209–51.

HWANG, S. W. (1960). Effects of ultra-low temperatures on the viability of selected fungus strains. *Mycologia*, **52**, 527–9.

HWANG, S. W. (1966). Long term preservation of fungus cultures with liquid nitrogen refrigeration. *Applied Microbiology*, **14**, 784–8.

HWANG, S. W. (1968a). Investigation of ultra-low temperature for fungal cultures. I. An evaluation of liquid nitrogen storage for preservation of selected fungal cultures. *Mycologia*, **60**, 613–21.

HWANG, S. W. (1968b). Investigation of ultra-low temperature for fungal cultures. II. Cryoprotection afforded by glycerol and dimethyl sulfoxide to eight selected fungal cultures. *Mycologia*, **60**, 622–6.

IIZUKA, H. and HASEGAWA, T. (Eds) (1970). *Culture Collections of Microorganisms*. Proceedings of the International Conference on Culture Collections, Tokyo, 1968. University Park Press, Baltimore, Maryland and Manchester, England.

MARTIN, S. M. (Ed.) (1963). *Culture Collections, Perspectives and Problems*. University of Toronto Press, Toronto.

MARTIN, S. M. and SKERMAN, V. B. D. (Eds) (1972). *World Directory of Collections of Cultures of Microorganisms*. John Wiley & Sons, New York, London, Sydney and Toronto.

MARX, D. H. and DANIEL, W. J. (1976). Maintaining cultures of ectomycorrhizal and plant pathogenic fungi in sterile water cold storage. *Canad. J. Microbiol.*, **22**, 338–41.

MELLOR, D. J. (1978). *Fundamentals of freeze drying*. Academic Press, London, New York and San Francisco.

OGATA, W. N. (1962). Preservation of *Neurospora* stock cultures with anhydrous silica gel. *Neurospora Newsletter*, **1**, 13.

OGATA, W. N. (1968). Report on the Fourth *Neurospora* Information Conference. *Neurospora Newsletter*, **13**, 6.

ONIONS, A. H. S. (1971). Preservation of fungi. In *Methods in Microbiology*. BOOTH, C. (Ed.). pp. 113–52, Academic Press, London and New York.

ONIONS, A. H. S. (1977). Storage of fungi by mineral oil and silica gel for use in the collection with limited resources. In *Proceedings of the Second International Conference on Culture Collections*. PESTANA DE CASTRO, DA SILVA, E. J., SKERMAN, V. B. D. and LEVERITT, W. W. (Eds). pp. 104–13.

PERKINS, D. D. (1962). Preservation of *Neurospora* stock cultures in anhydrous silica gel. *Canadian Journal of Microbiology*, **8**, 591–4.

PESTANA DE CASTRO, A. F., DA SILVA, E. J., SKERMAN, V. B. D. and LEVERITT, W. W. (Eds) (1977). *Proceedings of the Second International Conference on Culture Collections*. World Data Centre for Microorganisms, Queensland, Australia.

SIMMONS, E. G. (1963). Fungus cultures: conservation and taxonomic responsibility. In *Culture Collections, Perspectives and Problems*. MARTIN, S. M. (Ed.). pp. 100–10. Toronto Press, Toronto.

SMITH, R. S. (1967). Control of Tarsonemid mites in fungal cultures. *Mycologia*, **59**, 600–9.

SMITH, R. S. (1978). A new lid closure for fungal culture vessels giving complete protection against mite infestation and microbiological contamination. *Mycologia*, **70**, 499–507.

SNYDER, W. C. and HANSEN, H. M. (1946). Control of culture mites by cigarette paper barriers. *Mycologia*, **38**, 455–62.

THOM, C. (1930). *The Penicillia*. Bailliere, Tindall and Cox, London.

THOM, C. and RAPER, K. B. (1945). *A manual of the Aspergilli*. Williams & Wilkins, Baltimore.

WICKERHAM, L. J. and ANDREASEN, A. A. (1942). The lyophil process: its use in the preservation of yeasts. *Wallerstein Laboratories Communication*, No. 5. pp. 165–9.

Appendix I

Laboratory media for growth of fungi

Czapek (Dox) Agar

Sodium nitrate (NaNO$_3$)	*2.0 g*
Potassium dihydrogen phosphate (KH$_2$PO$_4$)	*1.0 g*
Magnesium sulphate (MgSO$_4$.7H$_2$O)	*0.5 g*
Potassium chloride (KCl)	*0.5 g*
Ferrous sulphate (FeSO$_4$.7H$_2$O)	*0.01 g*
Sucrose	*30.0 g*
Agar	*20.0 g*
Distilled water	*1.0 litre*

If glass distilled water is used trace elements must be added. Zinc sulphate (ZnSO$_4$.7H$_2$O), 0.01 g, and copper sulphate (CuSO$_4$.5H$_2$O), 0.005 g per litre are recommended. These trace elements can be prepared in concentrated solution and the appropriate volume added to the medium.

It is preferable to dissolve the KH$_2$PO$_4$ separately, especially if storing overnight, and mix the solutions immediately before preparing. The agar must be dissolved by boiling before sterilizing.

Autoclave   1.05 kgf/cm^2 (15 p.s.i.) for 20 minutes

Tap Water Agar

Agar	*15 g*
Tap water	*1 litre*

Dissolve agar in tap water by boiling.

Autoclave   1.05 kgf/cm^2 (15 p.s.i.) for 20 minutes

Various vegetable pieces may be added to this, such as sterile wheat straw, rice grains, etc., to induce sporulation.

Malt Extract

Malt extract	*20 g*
Agar	*20 g*
Water	*1 litre*

Dissolve malt extract and agar separately, mix and sterilize.

Autoclave   1.05 kgf/cm^2 (15 p.s.i.) for 20 minutes

pH is about 3–4 but may be adjusted with sodium hydroxide (NaOH) to 6.5

Propriety Malt Extract (without additions) from the chemist may be used; the variety is a matter of experience.

Variations of this media are possible.

+20% Sucrose, +40% Sucrose
The sugar should be added after boiling to avoid caramelization.

+ 10% Salt (NaCl)

+ Czapek's Salts

	Czapek's solution	*91%*
	Malt extract	*4%*
	Sucrose	*3%*
	Agar	*2%*

This is used for standard growth of *Penicillium* species.

Potato Based Media

(1) Potato Dextrose Agar (usually referred to as PDA)
Potato extract

Potato	*200 g*
Water	*1 litre*

Clean potatoes by scrubbing (do not peel), cut into approximately 12 mm cubes, rinse in running water, and boil until soft (approximately 1 hour) in 1 litre of water. Mash the potatoes, squeeze as much of the pulp as possible through a fine sieve, and make up to 1 litre. Instead of passing through the sieve the potato may be blended, in which case a thicker, opaque extract is produced. Old potatoes are preferable to new.

Medium

Potato extract	*1 litre*
Dextrose	*20 g*
Agar	*20 g*

Dissolve agar by boiling in extract, add dextrose, dissolve, make up to 1 litre and sterilize. If the medium is thick, agitate while dispensing.

Autoclave   1.05 kgf/cm² (15 p.s.i.) for 20 minutes

(2) Potato Sucrose Agar (for *Fusarium* species)
Potato extract

Potato	*1800 g*
Water	*4555 ml*

Peel and dice the potatoes, suspend in muslin and boil for 10 minutes. Discard the potato and use the liquid (which may be stored in the refrigerator).

Medium

Potato Extract	*500 ml*
Sucrose	*20 g*

(cont.)

> Agar   20 g
> Water   500 ml

Dissolve the agar in water and mix with the potato extract, adjust pH to 6.5 and sterilize.

Autoclave   1.05 kgf/cm² (15 p.s.i.) for 20 minutes

pH 6.5

(3) Potato Carrot Agar
>   Grated potato   20 g
>   Grated carrot   20 g
>   Agar   20 g
>   Water (tap)   1 litre

Boil grated potato and carrot for 1 hour in tap water. Strain (but do not press) through sieve, add agar and boil in water bath until agar dissolves.

Autoclave   1.05 kgf/cm² (15 p.s.i.) for 20 minutes

This is a starvation medium. Fungi produce very little mycelium, but usually spore well. It is good for cellulase destroying fungi such as *Chaetomium* and *Trichoderma* especially if a piece of sterile filter paper is placed on top of the medium. It is usually very clear and can be used for germination of spores in single spore culture.

Sabaraud's Dextrose Agar

> Dextrose (or Maltose)   40 g
> Peptone   10 g
> Agar   20 g
> Distilled water   1 litre

Mix the ingredients, boil to melt the agar, adjust pH to 5.6 and sterilize.

Autoclave 1.05 kgf/cm² (15 p.s.i.) for 10 minutes

This is the classical medium for dermatophytes, but the common moulds often look atypical on it.

Yeast Phosphate Soluble Starch (YPSS)

(Recipe of Emerson, R. (1941). *Lloydia*, **4**, 87.)
> Soluble starch   15.0 g
> Difco Yeast extract   4.0 g
> Potassium phosphate (K_2HPO_4)   1.0 g
> Magnesium sulphate ($MgSO_4.7H_2O$) 0.5 g
> Agar   20.0 g
> Water   1.0 litre

Dissolve the agar in ½ litre of water, dissolve the other ingredients in another ½ litre of water, mix and sterilize.

Autoclave   1.05 kgf/cm² (15 p.s.i.) for 20 minutes

Corn (Maize) Meal Agar (CMA)

Maize meal	*30 g*
Agar	*20 g*
Water	*1 litre*

(If maize meal is not available it can be prepared from dried maize using a coffee grinder or high speed blender.)

Heat the maize meal and water in a double saucepan (or waterbath) until boiling, stirring to prevent sticking, and simmer for 1 hour. Filter through muslin, discard the meal, add the agar to the filtrate, boil to dissolve, bottle and sterilize.

Autoclave   1.05 kgf/cm² (15 p.s.i.) for 20 minutes

This medium is particularly good for culturing *Phytophthora* and *Pythium* species, but pure water (glass distilled) and Japanese agar should be used, and 0.5 g of wheat germ oil added.

Oat Meal Agar (OA)

Oat meal	*30 g*
Agar	*20 g*
Water	*1 litre*

Prepare as for Corn Meal Agar.

V-8 Agar

'V8' vegetable juice or other vegetable juice	*200 ml*
Agar	*20 g*
Water	*800 ml*

Dissolve the agar in water and add to the 'V8' juice. Adjust pH to 6.0 with 10% sodium hydroxide (NaOH).

Autoclave   1.05 kgf/cm² (15 p.s.i.) for 20 minutes

pH 6.0

Rabbit Dung

Some fungi grow well on dung. The most convenient to use in the laboratory is rabbit dung.

Add 3 pellets to each 28 ml bottle, add tap water agar, and sterilize.

Autoclave   1.4 kgf/cm² (20 p.s.i.) for 20 minutes

The pellets can be sterilized separately and added to melted sterile tap water agar as required.

Soil

Half fill screw cap 28 ml bottles or test tubes with finely sieved loam, sterilize twice by autoclaving at 1.05 kgf/cm^2 (15 p.s.i.) for two 20 minute periods.

Vegetables

Various plugs of vegetables, stalks, leaves or grains (pre-cooked) may be used for growing fungi. They can be set in tap water agar or kept moist by a cotton wool pad at the base of the tube. They are often difficult to sterilize so autoclaving at 1.4 kgf/cm^2 (20 p.s.i.) for 20 minutes is recommended.

Thin tissues can be sterilized cold by leaving overnight in a screw cap honey jar with one or two drops of propylene oxide (this explosive should be handled carefully) ventilated in the morning by loosening the lid and used as required. Filter paper sterilized in this way is placed on the surface of tap water agar and the combination is a very useful medium to encourage sporulation. These cultures must be inspected for infections as sterilization may not be complete.

Antibacterial Media

Various antibacterial substances can be added to media in order to eliminate bacteria from fungus cultures.

(1)   Low pH Media
      Growth at low pH tends to eliminate bacteria, e.g. Czapek's agar is acid and inhibits bacteria.

(2)   Chloramphenicol
      (0.05 g per litre)
      Prepare the medium in the usual way. Just before sterilizing add 0.05 g chloramphenicol suspended in 10 ml of 95% alcohol and mix well.

(3)   Penicillin
      (1 phial = 6 mg (100 000 units))
      Add 1 phial to 200 ml sterile water and store in a refrigerator. Add 1 ml to 10 ml of cool medium when preparing plates. Greater concentrations are often preferable as it is better to err by using too much than too little.

(4)   Streptomycin
      (1 phial = 1 g (745 units per mg))
      Add 1 phial to 750 ml sterile water and use as for penicillin.

(5)   Mixtures of Penicillin and Streptomycin and other antibiotics
      At the Commonwealth Mycological Institute one phial of Crystomycin is added to 500 ml of cooled liquid agar when pouring plates. This has been found to be most satisfactory.
      (1 phial Crystomycin (Glaxo) = 300 mg (500 000 units) benzylpenicillin (sodium) BP + 500 mg streptomycin (as streptomycin sulphate BP))

Media for Yeast

Davenport (1980)* gives an outline for yeast media for all yeast studies, including detection, isolation, identification and classification. It is wise to first consult commercial catalogues for media products which are readily available for yeasts, then make a suitable choice, after considering the points given in Chapter 6 (p. 80). A general list of media is given here.

Media and supplier	Use
YM agar (Difco)	General (e.g. storage slopes and growth tests)
YM broth (Difco)	General (e.g. isolation and growth tests)
Malt extract agar (various) Wort agar (various) Malt and wort broths (various)	Isolation, purification and classification
WL (Difco and various)	All yeasts – isolation
WLD or acitidione agar (Difco and various)	Acitidione (cycloheximide) resistant yeasts Species and strain recognition during isolation and/or identification procedures
Potato dextrose agar Corn Meal agar	Identification – particularly for mycelium, pseudomycelium and sometimes ascospores

* DAVENPORT, R. R. (1980). An introduction to yeasts and yeast-like organisms. In *Yeasts* (SKINNER, F. A., PASSMORE, S. M. and DAVENPORT, R. R., Eds). Academic Press, London and New York.

Appendix II

Salt solutions for control of relative humidities

A short selection of saturated salt solutions suitable for controlling relative humidity in microbiological experimentation. Calculated from the data of Young (1967).* It is important that both solid and solution are present and that both the solution, and the air in contact with it, are at the same temperature.

Salt	Relative humidity (%) at various temperatures					
	10°C	15°C	20°C	25°C	30°C	35°C
NaOH.H_2O	–	7.0	7.0	7.0	6.7	6.3
LiCl.H_2O	–	–	11.3	11.3	11.3	11.2
CH_3.COOH.1.5H_2O	24.0	23.4	22.8	22.2	21.7	–
$MgCl_2$.6H_2O	33.6	33.3	33.0	32.7	32.4	32.1
KNO_2	50.9	50.0	49.1	48.2	47.3	46.4
$Mg(NO_3)_2$.6H_2O	67.3	66.3	65.4	64.4	63.5	62.5
$CuCl_2$.2H_2O	68.4	68.4	68.4	68.4	68.4	–
NaCl	75.4	75.3	75.2	75.1	75.0	74.9
$(NH_4)_2SO_4$	81.3	80.9	80.4	80.2	79.9	79.5
KCl	86.6	85.8	85.0	84.2	83.4	82.6
K_2CrO_4	–	–	86.8	86.5	86.2	85.9
$BaCl_2$.2H_2O	91.5	91.1	90.7	90.3	89.9	89.5
K_2SO_4	–	–	97.3	97.0	96.8	96.5

It should be noted that several hours are required for a solution to control the air above it. Also, if a material is being conditioned to a particular relative humidity over a saturated salt solution, the amount of material must be small and its minimum dimension very small. While a saturated solution is taking up water from the air (drying an exposed material) the relative humidity will tend to exceed the figures given. The reverse will apply if the solution is losing water (wetting an exposed material).

* YOUNG, J. F. (1967). Humidity control in the laboratory using salt solutions – a review. *J. appl. Chem.*, **17**, 241–4.

Appendix III

Glossary of common terms

Acervulus (dimin. of Lat. *acervus*, a heap)   A structure characteristic of the Melanconiales (see p. 101). Fungi of this order, when growing on the host plant, break through the cuticle of the host and produce a cushion-like mass of conidiophores, sometimes accompanied by stiff, sterile hyphae known as setae.

Acropetal (Gr. *acro*, topmost; *pet*, seek)   Describes a chain of conidia in which the youngest is at the apex.

Aleuriospore   A thick-walled terminal spore formed from the blown out end of the sporogenous cell. This term is tending to be dropped and such conidia are referred to as terminal thalloconidia.

Amerospore   One-celled spore; non-septate spore.

Ampulliform   Swollen or flask-like.

Anastomosis   Fusion between hyphal elements.

Anamorph   An asexual part of a fungus produced by mitosis (imperfect state).

Annellide, annellidic, annellophore (Lat. *annellus*, a little ring)   A sporogenous cell in which, after the first conidium is formed, a hyphal wall grows out through the conidial scar and produces a second conidium. A third is produced through the new scar. A chain of conidia is formed and the sporogenous cell has a series of rings at its apex.

Apiculate   Cells with points at one or both ends, rather like a lemon.

Apothecium (Gr. *apo*, from, away from; *theke*, a receptacle)   A cup-shaped or saucer-shaped structure on which there are numerous closely packed, cylindrical or club-shaped asci. The term signifies that it is not a true receptacle, since it is wide open.

Arthroconidium, arthrospore   A conidium produced from a thallic conidiogenous cell, by fragmentation or disarticulation of determinate fertile hyphae into chains of conidia, often cylindrical.

Ascus, pl. *asci* (Gr. *askos*, a skin bag)   A thin walled sac containing spores, characteristic of, and found only in, the class Ascomycetes. The number of spores in each sac is normally eight, but in certain cases a multiple of two. In the more simple members of the Ascomycetes the ascus is rounded with the spores packed tightly inside, in the more highly developed species it is cylindrical or club-shaped, with spores arranged in one or two rows.

Ascoma, ascocarp, etc.   Ascus-bearing structures.

Basidium (Gr. *basis*, a pedestal; *eidos*, like)   An organ characteristic of the class Basidiomycetes. In the larger Basidiomycetes (mushrooms and toadstools) it is a short cell, wider at the top than at the base, with four tiny projections (sterigmata), each of which bears a spore. The two nuclei of opposite 'sex' fuse in the basidium, the fusion being followed by two successive divisions to give four daughter nuclei, one passing into each developing spore (basidiospore).

Basipetal (Lat. *basi*, base; *pet*, seek)   A succession of conidia, with the youngest at the base of the chain.

Blastic, blastospores etc.   One of the main types of conidia. The blastic conidium is produced by the enlargement (blowing out) of a part of the conidiogenous cell before a septum is laid down.

Catenate, catenulate (Lat. *caten*, a chain)   Spores borne in a chain, end to end.

Chlamydospore (Gr. *chlamys*, a cloak)   A thick-walled resting spore formed by swelling and thickening of a single cell (thallic) in submerged vegetative hyphae, in aerial hyphae, or even in multiseptate conidia such as those of species of *Fusarium*.

Clavate   Club-shaped.

Cleistothecium   Ascocarp with no opening, or ostiole.

Collarette   Cup-shaped structure found at the apex of a phialide through which a succession of conidia emerge; frill left at the apex of a sporangiophore after rupture of the sporangium in some Mucorales.

Columella   Swollen tip of the sporangiophore projecting into the sporangium in some Mucorales.

Conidiogenous cell   The hypha which produces or becomes a conidium.

Conidiophore   The specialized hypha or cell on which, or as part of which, conidia are produced. Now used to describe the entire apparatus, while *stipe* is used to describe the main stalk which was called a conidiophore by Raper and Thom (1949) (see p. 22) and other authors.

Conidium, pl. *conidia* (Gr. *konis*, dust; *eidos*, like)   Asexual non-motile spore.

Coremium (Gr. *korema*, a besom) (= *Synnema*)   An erect, compact cluster of conidiophores. The latter may be of different lengths, the coremium being more or less cylindrical and clothed with spores throughout the greater part of its length, or may be of approximately equal length, in which case the coremium has a distinct stalk and a spore-bearing head.

Dematiaceous   Dark coloured or pigmented.

Denticle   Small tooth-like projection on which spores are borne.

Dichotomous   Branching into two more or less equal arms.

Dictyospore   Spore with both longitudinal and transverse septa.

Didymospore   Spore with one septum, i.e. consisting of two cells.

Distosepta (pseudosepta)   A cell appearing to be septate but not having true cross walls, as seen in *Helminthosporium, Drechslera* etc.

Echinulate   Spores, conidiophores or hyphal walls with spiny surfaces.

Enteroblastic   Blastic conidium produced from within the conidiogenous cell.

Eukaryote   An organism, plant, animal or microorganism, i.e. fungus, algae, protozoa etc., characterized by a membrane bound nucleus and the presence of other highly complex organelles such as mitochondria.

Fascicle   Bundle or tuft, usually of conidiophores, as in some *Penicillium* species.
Foot cell   Basal cell sometimes found supporting stipes of *Aspergillus*.
Funiculose   Term used to describe hyphae aggregated into ropes or bundles.
Fusiform   Conidia which are spindle-shaped or narrowing towards both ends.

Globose   Spherical.

Helicospore   Spirally coiled conidium, usually septate.
Holoblastic   A blastic conidium in which both the inner and outer walls of the conidiogenous cell swell out to form the new conidium (see p. 105).
Holomorph   The fungus in all its forms, imperfect and perfect, sexual and asexual.
Hyaline   Non-pigmented.
Hypha, pl. *hyphae* (Gr. *hyphe*, a tissue)   The individual filaments which constitute the vegetative part of a fungus.

Metula, pl. *metulae*   Apical branch of a conidiophore (complex) on which phialides (conidiogenous) cells are borne in *Penicillium* and *Aspergillus*.
Moniliaceous   Hyaline or brightly coloured.
Monopodial   Form of branching in which growth of main axis continues from the main axis.
Monoblastic   Blastic conidia produced at only one point on the conidiogenous cell.
Mycelium (Gr. *mykes*, a fungus; *elos*, a wort)   The mass of hyphae which forms the vegetative fungus.

Oidia   An old and rather loose term no longer in use, describing a series of very short cells, resembling a chain of beads; the chain fragmenting to form spores.

Penicillus   Broom-like arrangement of phialides, metulae and branches at the apex of the stipe as seen in *Penicillium*. In a wider use of the term the final conidiogenous cells need not be phialides.
Perithecium (Gr. *peri*, around; *theke*, a receptacle)   A globose or flask-shaped ascoma, closed at maturity except for a narrow passage (ostiole), through which the ascospores are liberated.
Phialide (Gr. *phial*, shallow cup)   A specialized conidiogenous cell from which conidia are produced in basipetal succession. The first conidium is blastic, leaving a collarette, and subsequent conidia are enteroblastic, being produced from within the colarette (see p. 106).
Phragmospore   Spore with two or more transverse septa.
Prokaryote   An organism with cells more complex than a virus particle but of relatively simple organization. The nucleus is represented by a body consisting largely of DNA but is not enclosed in a membrane.

Pycnidium   A globose, flask-shaped or irregular fruit body lined with conidiophores and resembling a perithecium. Conidia are liberated through an apical pore (ostiole). Pycnidia are usually readily visible to the naked eye.

Pyriform   Pear-shaped.

Rostrate   Having a beak or strongly drawn out at the apex.

Rostrum (Lat. *rostrum*, beak, snout, ships prow)   A beak-like process.

Sclerotium (Gr. *skleros*, hard)   Neither a spore nor a spore-bearing structure, but a compacted mass of mycelium, often very hard, varying in size (in mould fungi) from a fraction of a millimetre to several millimetres in diameter. Some of the larger fungi produce large sclerotia.

Scolecospore   Filiform, thread-like spore.

Septum   Cross wall in fungal hyphae or spores.

Seta   A stiff hair ornamenting vegetative hyphae of fruit bodies.

Sporangiole, sporangiolum   A small sporangium.

Sporangium (Gr. *angeion*, a vessel)   A closed receptacle, usually round or pear-shaped, borne on a stalk (*sporangiophore*) within which an indefinite number of asexual spores are produced by cleavage, produced, for example, by the Mucorales.

Sporodochium (Gr. *dochium*, a receptacle)   A cushion-shaped, tightly packed cluster of conidiophores, with conidia above the surface of the substrate. Rather like a coremium in which the stipes are too short to form a stalk.

Sporophore (Gr. *phoreo*, to bear)   A rather loose term used to describe any spore bearing structure.

Sporogenous cell   A cell which gives rise to one or more spores.

Staurospore   Forked or star-shaped (stellate) spore.

Sterigma, pl. *sterigmata*   Frequently applied to metulae and phialides of *Aspergillus* but correctly applied to the denticle on which basidiospores are produced.

Stroma, pl. *stromata* (Gr. *stroma*, a bed)   As seen in laboratory cultures, a gelatinous or leathery layer covering the surface of the culture medium, bearing spores on very short conidiophores or having perithecia or pycnidia embedded in it.

Sympodium   A kind of false axis, in which the conidiophore or sporangiophore forms a terminal spore or sporangium, puts out a branch which extends beyond the tip of the original sporophore, and then produces a second terminal spore or sporangium. This process is repeated, each new branch arising from the most recently formed branch.

Sympodula   A conidiophore showing compressed sympodial growth.

Synnema, pl. *synnemata* (Gr. *syn*, together, *nema*, a thread)   Much the same as coremium. The term is also used for associations of conidiophores of less definite form than true coremia. Conidiophores may be fused along their length or merely compacted.

Teleomorph   The perfect state or sexual morph involved in producing meiotically generated propagules.

Thallic, thallospores, etc.   Terms used to describe conidia produced by conversion of a whole cell or segment of fertile hypha, with no enlargement before the septa are laid down (see p. 104), (cf. blastic).

Thermophilic   Term used to describe fungi which have growth optima at high temperatures, e.g. 40–50°C, and which produce spores at 45°C and above.

Verticil   A whorl of 3 or more branches of conidiogenous cells.

Vesicle   The swollen apex of the stipe in *Aspergillus* or swollen part of a sporogenous cell in other fungi.

Zygospore   The thick-walled sexual spore produced by Zygomycetes.

Author Index

Subject and Organism Index

Page numbers in italics refer to diagnoses of genera and species, those in bold type to illustrations.
Readers are referred to the Glossary of common terms (p. 379) for definitions of technical terms.